A General Theory of Fluid Mechanics

Peiqing Liu

A General Theory of Fluid Mechanics

Science Press
Beijing

Springer

Peiqing Liu
Institute of Fluid Mechanics
Beihang University
Beijing, China

ISBN 978-981-33-6662-6   ISBN 978-981-33-6660-2   (eBook)
https://doi.org/10.1007/978-981-33-6660-2

Jointly published with Science Press
The print edition is not for sale in China (Mainland). Customers from China (Mainland) please order the print book from: Science Press.

This Springer imprint is published by the registered company Springer Nature Singapore Pte Ltd.
The registered company address is: 152 Beach Road, #21-01/04 Gateway East, Singapore 189721, Singapore

Preface

If modern civilization originated from the rise and development of the mechanical industry, the combination of fluid mechanics and the mechanical industry has played a decisive role. It can also be said that fluid mechanics is a bright pearl on the crown of modern mechanical industry. From the ancient hydraulic machinery to the birth of the most complex modern aero-engine, it is closely related to the development of fluid mechanics. It can also be said that fluid mechanics is involved in water turbines, steam turbines, gas turbines, expanders, wind turbines, pumps, fans, ventilators, compressors, hydraulic couplers, hydraulic torque converters, pneumatic tools, pneumatic motors, hydraulic motors, and various fluid transport and control equipment, all of which use the fluid as working fluid to convert energy. Therefore, there is no doubt that fluid mechanics is the most active professional basic subject in the mechanical industry, and it is also a professional basic course offered by the technological universities. In order to stimulate the learning interest of beginners and facilitate the understanding of the basic knowledge and development of the law of fluid mechanics, I tried a compilation mode that combines scientific knowledge with the human history of fluid mechanics in the form of biography and popular science. There are eight chapters in the book, namely, basic theories of fluid mechanics, aerodynamics, hydrodynamics, computational fluid dynamics, experimental fluid dynamics, wind tunnel and water tunnel equipment, flying mystery and aerodynamics principles. In the fundamentally theoretical part, readers are introduced to the development and application of theory; in the experimental part, they are mainly introduced to the principle of similarity, flow visualization, and measurement technology; in the wind tunnel and water tunnel equipment part, they are introduced to the development and application of wind tunnel and water tunnel equipment; in the flight mystery and aerodynamics principles part, they are mainly introduced to the flight cognition, flight principle, the function of aircraft components, etc.; in the personages of fluid mechanics part, they are introduced to the role and main contribution of famous scientists in the development of fluid dynamics, including Archimedes, Da Vinci, Galileo, Newton, Leibniz, Bernoulli, Euler, Lagrange, Helmholtz, Stokes, Reynolds, Mach, etc.

In the process of material selection and discussion, this book starts with the intuitive and understandable concepts, combines scientific knowledge with historical development, and introduces the development of fluid mechanics in a simple and understandable language.

This book has been reviewed by Prof. Wei Qingding who is from the College of engineering of Peking University and Prof. Zhu Keqin who is from the School of aerospace engineering of Tsinghua University. I would like to express my deep gratitude for the important modifications and suggestions.

The process of writing this book was greatly supported by the teachers and students of the Lu Shijia laboratory of Beihang University. I would like to express my heartfelt thanks to Dr. Liu Yuan, Dr. Ren Shuili, Dr. Zheng Yunlong, Dr. Guo Zhifei and Dr. Lu Weishuang from Lu Shijia laboratory for the translation of this book into English.

The theory of fluid mechanics is rigorous and widely used, which is difficult for the compilation of general introduction. This book covers three branches: theory, experiment, and numerical simulation. In addition, there are many industrial applications. Due to the limited space, this book cannot give a comprehensive introduction. Here, aerodynamics and hydrodynamics are mainly selected as representatives. Due to the limited knowledge and energy, the systematization of the book is far from enough. I have been deeply disturbed by the fact that mistakes cannot be avoided. I am very grateful to you for your criticism and correction.

Peiqing Liu

Beijing, China
September 2020

About This Book

This book is a general introduction to fluid mechanics in the form of biographies and popular science. Based on teaching perception and experience for many years, the author tries a compilation mode that combines natural science and human history, knowledge inheritance and cognition law, breaks the abstract and profound knowledge of fluid mechanics, and starts with intuitive and understandable physical concepts. The history of fluid mechanics and basic knowledge are organically combined in a way from shallow to deep and from the surface to the inside. The book includes the basic knowledge and development history of the basics of fluid mechanics, aerodynamics, hydrodynamics, computational fluid dynamics, experimental fluid dynamics, wind tunnel and water tunnel equipment, flying mystery and aerodynamics principles, personages of fluid mechanics, etc., in order to provide references for stimulating the interest of beginners and comprehensively understanding the development of fluid mechanics.

This book is suitable for all people who love fluid mechanics, including teachers, graduate students, and undergraduates of colleges and universities; technicians in various industries related to fluid mechanics; and science enthusiasts. Some of the contents in the book are also applicable to middle school students.

Contents

8   Introduction to Celebrities in Fluid Mechanics   581

1

Foundation of Fluid Mechanics

1.1 Combination of Early Development of Fluid Dynamics with Calculus

In 250 B.C., Archimedes (ancient Greeks, 287–212 B.C., as shown in Fig. 1.1), commissioned by King Syracuse of Sicily to examine the crown, studied the principle of force balance and proposed the famous hydrodynamic buoyancy theorem, which is also a part of hydrostatics. During this period, the achievements of Socrates, Aristotle, Plato, and other ancient Greek scientists mainly stayed at the philosophical level. At the mathematical level, Pythagoras put forward the concept that everything is a number and discovered the Pythagorean law. After A.D. until the Renaissance, society was dark and scientific development was slow. During the Renaissance (fourteenth to early seventeenth centuries A.D.), with the emergence of the new capitalism, the demand for handicraft and mechanical industries greatly promoted the development of mathematics and mechanics. During this period, the Italian scientist Galileo Galilei (1564–1642, as shown in Fig. 1.2) discovered the law of inertia of motion of objects and developed thermometers and telescopes. The Italian versatile scientist Leonardo Di Ser Piero Da Vinci (1452–1519 A.D., as shown in Fig. 1.3) published a series of qualitative cognitive results of flow, vortices, fluid mechanics, including the qualitative principle of bird flight, and even used vortices as an element of beauty in many of Leonardo's paintings (as shown in Fig. 1.4). In 1653, the French scientist B. Pascal (1623–1662) put forward the principle of hydrostatic pressure transfer (Pascal theorem) and made a hydraulic press. Later, he continued to carry out the atmospheric experiments by Galileo and the Italian scientist

© Science Press 2021
P. Liu, *A General Theory of Fluid Mechanics*,
https://doi.org/10.1007/978-981-33-6660-2_1

Fig. 1.1 Archimedes (287–212 B.C., Ancient Greek scholar)

Fig. 1.2 Galileo Galilei (1564–1642, Italian scientist)

E. Torricelli (1608–1647), and found that atmospheric pressure varies with altitude change (1643). These laid the foundation for the establishment of classical fluid mechanics theory.

But until the advent of calculus in the late seventeenth century, these qualitative human perceptions were fragmentary and unsystematic. It should be

Fig. 1.3 Leonardo Da Vinci (1452–1519 A.D., Italian versatile scientist)

said that only in the late seventeenth century, after the invention of calculus by the British scientist Newton (Isaac Newton, 1643–1727, as shown in Fig. 1.5) and the German scientist Gottfried Wilhelm Leibniz (1646–1716, as shown in Fig. 1.6), it laid a solid mathematical foundation for the development of fluid mechanics, and injected infinite vitality. According to historical records, Newton's "stream number concept" calculus was proposed in an unpublished essay written in 1666, while Leibniz's calculus was mentioned in unpublished manuscripts and correspondence in 1675; so they had independent inventive rights. Leibniz officially published his discovery of differential in 1684. Two years later, he published a study of integrals. The general calculus symbols are proposed by Leibniz. Studying Leibniz's manuscript, we also find that Leibniz and Newton created calculus from different ideas: Newton first had the concept of derivative and then integral to solve the problem of motion; Leibniz, in turn, influenced by his philosophy, had the concept of integral and then derivative. Newton only regarded calculus as a mathematical tool in physics research, while Leibniz realized that calculus would bring a revolution in mathematics. The dispute between Newton and Leibniz over the right to invent calculus has evolved historically into a confrontation between the British scientific community and the German scientific community, and even with the scientific community of the whole European continent. British mathematicians were reluctant to accept the research results of continental mathematicians for a long time. Their insistence on teaching using Newton's backward calculus symbols and outdated

Fig. 1.4 Leonardo Da Vinci painted "Old Man and Vortex" and "Turbulence"

concepts of mathematics made the study of mathematics in Britain stagnant for more than a century, and it was not until 1820 that they recognized the achievements of other countries on the European continent and rejoined the international mainstream.

Calculus introduces the viewpoint of development and change into mathematics (can be seen as dynamic mathematics), which can be said to be a thorough revolution of static mathematics, based on the limit of gradual approximation and infinite approximation, and is a process that can never

Fig. 1.5 Isaac Newton (1643–1727, British scientist)

Fig. 1.6 Gottfried Wilhelm Leibniz (1646–1716, German mathematician)

be reached but infinite approximation in philosophy. In 1686, Newton published the book "Mathematical Principles of Natural Philosophy". He put forward three theorems of gravitation and motion of objects, and expounded the laws of momentum and angular momentum, Cooling, and Newton's internal friction between the strata. Newton is a great scientist who organically combines the motion of objects with the concept of calculus. Under the influence of Newton, it can be said that the creation and development of fluid mechanics is the crystallization of the organic combination of calculus and flow phenomena, showing the great power produced by the perfect combination of mathematics and physics.

The relationship between mathematics and physics in fluid mechanics can be summarized in the following four sentences:

Mathematics is beautiful.

Physics is wonderful.

If we combine mathematics with physics,

Beauty will never end.

1.2   Methods of Describing Fluid Motion

By definition, a particle is a mass space point (material point) that is sufficiently large (composed of many molecules) at the microlevel and small enough to ignore its volume at the macro level. Take the air as an example. At sea level, the pressure is 101325 Pa, the temperature is 288.15 K, the space contains 2.7×10^{19} air molecules per cubic centimeter, and the average free path of molecules is 10^{-8} m. Under what conditions, the molecular flow satisfies the definition of particle continuous flow rather than discrete flow, which involves the relationship between the average free path of molecules and the characteristic scale of objects. The Denmark physicist M. Knudsen (1871–1949, as shown in Fig. 1.7) studied the theory of molecular motion and the phenomenon of low pressure in airflow. Knudsen number was proposed to judge the relative dispersion of molecules. Kn number was

Fig. 1.7   M. Knudsen (1871–1949, Denmark physicist)

Fig. 1.8 Qian Xuesen (1911–2009, Chinese scientist)

defined as the ratio of the average free path of molecules to the characteristic scale of objects. Qian Xuesen, the Chinese scientist (1911–2009, as shown in Fig. 1.8), put forward the condition of continuity of fluid motion by Kn number when he studied rarefied gas dynamics in 1946. In general, the average free path of air molecules is 10 nm. If the Knudsen number is less than 0.01, it is a continuous flow. That is to say, the macroscopic scale should be more than 1000 nm before it can be considered as a continuous flow. At this time, the fluid dynamics equation can be used to describe the fluid movement. When the Kn number is between 0.01 and 1, it is a slip flow. The viscous fluid motion equation with slip boundary conditions can be used to describe fluid motion. When Kn number is between 1.0 and 10, it is a transitional flow; when Kn number is greater than 10, it is a molecular flow. The Boltzmann equation of molecular motion is used to describe the fluid motion directly under the assumption of molecular discrete flow. That is to say, the molecular flow motion (discrete motion) is complete when the macroscale is less than 1 nm. Once in the continuous flow state, the impact of the collision and interpolation of individual molecules on the mainstream is almost negligible, as the elephant's behavior (object scale) depends on the random movement of individual ants (molecules) on the elephant.

Therefore, the hydrodynamic continuum hypothesis holds that fluid is composed of innumerable particles, which in any case fill the occupied space without voids. That is to say, the fluid particle and space point must satisfy the one-to-one relationship under any circumstances (motion and stationary), that is, each fluid particle can only occupy one space point at any time, but cannot occupy more than two space points (to ensure that the solution does not appear discontinuous); each space point can only be occupied by one fluid particle at any time, but not by more than two particles (to ensure that the solution does not have multiple values), so people will naturally introduce single-valued continuous differentiable functions into the analysis of fluid flow physical quantities. For the numerous fluid particles satisfying the continuity condition, when they move, how to correctly characterize the motion characteristics of each fluid particle must answer two basic questions. One is how to track and distinguish each fluid particle, and the other is how to describe the motion characteristics and changes in each fluid particle. This is the basic problem of fluid kinematics. According to the different viewpoints of observers, the motion of fluid particles can be described by the Lagrange method and the Euler method.

1. *Lagrange Method*

This method is also called the particle system method of fluid. It identifies and confirms all fluid particles (not space points), and then records the position coordinates of each particle at different times, so as to understand the overall flow behavior. Obviously, this method requires an observer to track every fluid particle at any time and anywhere, and record the particle movement process (directly measuring the position of particles at different times, leading to the concept of particle trajectory), so as to obtain the motion law of the overall flow. Whereas, the position coordinates of particles (a, b, c) at a stationary time or at an initial time t_0 are used as identifiers of fluid particles (so that the particle identification is not renamed, as shown in Fig. 1.9), at any time t, the spatial positions of particles (a, b, c) are $x(a, b, c, t)$, $y(a, b, c, t)$, $z(a, b, c, t)$, and the whole flow can be understood by tracking the whole process of all particles. Among them, the position record of any particle at different times is the direct measurement data, from which the velocity and acceleration data obtained by definition and law are indirect measurement data.

$$u = \frac{\partial x(a, b, c, t)}{\partial t}, v = \frac{\partial y(a, b, c, t)}{\partial t}, w = \frac{\partial z(a, b, c, t)}{\partial t}$$

$$a_x = \frac{\partial u(a, b, c, t)}{\partial t}, a_y = \frac{\partial v(a, b, c, t)}{\partial t}, a_z = \frac{\partial w(a, b, c, t)}{\partial t}$$

where u, v, and w, respectively, represent the velocity components in x, y, and z directions, and a_x, a_y, and a_z, respectively, represent the acceleration components in x, y, and z directions. For any fluid particle, the line

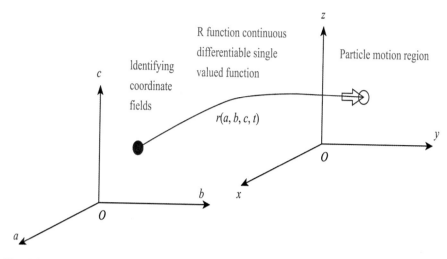

Fig. 1.9 Particle marking method

connecting its spatial position at different times is called the pathline of the particle (as shown in Fig. 1.10). The pathline equation can be expressed as

$$\frac{dx}{u} = \frac{dy}{v} = \frac{dz}{w} = dt$$

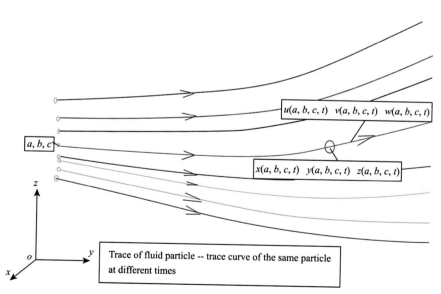

Fig. 1.10 Traces of different fluid particles characterized by the Lagrangian method

This method of staring at particles (which can be visually seen as the working way of "police tracking thieves") is a direct extension of the particle system method in theoretical mechanics. The reason is that it is a concept extension. Here, it refers to countless continuous particle systems, where it refers to countable discrete particle systems. Whether this extension from individual to general concepts is feasible or not is worth studying mathematically. This method is clear in concept and convenient for the direct generalization of physical laws. However, the disadvantage is that there are too many records, especially for the flow characteristics of only local areas, which is very inconvenient. For example, in the flood season every year, people only want to know the water regime of the Yangtze River in the Wuhan section (the water level at Wuhan Pass), but when using this method to describe it, we must make clear the origin of all water quality points passing through the Yangtze River in Wuhan section, track and record the flow process of each particle throughout the whole process, so as to depict the flow characteristics of the Yangtze River in Wuhan section. In fact, it is useless for many records of water quality points in the Yangtze River which are not located in the Wuhan section.

2. *Euler Method*

This method is also called the space point method or the flow field method. In order to avoid unnecessary data recording by the Lagrange method, Euler proposed that instead of identifying fluid particles, he changed the space points to identify the flow area (the relationship between space points and particles still satisfies the continuity hypothesis). The observer remained stationary relative to space points and recorded the speed of different particles passing through fixed space points at different times. The amount directly recorded by the observer is the particle velocity value passing through the space point at different times. For example, arranging an observer at each space point and recording the particle velocity value of each space point at each time can give a comprehensive understanding of the characteristics of the flow area investigated. Please note that although this method identifies spatial points, it still studies fluid particles, so it can be said that it is an unlabeled particle system method. For example, for any space point in the region under investigation, the velocity of fluid particles passing through the space point is directly recorded at time t as $u(x, y, z, t)$, $v(x, y, z, t)$, and $w(x, y, z, t)$. If the velocity of particles passing through all spatial points in the flow region is recorded, the flow field at any time in the region can be understood. This method does not need to identify the fluid particle information, but needs to record the velocity information of any fluid particle passing through a fixed space point. Therefore, it represents the spatial distribution of the fluid particle velocity at any time, so it is called the flow field method. The velocity records of fluid particles passing through any space point at any time are measured directly, and the acceleration values obtained by definitions and laws are measured indirectly. Imaginably speaking, this method can also be a "stand by and wait for rabbits" mode of work.

This method leads to the concept of streamline, i.e., a specific curve passing through any point at a certain time in the flow field. The velocity direction of the fluid particles at each point on the curve is parallel to the tangent direction of the curve at that point (as shown in Fig. 1.11). At some point, the streamline equation at any point in the flow field is

$$\frac{dx}{u} = \frac{dy}{v} = \frac{dz}{w}$$

Streamline is a curve reflecting the instantaneous velocity direction of the flow field. Compared with pathlines, streamlines are curves composed of different particles at the same time. According to the definition of streamlines, streamlines have the following properties:

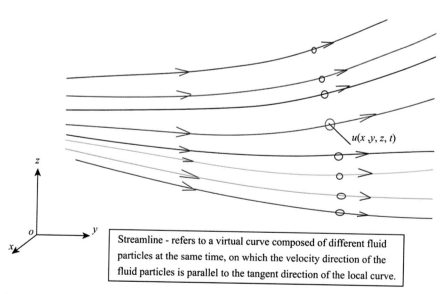

Streamline - refers to a virtual curve composed of different fluid particles at the same time, on which the velocity direction of the fluid particles is parallel to the tangent direction of the local curve.

Fig. 1.11 Streamlines in the flow field represented by the Euler method

(1) In steady flow, the traces of fluid particles coincide with streamlines. In unsteady flow, streamlines and traces do not coincide.
(2) In steady flow, streamline is a non-deviating curve of fluid particles.
(3) At constant points, streamlines cannot intersect, bifurcate, intersect, or turn, and streamlines can only be a smooth curve. That is to say, at the same time, a point can only pass through a streamline.
(4) Exceptions at singularities and zero velocities are not satisfied (3).

It should be pointed out that the velocity at a point in space essentially refers to the velocity at which t instantaneously occupies the fluid particle at that point. Mathematically, a space full of certain physical quantities is called a field, and the space occupied by fluid flow is called a flow field. If the physical quantity is velocity, it describes the velocity field. If it is pressure, it is called a pressure field. In high-speed flow, the density and temperature of the airflow also change with the flow, so there is a density field and a temperature field. These are all included in the concept of flow field.

When the Euler method is used to describe the flow field, the observer directly measures the velocity of the fluid particle through the space point. Then, if a fluid particle is tracked arbitrarily in a certain period, how can its velocity change and how to correctly express the acceleration of the particle motion in the Eulerian coordinate system? From this, the concept of Eulerian derivative is proposed, which is also called the body-dependent derivative in hydrodynamics. An example is given to illustrate the acceleration of locally tracking a fixed fluid particle. Suppose that at any time t, the velocity $u = (t, x, y, z)$ of the fluid particle occupying (x, y, z) space point, and at $t + \Delta t$ time, the tracked fluid particle moves to the space point $(x + \Delta x, y + \Delta y, z + \Delta z)$, and its velocity $u = u(t + \Delta t, x + \Delta x, y + \Delta y, z + \Delta z)$. According to the definition, the acceleration (the derivative of velocity) of the particle is

$$\frac{du}{dt} = \lim_{\Delta t \to 0} \frac{u(t + \Delta t, x + \Delta x, y + \Delta y, z + \Delta z) - u(t, x, y, z)}{\Delta t}$$
$$= \frac{\partial u}{\partial t} + u\frac{\partial u}{\partial x} + v\frac{\partial u}{\partial y} + w\frac{\partial u}{\partial z}$$

If we follow the motion of a fluid particle, the substantial derivative of the pressure is obtained.

$$\frac{dp}{dt} = \frac{\partial p}{\partial t} + u\frac{\partial p}{\partial x} + v\frac{\partial p}{\partial y} + w\frac{\partial p}{\partial z}$$

The general expression of the substantial derivative is

$$\frac{\mathrm{d}}{\mathrm{d}t} = \frac{\partial}{\partial t} + u\frac{\partial}{\partial x} + v\frac{\partial}{\partial y} + w\frac{\partial}{\partial z}$$

Note that the arbitrary derivative here is different from the total derivative in field theory. In field theory, the total derivative of a function u is

$$\frac{\mathrm{d}u}{\mathrm{d}t} = \frac{\partial u}{\partial t} + \frac{\mathrm{d}x}{\mathrm{d}t}\frac{\partial u}{\partial x} + \frac{\mathrm{d}y}{\mathrm{d}t}\frac{\partial u}{\partial y} + \frac{\mathrm{d}z}{\mathrm{d}t}\frac{\partial u}{\partial z}$$

If it is a substantial derivative, the specified fluid particles must be tracked. Because the coordinate increment satisfies the motion condition of the same particle, that is, $\mathrm{d}x = u\mathrm{d}t$, $\mathrm{d}y = v\mathrm{d}t$, $\mathrm{d}z = w\mathrm{d}t$. It can be seen from the above expression that the acceleration of any fluid particle in the Eulerian coordinate system consists of local acceleration and convective acceleration, the former depends on the unsteady velocity field and the latter on the non-uniformity of the velocity field. Since any physical theorem is for matter, the derivative of physical quantity following a fluid particle in the Eulerian coordinates refers to the substantial derivative.

The Lagrangian method describes fluid motion as follows: global tracking, full-course recording. The Euler method describes fluid motion as follows: local tracking and full region recording.

1.3 Establishment and Application of Differential Equations for Ideal Fluid Motion

In the eighteenth century, driven by the mechanical industry, classical mechanics entered the era of establishing system theory system and wide application under the support of calculus. During this period, the classical continuum mechanics system was formed based on the combination of the concept of calculus continuous differentiable function and the theory of particle system mechanics. The assumption of continuum based on the concept of particle system is the basis of introducing calculus into mechanics to establish the theoretical system.

Fig. 1.12 Daniel Bernoulli (1700~1782, Swiss mathematician and fluid dynamician)

In 1738, the Swiss scientist Daniel Bernoulli (1700–1782, as shown in Fig. 1.12) established the particle kinetic energy theorem along two sections of the same micro-element flow tube and derived the conservation equation of mechanical energy for an one-dimensional flow, which is the famous energy equation for a steady flow of ideal fluid (hereinafter referred to as the Bernoulli equation). In 1757, Leonhard Euler (1707–1783, as shown in Fig. 1.13), the Swiss mathematician, extended this equation to compressible flows. For the steady flow of ideal incompressible fluid, the total mechanical energy of fluid particles per unit weight along the same streamline is

Fig. 1.13 Leonhard Euler (1707~1783, Swiss mathematician and fluid dynamician)

conserved under the action of gravity force (the sum of potential energy, pressure energy, and kinetic energy of fluid particles per unit weight remains unchanged).

$$z + \frac{p}{\gamma} + \frac{V^2}{2g} = C$$

where z is the position of the fluid particle, p is the pressure of the fluid particle, V is the velocity of the fluid particle, γ is the volume weight of the fluid, g is the gravitational acceleration, and C is a constant. Without considering the mass force (the mass density of air is small, the influence of gravity can be neglected), the sum of pressure and energy of fluid particles per unit mass along the same streamline is constant.

$$\frac{p}{\rho} + \frac{V^2}{2} = C$$

The discovery of the Bernoulli equation correctly answers the contribution of suction on the upper wing to lift. Later wind tunnel tests show that for airfoils, the upper wing suction contributes about 60–70% of the total lift of the airfoil.

In 1752, the French scientist d'Alembert (1717–1783, as shown in Fig. 1.14) first expressed the field by the differential equation of fluid mechanics in his paper "A New Theory of Fluid Damping", and proposed the

Fig. 1.14 Jean le Rondd'Alembert (1717~1783, French mechanist)

d'Alembert Paradox of the steady flow of ideal fluid around arbitrary three-dimensional objects without drag. In 1753, Euler proposed the continuum hypothesis, and in 1755, he proposed the Euler method as a spatial point method to describe fluid motion. Based on the continuum hypothesis and the ideal fluid model, the differential equation of ideal fluid motion was established by Newton's second theorem, namely

$$\frac{du}{dt} = \frac{\partial u}{\partial t} + u\frac{\partial u}{\partial x} + v\frac{\partial u}{\partial y} + w\frac{\partial u}{\partial z} = f_x - \frac{1}{\rho}\frac{\partial p}{\partial x}$$

$$\frac{dv}{dt} = \frac{\partial v}{\partial t} + u\frac{\partial v}{\partial x} + v\frac{\partial v}{\partial y} + w\frac{\partial v}{\partial z} = f_y - \frac{1}{\rho}\frac{\partial p}{\partial y}$$

$$\frac{dw}{dt} = \frac{\partial w}{\partial t} + u\frac{\partial w}{\partial x} + v\frac{\partial w}{\partial y} + w\frac{\partial w}{\partial z} = f_z - \frac{1}{\rho}\frac{\partial p}{\partial z}$$

where u, v, and w are the velocity components of the particle, respectively; f_x, f_y, and f_z are the unit mass force acting on a particle, respectively. p is the velocity of the particle. The differential equations clearly show that the mass force acting on a fluid microelement and the pressure on the surface of the fluid microelement change the motion behavior of the fluid microelement. That is to say, if there is no pressure gradient along a certain direction without considering the mass force, the velocity of the fluid particle will remain unchanged along that direction. It can be written in vector form as

$$\frac{d\vec{V}}{dt} = \vec{f} - \frac{1}{\rho}\nabla p$$

For steady flow of incompressible fluid with potential mass force, the Bernoulli equation can be obtained by integrating the Euler equations along streamlines. Further studies show that the Bernoulli equation is satisfied not only along the same streamline but also along the same vortex line, potential flow field, and spiral flow.

In 1781, the French scientist Joseph-Louis Lagrange (1736–1783, as shown in Fig. 1.15) proposed the particle method for describing fluid motion and established the relationship between particle velocity and velocity potential function and flow function. On this basis, the conservation theorem of irrotational flow for ideal barotropic fluids with potential mass was established. In 1785, the French scientist Pierre-Simon Laplace (1749–1827, as shown in Fig. 1.16) established the Laplace equation based on the potential function. So far, the classical theoretical system of ideal hydrodynamics and irrotational flow has been basically established.

Fig. 1.15 Joseph-Louis Lagrange (1736~1813, French mathematician and fluid dynamician)

Fig. 1.16 Pierre-Simon Laplace (1749~1827, French mathematician and fluid mechanist)

In 1799, the Italian physicist G. B. Venturi (1746–1822) invented the famous Venturi flowmeter (as shown in Fig. 1.17) by experimenting with variable cross-sectional pipes. The pressure energy is transformed into kinetic energy through the contraction section, and then kinetic energy is transformed into pressure energy through the diffusion section. Venturi tube uses the combination of contraction and diffusion sections to measure the flow rate. Using the Bernoulli energy equation, as shown in Fig. 1.18, the energy

Fig. 1.17 Venturi flow tube

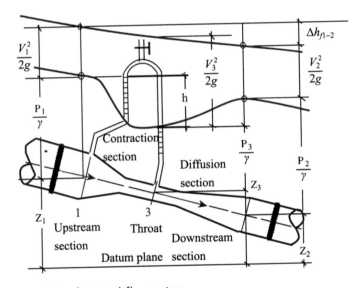

Fig. 1.18 Principle of Venturi flowmeter

equation and continuity equation between upstream section 1-1 and throat section 3-3 are established, and the flow calculation formula through the pipeline is obtained as follows:

$$V_3 = \frac{\mu}{\sqrt{1 - d_3^4/d_1^4}}\sqrt{2gh}, \quad Q = V_3 A_3$$

where d_1 is the diameter of the upstream section, d_3 is the diameter of throat section, Q is the flow through the pipeline, h is the head difference between the upstream section and throat section (measured by test), and μ is the Venturi flow coefficient (generally between 0.95 and 0.99).

In the nineteenth century, fluid mechanics focused on solving the problems and solutions of the theory of irrotational motion of the ideal fluid and established the theory of vortex motion of ideal fluid and the equation of viscous fluid mechanics. With the application of ideal fluid mechanics as the core, the ideal incompressible irrotational flow around different objects was solved. For example, the potential flow solutions around spheres, cylinders, and angular flows are obtained. Based on the potential flow superposition principle, the potential flow singularity solution was proposed. In 1813, French mathematician Augustin Louis Cauchy (1789–1857, as shown in Fig. 1.19) proposed the complex variable function. In 1850, the German mathematician Georg Friedrich Bernhard Riemann (1826–1866, as shown in Fig. 1.20) completed the single-valued condition for the complex variable function to be an analytic function. In 1868, Hermann Ludwig Ferdinand von Helmholtz (1821–1894, as shown in Fig. 1.21), the German hydromechanist, established a potential flow method of complex variable function based on flow function and potential function (as shown in Fig. 1.22). At the same time, Helmholtz put forward the velocity decomposition theorem of fluid mass in 1858, studied the swirling motion of ideal incompressible fluid under the action of force, and put forward three laws of Helmholtz's swirling motion, namely, the law of invariance of vorticity intensity along the vortex tube, the law of keeping vorticity tube, and the law of conservation of vorticity intensity. And the theory of vortex motion of the ideal fluid was established. In 1882, the British scientist W. J. M. Rankine (1820–1872, as

Fig. 1.19 Augustin Louis Cauchy (1789–1857, French mathematician and fluid mechanist)

Fig. 1.20 Georg Friedrich Bernhard Riemann (1826~1866, Ancient Greek mathematician)

Fig. 1.21 Hermann von Helmholtz (1821~1894, German physicist)

shown in Fig. 1.23) perfected the singularity superposition principle based on the ideal fluid theory, established the mathematical theory of free, forced, and combined vortices, and put forward the famous Rankine vortex model. In 1869, the Austrian physicist and philosopher Ludwig Edward Boltzmann (1844–1906, as shown in Fig. 1.24) extended Maxwell's velocity distribution law to the case of conservative force field and obtained Boltzmann's distribution law. In 1872, Boltzmann established the famous Boltzmann equation (also known as transport equation) to describe the statistical mechanics of

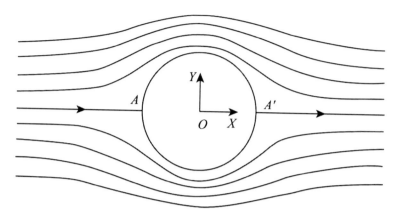

Fig. 1.22 Flow around a cylinder with an ideal fluid

Fig. 1.23 W. J. M. Rankine (1820~1872, British scientist)

Fig. 1.24 Ludwig Edward Boltzmann (1844~1906, Austrian physicist)

gas transition from non-equilibrium state to equilibrium state and explained the second law of thermodynamics from the statistical significance. The Boltzmann equation is an equation describing the motion of rarefied gases.

In 1872, the British scientist Rankine proposed the famous Rankine vortex model (as shown in Fig. 1.25) for the steady concentrated vortex and its induced flow field. The model established a combination model of a free vortex and forced vortex. The results show that the flow field in the vortex core is a circular cylinder with equal vorticity, and the flow field outside the vortex core is a free vortex-induced flow field. The obtained velocity and pressure fields are as follows:

In the vortex core, the flow field with equal vorticity (because the deformation rate is zero, in fact, it is also inviscid and rotational motion), and its circumferential velocity satisfies the rotation law of the rigid body around the spin axis, namely

$$u_\theta = \frac{\Gamma}{2\pi R^2} r$$

where u_θ is the velocity in the circumferential direction, R is the radius of the vortex core, and Γ is the vortex intensity (velocity circulation). The corresponding static pressure at arbitrary radius r is

$$p = p_c + \frac{1}{2}\rho u_\theta^2$$

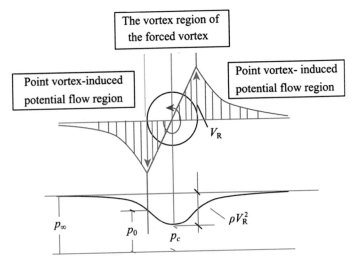

Fig. 1.25 Rankin eddy current model

where p_c is the static pressure at the center of the vortex core.

Outside the vortex core, there is no eddy current induced by point vortices (but because the deformation rate is not zero, it belongs to viscous potential flow), and the circumferential velocity at radius r is

$$u_\theta = \frac{\Gamma}{2\pi r}$$

The static pressure is

$$p = p_\infty - \frac{1}{2}\rho u_\theta^2$$

where p_∞ is the pressure at infinity. The difference between the outflow pressure and the pressure at the vortex center is

$$\Delta p = p_\infty - p_c = \rho u_\theta^2(R) = \rho V_R^2$$

Outside the vortex core, the viscous shear stress is

$$\tau_{r\theta} = 2\mu\gamma_{r\theta} = -\frac{\mu\Gamma}{\pi r^2}$$

On the boundary of the vortex core, the torque M_z and the power are, respectively,

$$M_z = \int_0^{2\pi} \tau_{r\theta} R d(R\theta) = -2\Gamma\mu$$

$$P_w = \frac{\mu\Gamma^2}{\pi R^2}$$

where ρ is the density of fluid and μ is the coefficient of hydrodynamic viscosity.

The solution obtained by this model is also the exact solution of N-S equation. For the two-dimensional flow field on the symmetrical plane, if the velocity field satisfies $u_\theta = f(r)$, $u_r = 0$, the above solution can be obtained by substituting the N-S equation system.

The Rankine vortex model provides a basis for understanding the formation mechanism of tornadoes, as shown in Figs. 1.26, 1.27, and 1.28. From the point of view of hydrodynamics, the tornado is a process of formation and

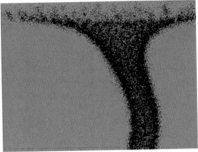

Fig. 1.26 Tornado

Fig. 1.27 Vortex

development of spatial concentrated vortices. It is not difficult to understand if it is the tail vortex of an airplane, because the root cause of vortex production is the motion of the airplane. But if there is no aircraft, it cannot be produced in ideal fluid flow and uniform flow field. However, when shear and convection occur in viscous fluids, there is no such conclusion. In fact, if wind shear occurs, there will be vortices. The question is whether these vortices can be concentrated, and if they can be concentrated, will they develop very strongly. Whether the distributed vorticity can be centralized depends on the angle between the direction of convective velocity and vorticity. If the direction of convective velocity is almost parallel to the vector direction of vorticity, it is possible for the vorticity flow to be combined by winding. If the velocity in the axis region is very high, the vorticity will become more and more concentrated, and eventually, the area of the vortex tube will become smaller. So did the cyclone. Tornadoes usually stand on top of the sky. Why? This is because in the horizontal convective wind shear zone (which can be either in the air or near the ground), there will be a region with small vorticity

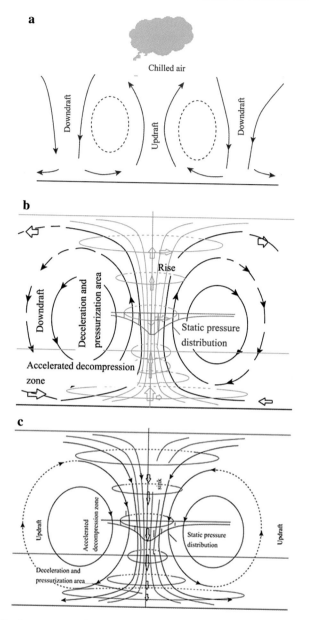

Fig. 1.28 **a** Surface atmospheric characteristics induced by thermal convection. **b** Tornado structure caused by updraft. **c** Subsidence convection-induced tornado structure

but a large distribution area. According to the Stokes formula,

$$\Gamma = \oint_L \vec{V} \cdot d\vec{s} = \iint_S 2\vec{\omega} \cdot d\vec{S} = \iint_S \nabla \times \vec{V} \cdot d\vec{S}$$

where Γ is the vorticity intensity (velocity circulation) passing through the contour L region; $\vec{V}$ is the velocity field; $\vec{\omega}$ is the rotating angular velocity of the fluid micro-cluster; $\nabla \times \vec{V} = 2\vec{\omega}$ is the vorticity of the fluid micro-cluster. Obviously, the vorticity integral value (vorticity intensity) of a large surface is very large. If you encounter strong vertical airflow (shown in Fig. 1.28a), such as strong updraft due to temperature difference (shown in Fig. 1.28b), or downdraft due to strong convection (shown in Fig. 1.28c), it will quickly wind up, with a smaller area and larger vorticity. It is possible to form a strong tornado.

From this point of view, the large-scale horizontal wind shear and vertical strong convection (both upward and downward) coupling will form a strong tornado.

1.4 Differential Equation of Viscous Fluid Motion and Vortex Transport Equation

Because there was no resistance in the flow around a cylinder with potential motion of ideal fluid, and people began to study the motion of viscous fluid. Based on Newton's law of internal friction (1687), the constitutive relationship between viscous stress and the deformation rate of the fluid microelement was established. On the basis of Euler's equation of ideal fluid motion in 1755, after the study by the French Engineer Claude-Louis Navier (1785–1836, as shown in Fig. 1.29) in 1822, the French scientist Simeon-Denis Poisson (1781–1840, as shown in Fig. 1.30) in 1829, and the French fluid mechanic Adhemar Jean Claude Barre DE Saint-Venant (1797–1886, as shown in Fig. 1.31) in 1843, finally in 1845, the British scientist George Gabriel Stokes (1819–1903, as shown in Fig. 1.32) proposed three relationships between stress and deformation rates at Trinity College, Cambridge University, and completed the Newtonian fluid viscous motion differential equation, namely the famous Navier–Stokes equation group, referred to as the N-S equation group. That is to say,

$$\frac{du}{dt} = \frac{\partial u}{\partial t} + u\frac{\partial u}{\partial x} + v\frac{\partial u}{\partial y} + w\frac{\partial u}{\partial z} = f_x - \frac{1}{\rho}\frac{\partial p}{\partial x} + v\Delta u$$

Fig. 1.29 Claude-Louis Navier (1785~1836, French mechanist)

Fig. 1.30 Simeon-denis Poisson (1781~1840, French scientist)

$$\frac{dv}{dt} = \frac{\partial v}{\partial t} + u\frac{\partial v}{\partial x} + v\frac{\partial v}{\partial y} + w\frac{\partial v}{\partial z} = f_y - \frac{1}{\rho}\frac{\partial p}{\partial y} + v\Delta v$$

$$\frac{dw}{dt} = \frac{\partial w}{\partial t} + u\frac{\partial w}{\partial x} + v\frac{\partial w}{\partial y} + w\frac{\partial w}{\partial z} = f_z - \frac{1}{\rho}\frac{\partial p}{\partial z} + v\Delta w$$

where u, v, and w are the velocity components of the particle, respectively; f_x, f_y, and f_z are the unit mass forces acting on the particle; p is the pressure acting on the particle; v is the viscous coefficient of fluid motion; and Δ is

Fig. 1.31 Adhemar Jean Claude Barre deSaint-Senant (1797~1886, French mechanist)

Fig. 1.32 George Gabriel Stokes (1819–1903, British mechanics and mathematician)

the Laplace operator. It can be written in vector form as

$$\frac{\mathrm{d}\vec{V}}{\mathrm{d}t} = \vec{f} - \frac{1}{\rho}\nabla p + v\Delta\vec{V}$$

This system of equations shows that the mass force, pressure difference force (surface normal force), and viscous force (surface tangential force) acting on the fluid microelement cause the acceleration of the fluid microelement, which is reflected in the viscous diffusion behavior of momentum in the motion equation. Note that there is no viscous dissipation here, and viscous dissipation can only occur in the energy equation. Comparing the Boltzmann equation with N-S equation, it is found that there is a certain relationship between them. In fact, N-S equation system is the hydrodynamic limit of the Boltzmann equation. So far, from 1755 to 1845, the Euler equations of ideal fluid motion were derived and the N-S equations of viscous fluid motion were derived. Through 90 years, mathematicians had made outstanding contributions to the establishment and derivation of the main equations of fluid mechanics. Thereafter, fluid mechanics began to enter the stage of solving and applying many flow problems.

For the steady flow of viscous fluid with only gravity and incompressible mass force, the Bernoulli equation similar to the ideal fluid can be obtained by integrating the N-S equations along the streamline, but there is an additional mechanical energy term in the energy equation which is lost by overcoming the viscous frictional force. That is to say,

$$z_1 + \frac{p_1}{\gamma} + \frac{V_1^2}{2g} = z_2 + \frac{p_2}{\gamma} + \frac{V_2^2}{2g} + \Delta h_{f1-2}$$

$$\Delta h_{f1-2} = \int_1^2 \frac{v}{g}[-\Delta u dx - \Delta v dy - \Delta w dz]$$

Compared with the Bernoulli equation of ideal fluid, the additional term on the right side of the formula above represents the mechanical energy consumed by a fluid particle per unit weight in overcoming viscous stress. This term can no longer be used by the mechanical motion of a fluid particle. Therefore, it is called the mechanical energy loss of a fluid particle per unit weight. This loss is related to the integral path (the shape of the streamline). The results show that in viscous fluids, the mechanical energy of fluid particles per unit weight per unit time along the same streamline always decreases along the flow direction (as shown in Fig. 1.33), and it is impossible to maintain conservation (in ideal fluids, the total mechanical energy is conserved without mechanical energy loss), and the fluid always flows from the place where the mechanical energy is large to the place where the mechanical energy is small.

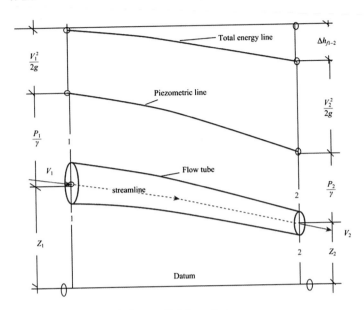

Fig. 1.33 Energy equation of viscous fluid motion

The familiar differential equation of vorticity transport (similar to the Helmholtz vorticity equation of ideal incompressible fluid) can be obtained by taking the curl of the N-S equation of motion of incompressible viscous fluid under the condition of potential mass force.

$$\frac{d\vec{\Omega}}{dt} = \left(\vec{\Omega} \cdot \nabla\right)\vec{V} + v\Delta\vec{\Omega}$$

where $\vec{\Omega} = \nabla \times \vec{V}$ is the vorticity of the flow field, and the vortex core region is as shown in Fig. 1.34. The left side of the equation represents the volume-dependent derivative of vorticity (or vorticity transport rate), the first term on the right represents the stretching and bending deformation of the vortex tube caused by the flow field heterogeneity, and the second term on the right represents the viscous diffusion of the vortex tube. If the viscous coefficient of fluid is zero, the Helmholtz vorticity equation of ideal incompressible fluid under the action of potential force can be obtained.

$$\frac{d\vec{\Omega}}{dt} = \left(\vec{\Omega} \cdot \nabla\right)\vec{V}$$

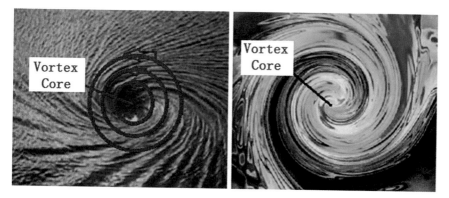

Fig. 1.34   Vortex core

Now, we will further discuss the physical significance of the above equations. For example, the terms contained in the equation

$$\frac{\partial \Omega_x}{\partial t} = \Omega_x \frac{\partial u}{\partial x} + \cdots$$

represents the change rate of the vorticity as a result of the axial tension deformation of the vortex tube (the axis of the vortex tube is stretched, $\frac{\partial u}{\partial x} > 0$, which increases the vorticity and decreases the cross section). Items contained in the equation

$$\frac{\partial \Omega_x}{\partial t} = \Omega_y \frac{\partial u}{\partial y} + \cdots$$

represent the change rate of the vorticity due to the shear action of the vortex tube. Items contained in the equation

$$\frac{\partial \Omega_x}{\partial t} = v \frac{\partial^2 \Omega_x}{\partial y^2} + \cdots$$

are a viscous diffusion term of vorticity.

1.5 Establishment and Application of Boundary Layer Theory

In the twentieth century, the machinery industry reached its peak and entered an era of all-round development and perfection, which undoubtedly promoted the comprehensive and rapid development of mechanics, and formed multi-disciplinary and multi-field research results, showing their own unique contents and directions in theory, experiment, and application. During this period, fluid mechanics was naturally divided into three branches: theoretical fluid mechanics, experimental fluid mechanics, and computational fluid mechanics. According to the study medium, it can be divided into hydrodynamics or aerodynamics. Under the guidance of basic theory, the complex flow problems related to viscous flow (such as laminar flow, turbulence, transition, jet, separation flow, wake, etc.) are mainly studied, and the problems of resistance and heat exchange of objects around flow are solved. In theory, since N-S equation was derived in 1845, people have been searching for its exact solution. However, because the system of equations is a non-linear system of second-order partial differential equations, the exact solution in a general sense is mathematically difficult. It is said that only 73 exact solutions of N-S have been found up to now, famous examples include the Couette flow (Couetteis French physicist at the end of the nineteenth century) produced by the dragging a flat plate under no pressure, Poiseuille flows (the fully developed laminar flow), produced by Poiseuilleis French physiologist (1799–1869, as shown in Fig. 1.35), the Stokes (1851) solution of the flow around a small Reynolds number sphere, etc. A lot of problems in practice can only be solved by the approximate method.

Since the French scientist D'Alembert put forward the D'Alembert paradox of steady flow of ideal fluid around an arbitrary three-dimensional object in 1752, people began to doubt the classical theory based on the ideal fluid model. By the first half of the nineteenth century, the study of ideal potential flow theory had gradually entered a perfect stage, and the classical hydrodynamics research was in a low ebb. Especially, the conclusion that there was no resistance to the flow around a cylinder was obtained using this model, which made people unable to do anything. Naturally, the N-S equation representing viscous fluid should be used to solve this problem. However, a difficult problem is how to deal with the effect of viscous flow around an object at the large Reynolds number. According to the accepted facts at that time, if the Reynolds number of incoming flow calculated by velocity and diameter of the cylinder is greater than 10^4, the influence of viscous effect

Fig. 1.35 Jean-Louis-Marie Poiseuille (1799–1869, French physiologist)

can be neglected, so we can return to the old proposition of flow around the ideal fluid. If we do not neglect the effect of viscosity, how to understand the concept of large Reynolds number? Besides, it was impossible to solve all N-S equations more accurately at that time. This problem had not been solved convincingly until 1904, when Ludwig Prandtl (1875–1953, as shown in Fig. 1.36), the world Master of fluid mechanics, put forward the famous boundary layer theory. It has been 152 years since the D'Alembert Question in 1752. It has been 59 years since N-S equations were derived in 1845. Now it seems to be a simple problem, that is, the relationship between global flow and local flow, which belongs to the problem of the size of the viscous region affected by the near wall, but at that time it was a big problem in the field of fluid mechanics. In 1904, Prandtl published a paper on the motion of small viscous fluids at the Third Annual Conference of International Mathematics in Heidelberg, Germany. He proposed the well-known concept of boundary layer (as shown in Figs. 1.37 and 1.38). The characteristics and governing equations of boundary layer flow with viscous effect on the surface of a body around a large Reynolds number are described in depth. The relationship between global flow and local flow is solved skillfully. That is to say, the incoming Reynolds number calculated by velocity and diameter of a cylinder can only characterize the overall flow characteristics, but cannot characterize the local flow behavior near the wall of the body around a flow (boundary layer flow). The Reynolds number of incoming flow can only

Fig. 1.36 Ludwig Prandtl (1875–1953, German mechanic and world Master of fluid mechanics)

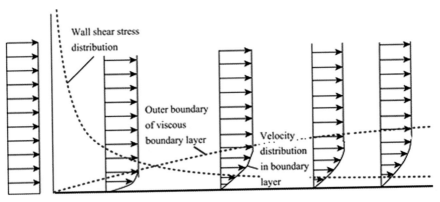

Fig. 1.37 Laminar boundary layer at zero pressure gradient

control the viscous effect on the flow outside the boundary layer, while the viscous effect in the boundary layer is determined by the flow characteristics in the boundary layer. On this basis, the separation and control of boundary layer (as shown in Figs. 1.39, 1.40, and 1.41) are proposed, and the matching relationship between viscous flow near the wall and inviscid outflow far from

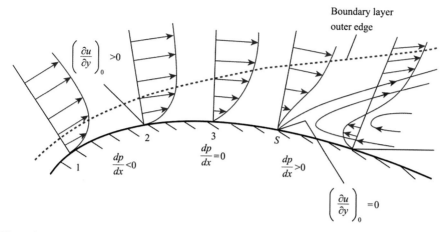

Fig. 1.38 Boundary layer development under variable pressure gradient

Fig. 1.39 Boundary layer separation around circular cylinder flow

the wall is found, thus finding a new way to solve the problem of viscous flow and playing an epoch-making milestone role.

Prandtl found through a large number of experiments that although the Re number of the whole flow is very large, the characteristics of the flow field are far from the ideal flow in the thin layer fluid near the surface, and there is a large velocity gradient along the normal direction, so the viscous force cannot be ignored. Prandtl calls the thin layer near the surface where the viscous force plays an important role in the boundary layer. The introduction of the concept of boundary layer opens a new way for people to take into

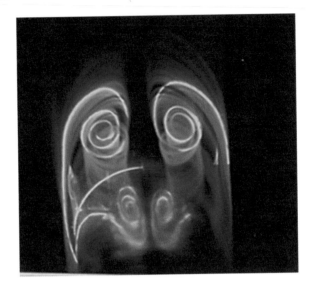

Fig. 1.40   Oblique cylindrical back vortex structure

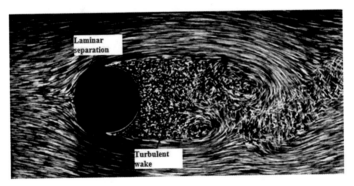

Fig. 1.41   Laminar boundary layer separation around a cylinder (cited from An Album of Fluid Motion, Re = 2000 for cylinder diameter calculation)

account the viscous effect. The basic zoning for the whole flow field is as follows:

(1) The whole flow area can be divided into an ideal fluid flow area (potential flow area) and a viscous fluid flow area (viscous flow area).
(2) In the ideal fluid flow region far away from the object, the effect of viscous flow can be neglected and treated according to potential flow theory.
(3) The viscous flow region is confined to the thin layer near the surface of the object, which is called the boundary layer. In this region, the

viscous stress cannot be neglected. It is the same order of magnitude as the inertial force, and the fluid particles are rotating. The thickness of the boundary layer can be estimated based on the assumption that the viscous force and inertial force in the boundary layer are of the same order of magnitude. Take the flow around a flat plate as an example. Let the velocity of the incoming flow be U, the length in the x-direction be L, and the thickness of the boundary layer be δ. In the boundary layer, the inertia force of the fluid micro mass is zero.

$$F_J = m\frac{du}{dt} \propto \rho L^2 \delta \frac{U}{T} = \rho L^2 \delta \frac{U}{L/U} = \rho L U^2 \delta$$

The viscous force of the fluid microelement is

$$F_\mu = \rho \nu A \frac{du}{dy} \propto \rho L^2 \nu \frac{U}{\delta} = \rho L^2 \nu \frac{U}{\delta}$$

Based on the assumption that the inertial force and the viscous force are of the same order of magnitude,

$$F_J \approx F_\mu, \rho L \delta U^2 \approx \rho L^2 \nu \frac{U}{\delta}$$

$$\frac{\delta}{L} \approx \frac{1}{\sqrt{Re_L}}, \quad Re_L = \frac{UL}{\nu}$$

It is shown that the ratio of the thickness of the boundary layer to the length of the plate L is inversely proportional to the square of the total Re_L number calculated by the previous flow velocity and the length of the plate. If the inflow velocity $U = 14.6$ m/s, the plate length $L = 1.0$ m, the air moving viscous coefficient $\nu = 1.46 \times 10^{-5}$ m²/s, $Re_L = 10^6$, the boundary layer thickness is in the order of millimeter, which is equivalent to 1/1000 of the plate length. Theoretical Solution $\delta \approx 5.0$ mm for Laminar Boundary Layer of Plate.

Accordingly, the two-dimensional laminar boundary layer governing equations derived by Prandtl (a simplified form of N-S equations) are as follows:

$$\frac{\partial u}{\partial x} + \frac{\partial v}{\partial y} = 0$$

$$\frac{\partial u}{\partial t} + u\frac{\partial u}{\partial x} + v\frac{\partial u}{\partial y} = f_x - \frac{1}{\rho}\frac{\partial p}{\partial x} + \nu\frac{\partial^2 u}{\partial y^2}$$

$$f_y - \frac{1}{\rho}\frac{\partial p}{\partial y} = 0$$

Compared with the original equations, the above equations are simplified in form, but their types have changed. The original equations are elliptic ones, but now they are parabolic ones. This system of equations seems simple, but still belongs to a non-linear partial differential equation system. Without other assumptions, the difficulty of solving the system is not much lower than that of the original system. For this reason, Prandtl introduced the second hypothesis, i.e., the similarity hypothesis of longitudinal velocity distribution, which can convert the solution of partial differential equations into the solution of ordinary differential equations, and can easily obtain the approximate solution of velocity distribution in the boundary layer. In addition, if the mass force is neglected $f_y = 0$, the pressure in the boundary layer remains unchanged along the normal direction from the third equation of the boundary layer equation system, and its value is equal to the pressure outside the boundary layer.

In 1908, Blasius, the German hydro mechanist, gave the boundary layer series solution of an unconfined gradient plate. In 1921, the American scientist Theodore von Karrman (1881–1963, as shown in Fig. 1.42) derived the momentum integral equation of the boundary layer. In 1921, Pohlhausen, the German scientist, established an approximate solution method based on the momentum integral equation. The influence of pressure gradient on the

Fig. 1.42 Theodore von Carmen (1881–1963, American scientist)

boundary layer was studied. In 1938, the British scientist Howard studied the problem of flow around the right angle and so on.

During this period, various viscous laminar boundary layer problems were approximated with the help of similarity assumption. For example, in 1929, the German scholar Tollmien obtained the Gaussian solution of the laminar wake of a flat plat;, in 1933, the German scholar Schlichting solved the similar solution of a circular laminar jet; and in 1945, the German scholar Mangler introduced a similar transformation to transform the axisymmetric laminar boundary layer problem into a plane boundary layer problem, and solved the flow around a cone.

The viscous flow and the similarity condition of a thin layer on the wall are the important basis for establishing the boundary layer theory of objects with high Reynolds number.

1.6 Laminar Flow Transition Phenomenon and Stability Theory

However, there are many flows in nature, which are not laminar flow, but turbulent flow, which is more complex and more urgent for practical application. Therefore, the study of turbulence has begun to attract great attention. There are two branches of research, one is the transition problem of layer loss stability and the other is the fully developed turbulence problem. For the first transition problem, as early as 1839, Hagen, the German scholar, found that the flow characteristics in a circular pipe are related to the velocity. In 1869, two different flow regimes were found to have different flow characteristics. In 1880, the British scholar Osborne Reynolds (1842–1912, as shown in Fig. 1.43) carried out a well-known transition experiment in a circular tube (as shown in Figs. 1.44 and 1.45). In 1883, the concept of laminar flow and turbulence was proposed, and a dimensionless number (hereinafter called Reynolds number) was used as a criterion to give the Reynolds number 2000 (now 2320). In 1872, William Froude (1810–1879, as shown in Fig. 1.46), the British scholar, observed that the resistance of a flat plate was proportional to the power of 1.85 of the velocity, rather than the power of the laminar flow. In 1914, Prandtl put forward the concept of a turbulent boundary layer when he studied the spherical drag. In 1924, the Dutch scholar Burgers studied the transition of boundary layer. In 1934, the American scholar Dryden gave the critical Reynolds number (calculated by the thickness of boundary layer) for the transition of the flat plate boundary layer, which was 2740. In 1946, he raised this number to 8700.

Fig. 1.43 Osborne Reynolds (1842–1912, British physicist)

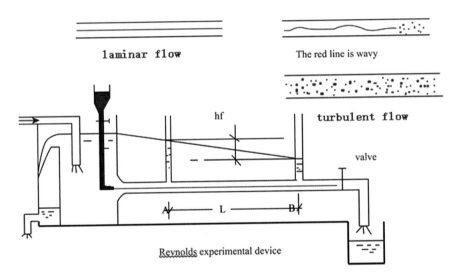

Fig. 1.44 Reynolds transition test device

In future studies, more attention has been paid to the development of disturbances in laminar flow, i.e., laminar flow stability. In 1880, Rayleigh (1842–1919, as shown in Fig. 1.47), the British scholar, studied the microwave perturbation problem with the inviscid effect. In 1907, Orr (German scholar) and Sommerfeld (German scholar) studied the evolution of the amplitude of perturbation wave motion with time, respectively, and proposed the famous perturbation stability equation, namely the Orr–Sommerfeld equation. Meanwhile, in 1897, Lorentz, the Dutch physicist (1853–1928, as shown in Fig. 1.48), put forward the equation of perturba-

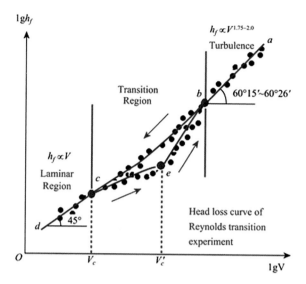

Fig. 1.45 Reynolds transition experimental results

Fig. 1.46 William Froude (1810–1879, British fluid dynamician)

tion energy and studied the evolution process of perturbation energy with time. Tollmien in 1935 and Lin Jiaqiao in 1945, the Chinese American mechanist, gave the critical Reynolds number of damped perturbations for Poiseuille flow between plates. However, the perturbation method is unsuccessful in studying the stability of Poiseuille flow in a circular tube. The stability theory does not give the physical mechanism of turbulent transition.

Fig. 1.47 Rayleigh (1842–1919, British physicist)

Fig. 1.48 Lorentz (1853–1928, Dutch physicist)

In 1960s, Kline, the American scholar, studied the transition phenomena in the boundary layer of a flat plate using hydrogen bubble technology (as shown in Fig. 1.49). It was found that the instability of the boundary layer begins with the two-dimensional T/S (Tollmien–Schlichting) wave instability, and then three-dimensional horseshoe vortices, such as stretching and deformation, fragmentation, ejection, and sweep, appear in turn (as shown in Fig. 1.50). This constitutes the basis of stability theory.

Layer loss is a stable turbulence, and one of the most obvious characteristics is the randomness of turbulence. It has been found that the stochastic characteristics of turbulence come not only from various disturbances and excitations of external boundary conditions, but also from the internal non-linear mechanism. The discovery of chaos has greatly impacted the deterministic theory. The deterministic system of equations is not like

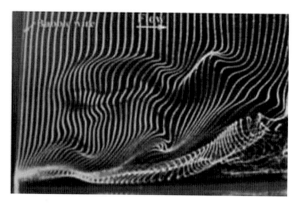

Fig. 1.49 Hydrogen bubbles show wall turbulence bursts

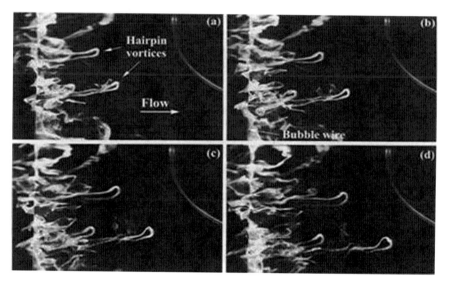

Fig. 1.50 Hairpin vortex structure displayed by hydrogen bubble flow

Laplace, a famous scientist said. It can decide everything in the future by giving the conditions of definite solutions, but the deterministic system can produce uncertain results. Chaos makes deterministic theory and stochastic theory link organically, which makes us more convinced that deterministic Navier–Stokes equations can be used to describe turbulence (that is, a dissipative system is affected by non-linear inertia force, under certain conditions, multiple non-linear bifurcations may occur and eventually become chaotic structures). A word is used to describe the transition process of laminar flow.

Who can block the splitting and breaking up?

Don't forget the way back in troubled times.

A sudden splitting of the great eddy.

Order and disorder always go hand in hand.

   Note: Who can block the splitting and breaking up? It means that when Re reaches a certain value, the laminar flow will turn into turbulence, which is determined by internal instability and cannot be blocked. Don't forget the way back in troubled times—it means that average motion after turbulence is distinguishable (controllable). A sudden splitting of the large eddy means that large eddies suddenly break up into small eddies. Order and disorder always go hand in hand—indicates the coherent structure of large eddies in a turbulent flow field.

1.7 Turbulence Phenomenon and Its Characteristics

As far as turbulence is concerned, Da Vinci, the Renaissance Italian all-round scientist, was the first to observe in detail. He observed the eddies and turbulences qualitatively on the beach and recorded the flow structure of the eddies and turbulences in his notes. In a famous turbulence painting, he wrote that the clouds were torn apart by the strong wind, the sand particles rose from the beach, and the trees stooped down. The splitting and fragmentation of turbulence, entrainment of turbulent eddies, and wall shear are clearly depicted. Since Reynolds' transition experiment in 1880, Reynolds put forward the concept of time mean in 1883, which considered that the instantaneous motion of turbulence consisted of time mean motion and pulsating motion, but Reynolds called turbulence tortuous motion at that time. In 1895, Reynolds put forward the Reynolds equations describing the time-averaged motion of N-S equation by assuming that the instantaneous motion of turbulence satisfies Navier–Stokes equations. From then on, turbulence research began to embark on the road of closed turbulence equation (In fact, whether the instantaneous physical quantity of motion satisfies the N-S equations has been debated at the beginning. The main concern is whether Newton's law of internal friction, which characterizes the constitutive relationship between stress and deformation rate in fluid motion, is applicable to instantaneous turbulence. In addition, N-S equations require that physical quantities be continuous differentiable functions, but in fact, the

Fig. 1.51 G. I. Taylor (1886–1975, British mechanist)

instantaneous physical quantities cannot be continuous differentiable from the measurement results, at most a continuous function).

In 1937, Taylor (G. I. Taylor, 1886–1975, as shown in Fig. 1.51) and Von Karman considered turbulence to be an irregular motion, which generally occurs in fluids when fluids flow through solid surfaces or adjacent similar fluids. In 1959, J. O. Hinze, the Dutch scholar, believed that turbulence was an irregular flow state, but its various physical quantities varied randomly with time and spatial coordinates, so that different statistical averages could be distinguished. Zhou Peiyuan, the Chinese scholar, believes that turbulence is an irregular eddy motion. The general textbook defines turbulence as an irregular random motion that is disorderly and mixed with each other. At present, it is generally accepted that turbulence is composed of vortices of different sizes and frequencies, which make the changes in physical quantities in time and space show irregular randomness. In the study of turbulence, the engineering turbulence method represented by L. Prandtl and the turbulence statistical theory represented by G. I. Taylor have been formed. With the improvement of computational techniques in recent decades, the numerical study of turbulence has been developed rapidly.

Compared with the definition of turbulence, the basic characteristics of turbulence can be easily expressed as follows:

(1) Eddy in turbulence

Turbulence is accompanied by large and small vortices. Vortex is the main cause of turbulence fluctuation. It is generally believed that large eddies produce large fluctuations and small eddies produce small fluctuations during

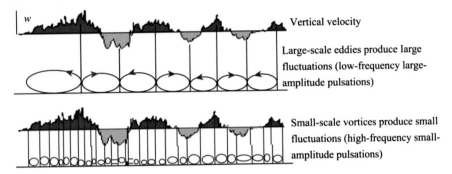

Fig. 1.52 Turbulent velocity fluctuation and vortex structure

a change in physical quantities. If small eddies exist in large eddies, small fluctuations will occur in large fluctuations (as shown in Fig. 1.52). Because the direction of velocity around these vortices is relative (opposite), large shear stress will be produced.

(2) Irregularity of turbulence

The motion of fluid particles in turbulence is random and disorderly. However, because turbulent flow contains vortices of different sizes, there is no characteristic scale in theory, so this random walk must be accompanied by transitions of various scales.

(3) Random Behavior of Turbulence

The physical quantities of particles in turbulent flow field are random variables of time and space, and their statistical average values obey certain regularity. In recent years, with the advent of fractal, chaotic science and the rapid development of non-linear mechanics, people have a new understanding of this randomness.

(4) Diffusion of turbulence

Due to the fluctuation and mixing of fluid particles, the diffusion of momentum, energy, heat, mass, and concentration in turbulence is greatly increased, which is obviously larger than that in laminar flow.

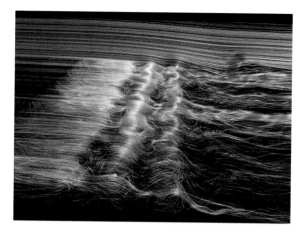

Fig. 1.53 Quasi-ordered structure of flow around the back step

(5) Dissipation of turbulent energy

Small-scale vortices in turbulence will dissipate large turbulent energy by shearing, which is caused by viscous flow. This is because the dissipation caused by small-scale vortices (shown in Fig. 1.53) is much larger than that caused by laminar viscous friction.

(6) Coherent Structure of Turbulence

Pulsations in turbulence are not completely irregular random motions, but still have detectable ordered motions in seemingly irregular motions, which play a dominant role in the generation and development of shear turbulent pulsations (as shown in Figs. 1.54 and 1.55). For example, the discovery of coherent structures in free shear turbulence (turbulent mixing layer, far-field turbulent jet, turbulent wake, etc.) clearly depicts the mixing and entrainment of coherent large-scale eddies in these turbulences (as shown in Fig. 1.56). The discovery of band structure in wall shear turbulence reveals the mechanism of turbulence generation near the wall (as shown in Figs. 1.57 and 1.58).

Fig. 1.54 Turbulent large eddy structure

Fig. 1.55 Turbulent large eddy structure

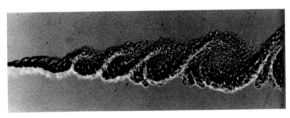

Fig. 1.56 Large-scale coherent structure (cited from An Album of Fluid Motion)

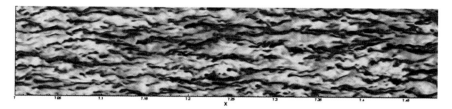

Fig. 1.57 Cloud map of the equivalent surface distribution of turbulent velocity near the wall

(7) Intermittency of turbulence

It was first found that the intermittency of turbulence occurs in the turbulent and non-turbulent boundary areas, such as the outer region of the turbulent boundary layer and the entrainment region of the turbulent jet, where turbulence and non-turbulence occur alternately. However, recent studies have shown that even in the interior of turbulence, it is intermittent because the energy of large eddies will eventually cascade to the small eddies with viscous dominant role in the process of splitting and fragmentation of turbulent vortices, which occupy only a small area in the space field. So the intermittence of turbulence is universal and strange.

(8) The concept of time-averaged decomposition of turbulence

Considering the randomness of turbulence, Reynolds first decomposed the instantaneous motion into the sum of time-averaged motion (describing the average trend of flow) and fluctuating motion (the degree of deviation from time-averaged motion) in 1895. Later, people put forward the methods of spatial decomposition and statistical decomposition.

Time decomposition (Reynolds concept of time mean)

If the turbulent motion is a stationary random process, the instantaneous velocity u at any point in the turbulent flow field can be decomposed into time-averaged velocity + fluctuating velocity. For non-stationary stochastic processes, the time-averaged decomposition method cannot be used strictly, but if the characteristic time of the time-averaged motion is much longer than that of the pulsating motion, and when the mean time T is much less than that of the time-averaged motion and much larger than that of the pulsating motion, the time-averaged decomposition still holds approximately.

Spatial Decomposition Method (Spatial Average Method)

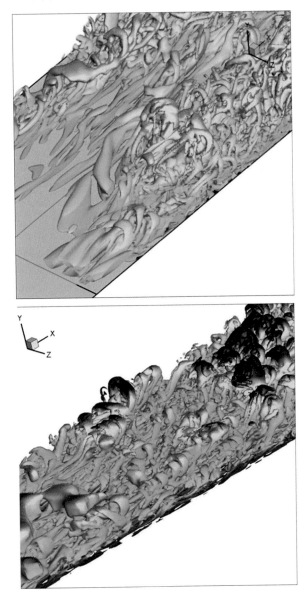

Fig. 1.58 Quasi-ordered structure of the near-wall region

If the turbulent flow field is a random field with spatial homogeneity, the instantaneous motion of turbulence can be decomposed into spatial mean motion + pulsating motion by using the spatial averaging method.

Ensemble averaging (decomposition in probability sense)

If the turbulent motion is neither stationary in time nor uniform in space, we can decompose the instantaneous motion of turbulence into statistical mean motion and pulsating motion in probability sense.

Although the above three decomposition methods are proposed for different turbulent flow fields, they are statistically equivalent under certain conditions. According to the Ergodic Theorem of probability theory, all possible values of a random variable appearing in repeated experiments will occur many times in a long time (or in a considerable space), and the probability of appearing is the same. Therefore, for a time-stable and space-uniform turbulent flow field, the average values of each physical quantity obtained by the above three decomposition methods are equal.

1.8   Statistical Theory of Turbulence

In the statistical theory of turbulence, the British meteorologist Richardson (L. F. Richardson, 1881–1953, as shown in Fig. 1.59) put forward the energy cascade theory of turbulence in 1922. Large-scale eddies obtain energy from basic (time-averaged or average) flows through turbulent shear, and then through viscous dissipation and dispersion (instability) processes, these large vortices are cascaded into small eddies of different scales (as shown in Fig. 1.60), and the energy is transferred to small-scale vortices step by step

Fig. 1.59 L.F. Richardson (1881–1953, British meteorologist)

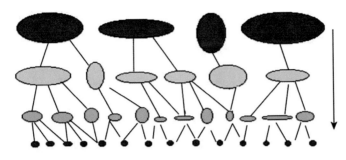

Large-scale vortex
turbulence energy
input area

Energy transfer
region of medium
scale eddy turbulence

Energy dissipation
region of small scale
eddy turbulence

Fig. 1.60 Cascade Viewpoint of Turbulent Vortex

during the splitting and breaking process until viscous dissipation is achieved. It is described in a famous poem.

Big whirls have little whirls,

Which feed on their velocity.

Little whirls have smaller whirls,

And so on to viscosity.

In 1935, the British scientist Taylor (1886–1975, as shown in Fig. 1.51) put forward the theory of homogeneous isotropic turbulence, gave a series of important concepts, established a one-dimensional energy spectrum relationship, and proposed the hypothesis of frozen turbulence. In 1938, based on the two-point velocity correlation function, the American scientists Carmen (shown in Fig. 1.42) and Howarth derived the dynamic equation of the isotropic turbulent structure function, the famous K-H equation. In 1953, the British scientist G. Batchelor (1920–2000, as shown in Fig. 1.61) further studied the theory of homogeneous isotropic turbulence. In 1941, Kolmogorov (1903–1987, as shown in Fig. 1.62), the Russian statistician, put forward the theory of locally homogeneous isotropy and derived the −5/3 law of the spectral density distribution of turbulent structure functions (as shown in Figs. 1.63, 1.64, 1.65, and 1.66). In 1949, Batchelor and Townsend discovered the intermittent nature of turbulence. In 1967, the American scientist Kline proposed the coherent structure of turbulence. In 1991, Robinson plotted the burst pattern of the turbulent boundary layer.

For homogeneous isotropic turbulence (as shown in Figs. 1.67, 1.68, 1.69, and 1.70), this is an ideal model for small-scale turbulence-induced flow field proposed by Taylor. But the shape of these small-scale turbulent eddies is still unknown. How do they exist in turbulent flow field? Are their vortices the

Fig. 1.61 George Keith Batchelor (1920–2000, British mechanist)

Fig. 1.62 Kolmogorov (1903–1987, Russian Master of Statistics)

end of turbulence? If there are small-scale eddies, does homogeneous isotropic turbulence mean that the flow field induced by these small-scale vortices is homogeneous isotropic? But it is not necessarily the turbulent eddies themselves, because the real visible vortices are difficult to be isotropic. If there are no small-scale vortices, there are two kinds of turbulence: eddy turbulence (turbulence) and non-eddy turbulence (turbulence)? What role does homogeneous isotropic turbulence play in turbulence research? Is it dissipative? What is the mechanism? What are the differences and similarities with the dissipation mechanism of fluid viscosity? These problems need further clarification

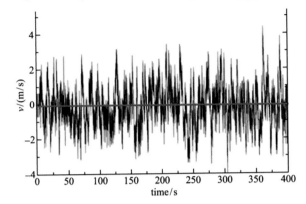

Fig. 1.63 Turbulent fluctuating velocity processes with normal probability distribution

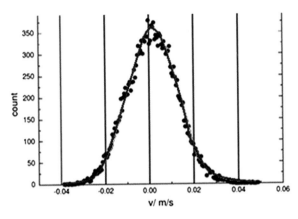

Fig. 1.64 Normal probability distribution of turbulent fluctuating velocity

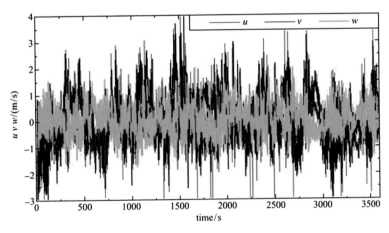

Fig. 1.65 Uniform and isotropic turbulence (fluctuating velocity components along three coordinate axes)

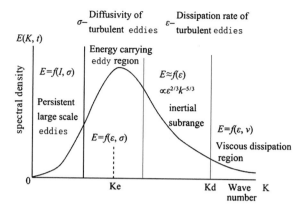

Fig. 1.66 Three-dimensional spectral density distribution of isotropic turbulence

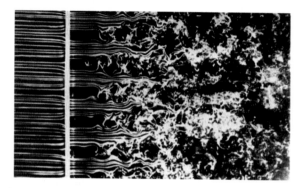

Fig. 1.67 Turbulence structure after grid (cited from An Album of Fluid Motion)

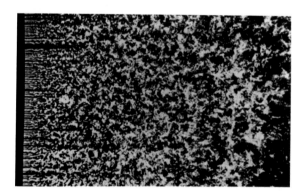

Fig. 1.68 Homogeneous isotropic turbulence after grid (cited from An Album of Fluid Motion)

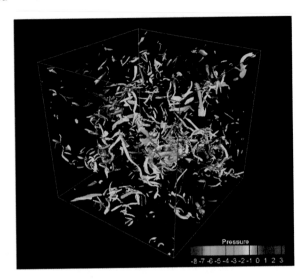

Fig. 1.69 Small-scale turbulent vortex structure (ltcs.pku.edu.cn)

Fig. 1.70 Uniform turbulence structure in the atmosphere

from the cognitive level. For turbulent eddies, it gives people the feeling that, like Confucius's treatment of Lao Tzu's conversation with him, there is a kind of magic dragon that cannot see the end (visible big eddies, invisible small eddies). Turbulence in turbulence is also an abstract concept. Like "benevolence" in Confucianism and "Tao" in Taoism, it often gives people the feeling that it is both real and vague, and sometimes it can only be understood and cannot be said.

1.9 Engineering Turbulence Theory

For practical engineering turbulence theory, Prandtl is the representative of semi-empirical theory based on phenomenology and turbulence model theory developed later, which plays an important role in solving turbulence engineering calculation. Based on the phenomenological principle, in 1877, Joseph Valentin Boussinesq (1842–1929, as shown in Fig. 1.71) first compared the additional shear stress (later known as the Reynolds stress) caused by turbulent fluctuation with the viscous stress, put forward the famous eddy viscous hypothesis, and established the analogy relationship between the Reynolds stress and time-averaged velocity gradient. Although the concept of eddy viscosity appeared earlier than the Reynolds equations, it laid the foundation for later engineering turbulence. For simple near-wall time-averaged binary flows (as shown in Figs. 1.72 and 1.73), Turbulent shear stress (Reynolds stress) can be expressed as follows:

$$\tau_t = -\rho \overline{u'v'} = \rho v_t \frac{\partial u}{\partial y}$$

Fig. 1.71 Joseph Valentin Boussinesq (1842–1929, French scientist)

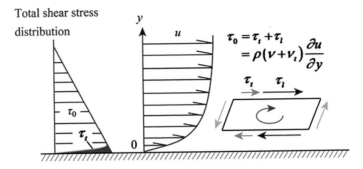

Fig. 1.72 Near-wall shear turbulence

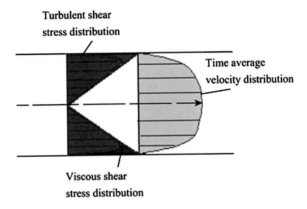

Fig. 1.73 Pipeline velocity and stress distribution

In the formula, v_t is the Turbulent or Eddy Viscosity. In contrast, the viscous shear stress produced by time-averaged flow is

$$\tau_l = \rho v \frac{\partial u}{\partial y}$$

The total shear stress acting on the stratosphere is

$$\tau_0 = \tau_t + \tau_l = \rho(v + v_t)\frac{\partial u}{\partial y}$$

Compared with the molecular viscous coefficient v, the eddy viscous coefficient v_t is not a physical property of the fluid, but a function of the turbulent motion state. In this way, the closure of the turbulence problem can be attributed to how to determine the size and distribution of v_t. At first, Boussinesq thought that v_t was a constant. Later, it was found that v_t not only had different values for different flow problems, but also had different values for the same flow problem in different regions at different times. According to turbulent motion characteristics, v_t could change obviously in the flow field. According to the results of dimensional analysis and turbulence research, the characteristic length scale and characteristic velocity scale of eddy viscous v_t with loaded energy are determined, namely

$$v_t \propto l_t V_t$$

l_t—Energy – carrying Vortex Length Scale

V_t—Velocity Scale of Energy – carrying Vortex
    Ratio of turbulent stress to viscous stress

$$\frac{\tau_t}{\tau_l} = \frac{-\rho \overline{u'v'}}{\rho v \frac{\partial u}{\partial y}} = \frac{\rho v_t \frac{\partial u}{\partial y}}{\rho v \frac{\partial u}{\partial y}} = \frac{v_t}{v} = \frac{l_t V_t}{v} = \mathrm{Re}_t$$

In the formula, Re_t represents the Reynolds number of large-scale turbulent eddy motion characteristics, generally $\mathrm{Re}_t = 10^3 \sim 10^5$.

In 1925, based on the analogy of molecular motion theory, Prandtl proposed the theory of mixed length. On the basis of the experimental results of resistance in 1932 by Nikuradse, the German scholar, the problem of time-averaged velocity distribution and resistance loss in pipe turbulence was solved, and the well-known formula of logarithmic velocity distribution was derived. The semi-empirical and semi-theoretical solution of resistance coefficient along the resistance loss formula proposed by Darcy, the French engineer, and Weisbach, the German scholar, in 1858 was given.

According to Prandtl's mixing length theory, for shear turbulence, Prandtl holds that the characteristic velocity V_t of turbulent eddies is proportional to the product of the time-averaged velocity gradient and mixing length (the average scale of free mixing of fluid particles under the action of turbulent vortices, which is the same order of magnitude as the average scale of turbulent vortices), i.e.,

$$V_t \propto l_m \left| \frac{\partial u}{\partial y} \right|$$

Using the above formula and absorbing the proportional coefficient into the mixing length l_m, we can get the result.

$$\tau_t = -\overline{\rho u'v'} = \rho l_m^2 \frac{\partial u}{\partial y}\left|\frac{\partial u}{\partial y}\right|, \quad v_t = l_m^2 \left|\frac{\partial u}{\partial y}\right|$$

In near-wall turbulence (as shown in Fig. 1.74), the fluctuating velocity near the wall is very small and the turbulent shear stress is very small, but the velocity gradient is very large, the viscous shear stress plays a leading role, and the velocity distribution is linear. This layer is called the viscous bottom layer. In the outer region of the viscous bottom layer, the turbulent shear stress plays a dominant role, and the velocity distribution conforms to logarithmic or power distribution. The transition zone is between the turbulent core region and the viscous bottom region. The viscous bottom is neither laminar nor turbulent. There are turbulent spots in this layer. The viscous bottom thickness and wall roughness directly affect the loss along the course. In the near-wall turbulent region, assuming that the turbulent shear stress is approximately equal to the wall shear stress τ_w and the mixing length is proportional to the distance y from the particle to the wall, i.e., $l_m = ky$ (k is Carmen constant ≈ 0.4), the following results are obtained

$$\frac{\tau_w}{\rho} = k^2 y^2 \left(\frac{du}{dy}\right)^2$$

The famous logarithmic distribution curve of near-wall time average velocity is obtained by integrating the above formula.

$$\frac{u}{u^*} = \frac{1}{k}\ln\frac{u^* y}{v} + C$$

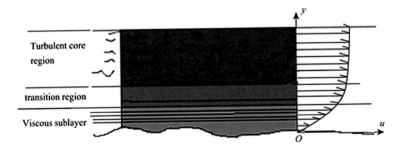

Fig. 1.74 Near-wall turbulence structure

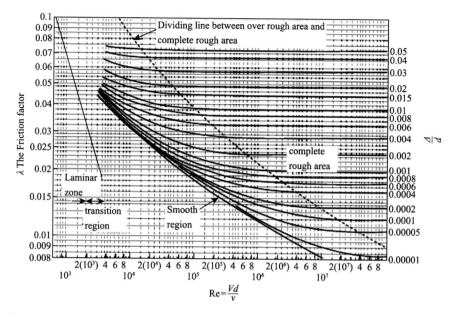

Fig. 1.75 Modi diagram (resistance coefficient relation curve)

where C is constant and $u^* = \sqrt{\frac{\tau_w}{\rho}}$ is the friction velocity.

Similar to the mixing length theory, the similarity theory proposed by von Carmen in 1931 and the vorticity transport theory proposed by Taylor in 1932 are similar. Later, because the Nicolas test curve was artificially roughened, it was not suitable for the natural rough commercial pipeline. Now the commonly used curve in engineering was drawn by Moody, the American engineer, in 1944 (called Moody diagram, as shown in Fig. 1.75). The graph borrows the formula of resistance coefficient along the course proposed by the British physician Leonard Colebrook (1883–1967, as shown in Fig. 1.76) and the American scholar White in 1939.

1.10 Turbulence Model

The turbulence model theory was developed based on the above semi-empirical theory. In the 1930s, although the mixed-length theory successfully solved some common problems of time-averaged velocity distribution in strong shear turbulence, such as turbulent boundary layer, turbulent jet, turbulent wake, and pipe turbulence. However, the mixing length theory only considers the relationship between eddy viscosity and time averages and does not consider the diffusion and transport of turbulent momentum. The

Fig. 1.76 Leonard Colebrook (1883–1967, British physician)

empirical constants of mixing length theory are too limited to be applied to complex separated flows. Therefore, in 1940, Mr. Zhou Peiyuan (1902–1993, as shown in Fig. 1.77), the famous mechanic in China, first deduced the transport differential equation of Reynolds stress, which made it possible to

Fig. 1.77 Zhou Peiyuan (1902–1993, Chinese Fluid mechanist)

close the Reynolds equation by introducing the transport equation of turbulent momentum. Kolmogorov (1942) and Prandtl (1945) first proposed a closed model (later called One-Equation Model) by introducing a turbulent kinetic energy equation, based on the eddy viscosity hypothesis. In 1967, the British scientist Peter Bradshaw (as shown in Fig. 1.78) proposed that the shear stress transport equation should be used to close the strong shear flow. The characteristic length scale of one-equation model needs to be determined experientially, which is strongly dependent on flow. For this reason, Kolmogorov (1942), Prandtl (1945), Zhou Peiyuan (1945), Rotta (1951), Spalding (1967), and Launder (1967) proposed closing the medium-length scale of one-equation model with differential transport equation and established a two-equation model. The most commonly used two-equation model is to characterize turbulent kinetic energy and turbulent energy dissipation rate. Turbulent kinetic energy K per unit mass is used to characterize the characteristic velocity scale of energy-carrying vortices, and dissipation rate ε per unit mass is used to characterize the characteristic length scale of energy-carrying vortices. Vortex viscous v_t is regarded as a function of K and ε, which can be expressed by dimensional analysis.

$$v_t = C_\mu \frac{K^2}{\varepsilon}$$

Fig. 1.78 Peter Bradshaw (British scientist)

where C_μ is an empirical constant. Turbulent kinetic energy K and turbulent energy dissipation rate ε are closed by the transport equation. For the standard $K - \varepsilon$ turbulence model,

The turbulent kinetic energy K equation is

$$\frac{\partial K}{\partial t} + u_j \frac{\partial K}{\partial x_j} = \frac{\partial}{\partial x_j}\left[\left(\frac{v_t}{\sigma_K} + v\right)\frac{\partial K}{\partial x_j}\right] + P - \varepsilon$$

The turbulent kinetic energy dissipation rate equation ε is

$$\frac{\partial \varepsilon}{\partial t} + u_j \frac{\partial \varepsilon}{\partial x_j} = \frac{\partial}{\partial x_j}\left[\left(\frac{v_t}{\sigma_\varepsilon} + v\right)\frac{\partial \varepsilon}{\partial x_j}\right] + C_{\varepsilon 1}\frac{\varepsilon}{K}P - C_{\varepsilon 2}\frac{\varepsilon^2}{K}$$

where $P = -\overline{u'_i u'_j}\frac{\partial u_i}{\partial x_j}$ is the turbulent kinetic energy generation term. The empirical constants in the model should be determined by experiments. At present, most scholars recommend the values of the constants as follows: $C_\mu = 0.07 \sim 0.09$, $\sigma_K = 1.0$, $\sigma_\varepsilon = 1.3$, $C_{\varepsilon 1} = 1.41 \sim 1.45$, $C_{\varepsilon 2} = 1.9 \sim 1.92$. $K \sim \varepsilon$ model has been widely used in turbulence engineering calculation, and many successful examples have been obtained, such as various turbulent jets (shown in Fig. 1.79), sudden expansion separation flows, and other strong shear flow problems.

Later, the turbulent stress transport equation model and its simplified algebraic stress model were developed (W. Rodi, 1972). On this basis, several first and second equation models with different turbulence characteristics have

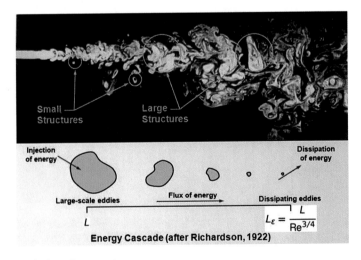

Fig. 1.79 Turbulent jet structure

been established. At present, commonly used are K-ε two-equation models, SST K-ω two-equation models, SA one-equation model, and so on. (These models play an important role in the numerical simulation of flow field, as shown in Fig. 1.80.)

A general relation used in the establishment of turbulent momentum transport equation is as follows:

The transport rate of any turbulent momentum is equal to the diffusion term, the production term, the dissipation term, and the additional term.

$$\frac{d()}{dt} = \frac{\partial}{\partial x_i} + \text{Production term} - \text{Dissipation term} + \text{Additional item}$$

Although turbulence models play an important role in engineering turbulence calculation (as shown in Figs. 1.81, 1.82, and 1.83), careful people

(a) SST   (b) SA   (c) SADES

Fig. 1.80 Turbulent jet structure

Fig. 1.81 Numerical simulation of flow around a square cylinder

Fig. 1.82   Aircraft landing gear winding

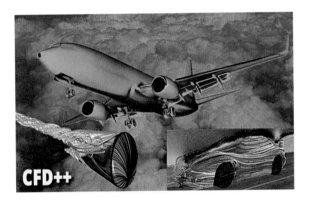

Fig. 1.83   CFD simulated flow field

will find that it is quite difficult to predict various complex flows accurately through a unified model, and that complex turbulence models are not necessarily better than simple ones. Different types of flow adapt to different turbulence modes, depending on the type of flow for which the model is established.

1.11 Turbulence Advanced Numerical Simulation Technology

Turbulence is a complex multi-scale and multi-level flow phenomenon. The prediction and control of turbulence are closely related to people's cognitive level and understanding level. For example, when only time-averaged changes are needed, the Reynolds time-averaged equations (RANS) can be used to solve them; if large-scale turbulence structure is needed, LEM (Large Eddy Simulation) is developed; and if all information of turbulent flow field is needed, the direct numerical simulation technique of full-scale turbulent motion must be developed from the instantaneous N-S equation. The three simulation techniques have different resolutions to the flow field and different turbulence scales. Generally speaking, direct numerical simulation requires the simulation of turbulence components at all scales, ranging from the minimum scale to the dissipative scale, which is equivalent to the Re number of the grid approaching 1.0. In Reynolds time-averaged method, turbulence fluctuation components are closed by a statistical turbulence model, and the grid scale of numerical simulation can be determined by the nature of the time-averaged flow. Large eddy simulation (LES) technology can simulate large-scale turbulence component because the grid scale is above the inertial sub-region and the dissipative scale component is replaced by the modeling equation. Direct numerical simulation technology was developed in the 1970s. Orzag and Patterson (1972) were the first to use direct numerical simulation to calculate isotropic turbulence with only 32^3 meshes and the corresponding Reynolds number is $Re_\lambda = 35$. On the contrary, the experimental measurements can only obtain limited flow field information, including finite scale turbulence components. For example, the vorticity distribution in the turbulent flow field is difficult to measure. So far, the development and evolution of turbulent vorticity structure can only be quantitatively observed by flow visualization or numerical simulation. Direct numerical simulation is an effective tool for studying the mechanism of turbulence and controlling turbulence (as shown in Fig. 1.84). The database of direct numerical simulation can also be used to evaluate the existing turbulence model, and then study the way to improve the turbulence model.

It has been found that in turbulent motion, besides many small-scale vortices with strong randomness, there are also some large-scale vortices with good organization. They have a relatively regular vortices' structure. Their shape and scale are of universal significance for the same type of turbulent motion. They play an important role in the turbulent Reynolds stress and the turbulent transport process of various physical quantities. For this reason, the

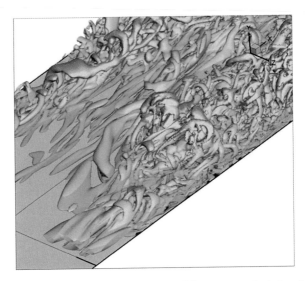

Fig. 1.84 Transition quasi-sequential structure (direct numerical simulation)

main idea of large eddy simulation (LES) is to abandon the numerical simulation of turbulent eddies in full scale, instead of using N-S equation to directly simulate the large eddy motion which is larger than the grid scale, while the effect of small eddy motion which is smaller than the grid scale on the large-scale eddy motion is simulated by establishing a general model. In a sense, large eddy simulation is a compromise between direct numerical simulation and general model theory. The model used to simulate the effect of small eddy motion on large-scale motion is called the Subgrid Scale Model, which is essentially the dissipation model of small-scale eddy. Large eddy simulation (LES) method was first proposed by Smagorinsky (1963), the American meteorologist, in the study of global weather forecasting. Then in 1970, Deardoff, the meteorologist, first applied this method for the calculation of turbulence in the channel. In the 1970s, Ferziger introduced the concepts of turbulent kinetic energy and dissipation rate like the time-averaged method, and corrected Smagorinsky's sublattice model by considering the effect of vorticity scale on eddy viscous coefficient. In 1991, Germano et al. proposed a dynamic model, which made the Smagorinsky model have a self-calibration effect. Su Mingde, the Chinese scholar, proposed an algebraic stress model in large eddy simulation in 1986 and used it to calculate turbulent flows in straight and curved channels.

The study of turbulence has been an unsolved problem for 136 years since Reynolds' transition test in 1880. At present, there are many branches in the basic theory of turbulence, such as turbulence stability theory, turbulence statistical theory, turbulence model theory, turbulence experiment, coherent structure of shear turbulence, large eddy simulation, and direct numerical simulation. Among them, the turbulence model is the most representative one, but its time-averaged operation erases all the details of the fluctuating motion, losing a lot of important information contained in the fluctuating motion, and all kinds of turbulence models have certain limitations. The rapid development of modern computer technology has provided people with a new way to solve turbulence problems. It is recognized that large eddy simulation and direct numerical simulation are promising. However, due to the limitation of computer speed and capacity, direct numerical simulation is still limited to low Reynolds number flow, and it is impossible to complete numerical simulation for high Reynolds number at present. Large eddy simulation (LES) is a compromise between direct numerical simulation and turbulence model theory. Because of its low computational cost and high computational accuracy, LES has attracted much attention. Chinese scholars have made remarkable achievements in turbulence research in recent decades, which are not listed here because of the limitation of space.

1.12 Multi-scale Discussions of Turbulent Eddies

1. *Multi-scale Structure of Turbulent Eddies*

Turbulence is a kind of complex flow phenomenon in nature. Since the British scientist Renault completed the laminar flow transition test in 1880, many basic problems of turbulence have not been solved. However, after 136 years of exploration, considerable achievements have been made in the study of turbulence. Especially with the rapid development of computer and modern measurement technology, the recognition of turbulence has gone from simple time-averaged level to turbulent structure level of different scales. Since the 1950s, with the discovery of turbulent coherent structure, it is generally believed that turbulence is not determined by the randomness of small-scale eddies, but by the existence of large-scale eddies. Turbulence is a complex flow phenomenon consisting of eddies of different scales and frequencies. The scale of the largest eddies is the same order of magnitude as the characteristic size of the flow region, and the scale of the smallest eddies is the same as the viscous scale of the fluid, which makes the turbulence become a multi-scale complex flow phenomenon. In the process of fluid motion, turbulent eddies of different scales interact and evolve, and there are random small-scale eddies and coherent structures of large-scale eddies, as shown in Figs. 1.85, 1.86 and 1.87.

2. *Evolution Characteristics of Turbulent Eddy Scale*

During the development of turbulence, the scale of turbulent eddies is not only wide but also evolves from large-scale eddies to small-scale eddies over time. The scale evolution may be gradual or catastrophic. For example, in 1922, the British meteorologist L. F. Richardson (1881–1953) proposed a theory of gradual evolution of turbulent scale, namely the energy cascade theory of turbulence, as shown in Fig. 1.60. The theory shows that large-scale eddies obtain energy from time-averaged flows through turbulent shear, and then through viscous dissipation and dispersion processes, these large-scale eddies are cascaded into small eddies of different scales, and the energy is transferred to smaller scale eddies step by step during the splitting and fragmentation of eddies until viscous dissipation is achieved.

But is there a sudden change in the scale evolution of turbulent eddies? This problem seems to have never been reported. Recently, the author observed the decay process of the wing wake vortex in the towing flume of

Fig. 1.85 Vincent Willem van Gogh (1853–1890), a Dutch post-Impressionist painter, created Star Sky in 1889 (like large-scale eddies)

Fig. 1.86 Large eddy structure in atmospheric turbulence

Fig. 1.87 Large eddy structure in turbulent flow

the Lu Shijia Laboratory of Beijing University of Aeronautics and Astronautics. The results show that the relatively slow decay process of the wing wake vortices is produced by large-scale and small-scale vortices, while the medium-scale vortices evolve rapidly, and there is hardly a gradual process of scale evolution. It can be divided into three stages: (1) The slow decay period of large-scale eddies, in which large-scale eddies are mainly affected by convection and diffusion, and their dissipation is weak. They are in the process of mutual induction and winding of large-scale eddies, and their decay is relatively slow, as shown in Fig. 1.88; (2) During the rapid evolution period of

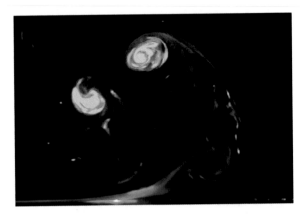

Fig. 1.88 Large-scale vortices (two co-directional vortices and one near-wall vortices coiling, convection, and diffusion dominate)

mesoscale eddies, we can see that only when the large-scale eddies cannot be maintained and break up rapidly, the evolution process of mesoscale eddies will occur. The shortest period in the whole process is the fast decay period, as shown in Fig. 1.89. (3) The small-scale eddy dissipation period belongs to the later stage of turbulent decay, as shown in Fig. 1.90. During this period, the small-scale eddies are mainly affected by viscous diffusion and dissipation, and convection is very weak. It belongs to the slow dissipation period of the small-scale eddies. The whole process takes a long time. The energy of the turbulent eddies is mainly dissipated by viscous energy in the first-scale eddies. These need further experimental verification.

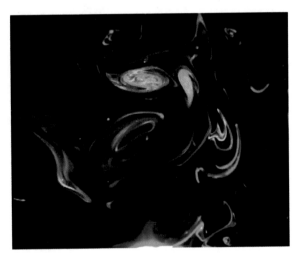

Fig. 1.89 Mesoscale eddies (large eddies rapidly broken, mesoscale eddies transition)

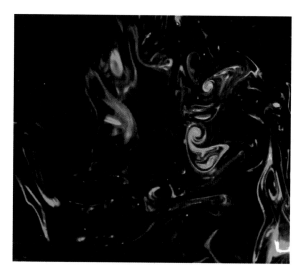

Fig. 1.90 Small-scale eddies (turbulent decay period, viscous diffusion, and dissipation dominate)

3. *Large and Small Scales of Turbulent Eddies*

Although the scale variation of turbulent eddies is very wide, large-scale vortices and small-scale eddies often control the steady-state turbulent structure. The former plays a role in the turbulent eddies, while the latter dissipates the turbulent eddies. Therefore, in the establishment of turbulence models, not all scales of turbulent eddies need to be modeled, but only those scales of turbulent structures that control the dynamic equation. For this reason, special attention has been paid to two scales of eddy structures. One is the large eddies interacting with time-averaged flow, which continuously extracts energy from time-averaged flow energy to maintain turbulent fluctuating motion through the action of time-averaged shear motion. Another scale of vortices is dissipative eddies, which are very small in scale and dissipate turbulent fluctuating kinetic energy through the viscous flow. These dissipative vortices are also considered to be the smallest scale vortices to maintain the macroscopic motion of turbulence, because smaller scale eddies cannot be sustained under the action of strong viscous dissipation.

For large-scale turbulent eddies, the length scale can be characterized by an integral scale, because the integral scale represents a region with a strong correlation of fluctuating velocities. The integral scale is characterized by l_t, and the velocity scale is characterized by the square root of turbulent kinetic energy K per unit mass, $V_t = \sqrt{K}$. Large eddies of this kind of scale are also called energy-containing eddies. For dissipative scale vortices, Kolmogorov considers that their length and velocity scales are determined by the viscous coefficient of fluid motion v and the dissipation rate of turbulent energy ε. In the dissipative vorticity scale, assuming that the length scale is η and the velocity scale is v, the particle fluctuation inertia force is considered to be the same order of magnitude as the viscous force due to the viscous restriction, that is, the particle fluctuation inertia force is equal to the viscous force.

$$\mathrm{Re}\eta = \frac{v\eta}{\upsilon} \approx 1.0$$

Through dimensional analysis, we can get the result.

$$\eta = \left(\frac{\upsilon^3}{\varepsilon}\right)^{1/4} \quad \mathrm{v} = (\upsilon\varepsilon)^{1/4} \quad \tau = \left(\frac{\upsilon}{\varepsilon}\right)^{1/2}$$

Among them, τ is the time scale of dissipative eddies. These scales are also called the Kolmogorov microscales. The dissipation rate of turbulent energy dissipation ε is expressed in microscale terms.

$$\varepsilon \approx \frac{\mathrm{v}^3}{\eta}$$

Now we look at the relationship between large scale and microscale. According to turbulent kinetic energy transport equation,

$$\frac{\partial K}{\partial t} + u_j \frac{\partial K}{\partial x_j} = \frac{\partial}{\partial x_j}\left[-\frac{\overline{u_i' u_i'}}{2}u_j' - \frac{\overline{p' u_j'}}{\rho} + v\frac{\partial K}{\partial x_j}\right] - \overline{u_i' u_j'}\frac{\partial u_i}{\partial x_j} - \varepsilon$$

In shear turbulence, the turbulence in the local equilibrium state should be kept in the order of magnitude as follows:

$$-\overline{u_i' u_j'}\frac{\partial u_i}{\partial x_j} = \varepsilon$$

It is estimated that Reynolds stress $-\overline{u_i' u_j'}$ is mainly determined by large-scale eddies, while it is $\overline{u_i' u_j'} \approx V_t^2$. The time-averaged velocity gradient interacts with large-scale vortices to generate turbulent kinetic energy, so the time-averaged velocity gradient can be represented by large-scale eddies, namely

$$\frac{\partial u_i}{\partial x_j} \approx \frac{V_t}{l_t}$$

In this case, the dissipation rate of turbulent kinetic energy can be expressed as follows in large eddy scale.

$$\varepsilon \approx \frac{V_t^3}{l_t}$$

The relationship between the length scale of large eddy and the length scale of small eddy can be obtained by substituting the large eddy scale expression of ε into the Kolmogorov microscale.

$$\frac{l_t}{\eta} \approx \left(\frac{V_t l_t}{\nu}\right)^{3/4} = Re_t^{3/4}$$

where Re_t is the turbulent Reynolds number represented by the energy-carrying eddy scale. Similarly, the relationship between the velocity scale of large eddies and the velocity scale of small eddies is as follows:

$$\frac{V_t}{\nu} \approx \left(\frac{V_t l_t}{\nu}\right)^{1/4} = Re_t^{1/4}$$

It is shown that the ratio of large eddies to small eddies is Re_t function, and the scale width between them is larger with the increase of Re_t.

For example, for $V_t = 1.46$ m/s, $l_t = 10$ mm, and thus $Re_t = 1000$, then

$$\frac{l_t}{\eta} \approx Re_t^{3/4} = 178, \quad \frac{V_t}{\nu} \approx Re_t^{1/4} = 5.6$$

Currently, the length scale of dissipative vortices is $\eta = 0.056$ mm $= 56$ μm, which is 56 times the minimum macroscale of 1 μ to maintain air continuous flow, which indicates that turbulence satisfies the condition of macro continuous flow of particles. The velocity scale of dissipative vortices is $\nu = 0.26$ m/s and the dissipation rate is $\varepsilon = 311$ m²/s³. If $V_t = 1.46$ m/s, $l_t = 1$ mm, $Re_t = 100$, the length scale of dissipative vortices is $\eta = 0.0316$ mm $= 31.6$ μm, the velocity scale of dissipative vortices is

0.46 m/s, and the dissipative rate is $\varepsilon = 3112$ m^2/s^3; if $V_t = 1.46$ m/s, $l_t = 0.1$ mm, Re$t = 10$, the length scale of dissipative vortices is $\eta = 0.0178$ mm $= 17.8$ μm, the velocity scale of dissipative vortices is 0.82 m/s, and the dissipative rate is $\varepsilon = 31121$ m^2/s^3. If $V_t = 1.46$ m/s, $l_t = 0.01$ mm, Re$t = 1$, the length scale of dissipative vortices is $\eta = 0.01$ mm $= 10$ μm, the velocity scale of dissipative vortices is 1.46 m/s, and the dissipative rate is $\varepsilon = 311214$ m^2/s^3. It can be seen that the scale of the minimum vortices is 10 times larger than that of the minimum macro-scale of 1 μ, which means that the turbulence satisfies the continuity condition and that the turbulence is the result of Macro-motion of particles.

If the minimum length scale of dissipative vortices (the requirement of continuity) is taken as $\eta = 1$ μm, and from $\eta v / v = 1$, the maximum velocity scale of dissipative vortices is $v = 14.6$ m/s, the minimum time scale $\tau = 6.8 \times 10^{-8}$ s, and the maximum dissipation rate of dissipative vortices is $\varepsilon = 3.1 \times 10^9$ m^2/s^3. For ease of comparison, the relationship between the scale ratio of energy-carrying vortices and dissipative vortices and turbulent Reynolds number is given in Fig. 1.91, the scale relationship between energy-carrying vortices and dissipative vortices in Fig. 1.92, and the relationship between the scale of energy-carrying vortices and dissipative rate in Fig. 1.93.

According to the definition of turbulent energy dissipation rate,

$$\varepsilon = v \, \overline{\frac{\partial u'_i}{\partial x_j} \frac{\partial u'_i}{\partial x_j}} \approx \frac{V_t^3}{l_t}$$

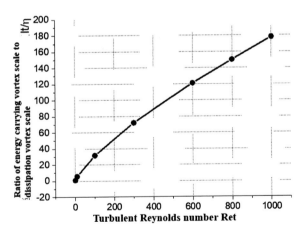

Fig. 1.91 Relation between scale ratio of energy-carrying eddies and dissipative eddies and turbulent Reynolds number

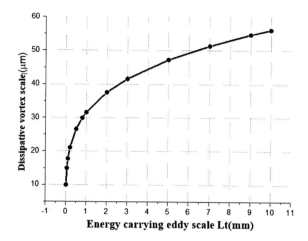

Fig. 1.92 Scale relation between energy-carrying eddies and dissipative eddies (*Vt* = 1.46 m/s)

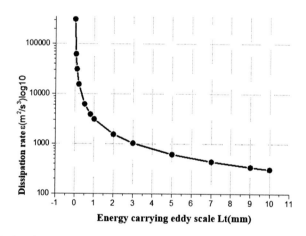

Fig. 1.93 Relation between scale of energy-carrying eddies and dissipation rate ε(*Vt* = 1.46 m/s)

The scale of the fluctuating velocity gradient can be expressed as

$$\left[\frac{\partial u_i'}{\partial x_j}\right] \approx \frac{V_t}{l_t} \mathrm{Re}t^{1/2} = \frac{\mathrm{v}}{\eta}$$

It is shown that the fluctuating velocity gradient can be characterized by Kolmogorov's microscale, which is consistent with the concept that turbulent kinetic energy is dissipated by small-scale eddies. This shows that the fluctuating velocity gradient in the dissipation rate is determined by small-scale eddies. According to the above analysis, the expression of fluctuating velocity gradient and mean velocity gradient is obtained as follows:

$$\left[\frac{\partial u_i'}{\partial x_j}\right] \approx \frac{V_t}{l_t} Re_t^{1/2}, \left[\frac{\partial u_i}{\partial x_j}\right] \approx \frac{V_t}{l_t}, \left[\frac{\partial u_i'}{\partial x_j}\right] \approx \left[\frac{\partial u_i}{\partial x_j}\right] Re_t^{1/2}$$

2

Aerodynamics

2.1 Development of Aerodynamics

Aerodynamics is the study of the law of air motion and force when there is relative motion between the object and air, that is, when the object moves in the air or the object does not move, but air moves around the object. Traditional aerodynamics refers to the aerodynamics of aircraft, especially the aerodynamics of ordinary aircraft. The component force acting on a moving aircraft perpendicular to the direction of flight is known as the lift for the aircraft, which makes the aircraft "hold" in the air. The component force acting on an aircraft parallel to the direction of flight plays a resistance role in the flight of an aircraft.

In the study of aerodynamics, the motion of an aircraft passing through the air is often equivalent to the motion of an aircraft bypassing the aircraft without moving air according to the principle of relative flight. The principle of relative flight refers to the situation when an aircraft moves along a straight line at a constant speed in static air. The relative motion law states that the mutual forces between the aircraft and the air are equivalent to the situation when the aircraft is fixed but the air flows over the aircraft with the same speed in the opposite direction, as shown in Fig. 2.1.

The principle of relative flight, which is the basic principle of aerodynamics experiment, provides convenience for aerodynamics research. In the experimental study, the aircraft model can be fixed, and the straight and uniform airflow around the model can be manufactured manually, so as to observe the flow phenomenon, measure the aerodynamic forces on the model, conduct the experimental aerodynamic study, and make the airflow easier to achieve

© Science Press 2021
P. Liu, *A General Theory of Fluid Mechanics*,
https://doi.org/10.1007/978-981-33-6660-2_2

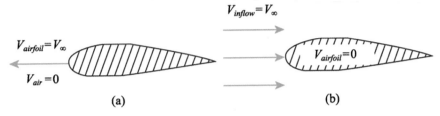

Fig. 2.1 Relative flight principle for aircraft

in the wind tunnel test than to make the model move in still air (as shown in Figs. 2.2 and 2.3).

As early as the Italian Renaissance, the Italian scientist Leonardo da Vinci (1452–1519, as shown in Fig. 1.3) studied the flight principle of birds and

Fig. 2.2 Wind tunnel test of scaled model of an aircraft

Fig. 2.3 Wind tunnel test with a prototype car

gave some qualitative concepts. Later, famous scientists such as Kelly in England, Langley in America, and Otto Lilienthal in Germany made further and deeper research. George Kelly (1773–1857, as shown in Fig. 2.4) is known as the father of classical aerodynamics. He has done a lot of research on the flying principle of birds. By observing the wing area, bird weight, and flying speed, he estimated the relationship between speed, wing area, and lift and proposed that the propulsion force and lift surface should be considered separately. Samuel Pierpont Langley (1834–1906, as shown in Fig. 2.5) proposed the calculation formula of wing lift. Otto Lilienthal (1848–1896), the German engineer and glider, began to manufacture gliders. He was one of the pioneers in the manufacture and practice of fixed-wing gliders. He flew more than 2000 times near Berlin, and accumulated rich data, which provided valuable experience for the later realization of power flight by the American Wright brothers.

In the twentieth century, human beings have established a complete scientific system of aerodynamics, which has been developing vigorously. American Wright brothers are two engineers with practical experience, theoretical knowledge, imagination, and vision. On December 27, 1903, Orville Wright piloted them to design and manufacture the first successful test flight of "pilot one". This is the first powered, manned, sustained, stable, and controllable aircraft in human history, which ushered in a new era of

Fig. 2.4 George Kelly (1773–1857, British Aerodynamics)

Fig. 2.5 Samuel Pierpont Langley (1834–1906, American scientist)

power flight. Since then, the development of aircraft has promoted the rapid development of aerodynamics. In 1906, Joukowky (1847–1921, as shown in Fig. 2.6) published the famous lift formula, laid the foundation of the two-dimensional wing theory, and proposed the airfoil named after him. From

Fig. 2.6 N. Joukowsky (1847–1921, Russian scientist)

1918 to 1919, Ludwig Prandtl (1875–1953, as shown in Fig. 1.36) proposed the theory of lift line for wings with a large aspect ratio. From the 1920s to the 1930s, the theory and experiment of aerodynamics developed rapidly. Many low-speed wind tunnels were built to carry out a large number of experiments on the development of various kinds of aircraft, which greatly improved the aerodynamic shape of the aircraft, and realized the increase in the aircraft speed from 50 m/s to 170 m/s when the increase in the aircraft power was small. In 1925, Ackeret derived the theory of supersonic linearization of airfoils. In 1927, H. Glauert and Prandtl put forward the similarity rule of subsonic three-dimensional wings. In 1939, Gothert put forward the similarity rule of subsonic three-dimensional wings. In 1944, von Karman (as shown in 1.42) and Qian Xuesen (as shown in Fig. 1.8) adopted the speed chart method and proposed a more accurate subsonic similarity rule than that of Glauert. In 1946, Qian Xuesen first proposed a hypersonic similarity rate. From the 1930s to the 1940s, humans built a number of supersonic wind tunnels, which made the aircraft break through the "sound barrier" at the end of the 1940s, and then brake through the "thermal barrier" in the 1950s, realizing supersonic flight and artificial satellites.

Since the 1950s, with the emergence and development of computers, computational aerodynamics has developed rapidly. Theory, experiment, and calculation have become an indispensable part of aircraft design (as shown in Fig. 2.7).

According to the velocity of the flow, the aerodynamics of aircraft can be divided into several categories. The low-speed problem is called low-speed aerodynamics, and the high-speed problem is called high-speed aerodynamics (as shown in Fig. 2.8). In the high-speed range, according to the important reference quantity of sound speed, it can be divided into several parts: the

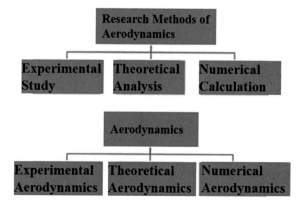

Fig. 2.7 Aerodynamics research methods

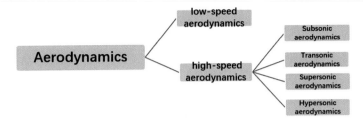

Fig. 2.8 Aerodynamic classification

study of flight speed lower than sound speed is called subsonic aerodynamics, the study of flight speed higher than sound speed is called supersonic aerodynamics, and the study of flight speed around sound speed is called transonic aerodynamics. When intercontinental missiles and spacecraft return to the atmosphere, their speed will be more than ten times of the sound speed in a short time at first. This kind of flight is called hypersonic flight. Generally speaking, the hypersonic flight is defined as a flight with a speed greater than five times the sound speed, and the one dealing with this problem is called hypersonic aerodynamics. In addition, we also study the rarefied gas mechanics in the outer atmosphere, magnetohydrodynamics of ionization in the outer atmosphere, etc.

2.2 Low-Speed Airfoil Flow

First of all, humans have realized the dream of flying by air buoyancy based on balloons, but it is not enough to fly only by balloons. Birds and many aircraft do not fly by air buoyancy, but by the relative motion of air around them. So people gradually realize that when the air around the wing has relative motion, it can produce lift, and the lift is much greater than the buoyancy. The second is that the object moving in the air must overcome air resistance, so the concept of aircraft dynamics is introduced. The third is that the concept of force multiplying distance, i.e., the concept of moment, is introduced when the aircraft is able to maintain stable flight in the air. So these three concepts lay the foundation for us to realize real flight (as shown in Fig. 2.9).

Aircraft research starts from imitating birds, which is called imitation birds. Later, people separated the power from the lift, and then there appeared fixed-wing aircraft. The so-called fixed wing is that the lift surface is fixed, so the earliest fixed wing is the glider. A glider is an unpowered aircraft, which generates lift through its wings. In addition, when the aircraft is equipped with an

1、 **Understanding of force**

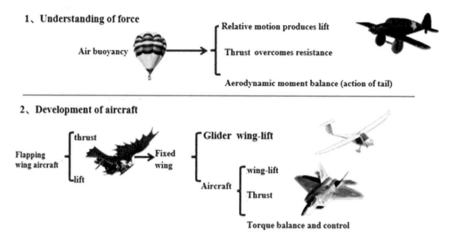

Air buoyancy → ⌈ Relative motion produces lift

Thrust overcomes resistance

⌊ Aerodynamic moment balance (action of tail)

2、 **Development of aircraft**

Flapping wing aircraft ⌈thrust / ⌊lift → Fixed wing ⌈ Glider wing-lift

Aircraft ⌈ wing-lift / Thrust

⌊ Torque balance and control

Fig. 2.9 The role of aerodynamics

engine, it becomes a real aircraft with power. If we balance and control the power, lift, resistance, and gravity of the aircraft, plus the navigation system, the aircraft will be able to flap its wings and fly high, which will realize the dream of human flight. People can travel far by aircraft. Now the passenger aircraft is fast, safe, and comfortable, so the aircraft is one of the greatest scientific and technological achievements of the twentieth century (as shown in Fig. 2.10). With the aircraft, I'm afraid that Confucius's original saying "don't leave your parents too far" would have to be changed to "don't leave your parents too long". That is, don't leave your parents for a long time, and don't worry about the distance. Will it take 13 h to fly to America? Therefore,

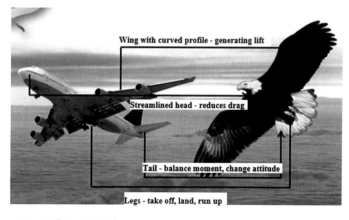

Wing with curved profile - generating lift

Streamlined head - reduces drag

Tail - balance moment, change attitude

Legs - take off, land, run up

Fig. 2.10 Layout of an Aircraft

the emergence of the aircraft makes the distance between people's living place appear much shorter.

How a real aircraft can fly depends on the understanding of birds. In terms of flight, it can be said that birds are the ancestors of human beings. No matter what kind of aircraft they build, they need to learn from birds. Because birds have enough flight experience in the air, they have evolved over a long period of time. Next, we analyze the flight principle of birds to explore the mystery of the lift generated by airfoils.

For a bird's skeleton and wing (as shown in Fig. 2.11), a section is cut on the wing along the airflow direction. We will find that the front of this section is the wing bone, and then the feathers are connected by muscles; so after cutting, it is equivalent to a round head in the front and a thin feather at the back, with a certain degree of bending, and the section is streamlined. Considering only the flow around this unique section shape, aerodynamics is called airfoil flow (as shown in Fig. 2.12).

At the airflow around an airfoil, the aerodynamic force will act on the airfoil. The component force perpendicular to the inflow direction is the airfoil lift, and the component force parallel to the inflow direction is the airfoil drag. Their magnitudes are not only related to the inflow speed, airfoil geometry, and size but also to the angle of attack between the airfoil and the inflow direction. In 1686, Newton proposed in the application of mechanical principles and deductive methods that the force on a moving object in the air

Fig. 2.11 Bird wings and section

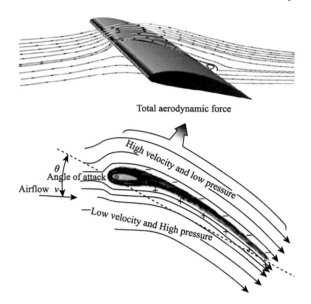

Fig. 2.12 The flow around an Airfoil

is directly proportional to the square of its moving speed, the characteristic area of the object, and the density of the air. According to the principle of force and reaction force, Newton put forward the so-called "Skipping Stone Theory". He believed that the lift on the airfoil was the result of the impact of the lower wing on the airflow, independent of the upper wing (as shown in Fig. 2.13). Now it seems wrong.

In 1738, Bernoulli, the Swiss scientist, derived an ideal fluid energy equation and established a quantitative relationship between air pressure and velocity, which provided a theoretical basis for a correct understanding of lift. Especially, according to the energy theorem, the lift on an airfoil is not only related to the air uplift force acting on the lower wing surface but also to the suction on the upper wing surface (as shown in Fig. 2.14). Later, wind

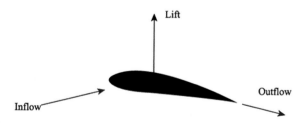

Fig. 2.13 Newton's floating stone theory

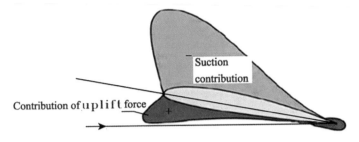

Fig. 2.14 Pressure distribution of upper and lower airfoil surfaces

tunnel tests confirmed that the suction of the upper wing surface accounts for about 60–70% of the total lift of the airfoil.

In 1902, the German mathematician Martin Wihelm Kutta (1867–1944, as shown in Fig. 2.15) and in 1906, the Russian physicist N. Joukowsky (1847–1921, as shown in Fig. 2.6) extended the calculation formula of the lift for the flow around a cylinder with velocity circulation to the flow around a body of any shape and proposed that the lift will be generated for any object as long as there is a velocity circulation. The direction of lift is that the flow direction rotates 90° according to the reverse circulation. Later, it is called

Fig. 2.15 Martin Wihelm Kutta (1867–1944, German mathematician)

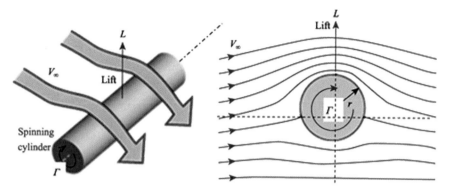

Fig. 2.16 The law of lifting circulation of Kutta–Joukowsky

Kutta and Joukowsky's law of lift circulation (as shown in Fig. 2.16), namely

$$L = \rho V_\infty \Gamma$$

where L is the lift acting on the object, ρ is the air density of the inflow, V_∞ is the velocity of the inflow, and Γ is the velocity circulation around the object.

When different circulation values bypass the airfoil, there may be three different flow pictures of the rear stagnation point located on the upper wing surface, the lower wing surface, and the trailing edge point. When the rear stagnation point is located on the upper and lower wing surfaces, the airflow should bypass the trailing edge of the tip. According to the potential flow theory, there will be infinite velocity and negative pressure there, which is physically impossible. Therefore, the possible flow picture in physics is that the stagnation point coincides with the trailing edge point, or the airflow smoothly flows through the trailing edge of the airfoil from the upper and lower wing surfaces, and the velocity value of the trailing edge remains limited. The flow experiment also confirms this analysis. Kutta and Joukowsky use this condition to give the unique condition for determining the attached circulation.

According to Kelvin's law of conservation, for an ideal incompressible fluid, under the action of a potential force, the velocity circulation around the closed circumference composed of the same fluid particles does not change with time, i.e., $d/dt = 0$. All airfoils accelerate from static state to steady state. According to the law of vortex conservation, the circulation of velocity caused by airfoil motion should be zero everywhere as in the static state, but a non-zero circulation value is obtained by the Kutta condition, which is a contradiction. How to understand the physical cause of circulation?

When the airfoil starts at the beginning, because the viscous boundary layer has not been formed on the airfoil surface, the velocity circulation around the airfoil is zero, and the rear stagnation point is not at the trailing edge, but at some point on the upper airfoil surface, the air will flow around the rear edge to the upper airfoil surface. With the development of time, the boundary layer on the wing surface is formed. When the air on the lower wing surface flows around the rear edge, it will form a large velocity and a low pressure. There is a large counter pressure gradient from the rear edge point to the rear stagnation point, which causes the boundary layer to separate, resulting in a counter-clockwise circulation, which is called the starting vortex (as shown in Fig. 2.17). The starting vortices are separated from the trailing edge of the airfoil and flow downstream with the airflow, and the closed fluid line also moves with the airflow, but it always surrounds the airfoil and the starting vortices. According to the law of vorticity retention, there must be a counter-clockwise velocity circulation around the airfoil, so that the total circulation around the closed fluid line is zero. This way, the position of the rear stagnation point of the airfoil moves backward. As long as the rear stagnation point has not moved to the trailing edge point, there is a continuous anticlockwise vortex shedding from the trailing edge of the airfoil, so the circulation around the airfoil continues to increase until the airflow leaves smoothly from the trailing edge point (the rear stagnation point moves to the trailing edge, as shown in Fig. 2.18), forming the final attachment vortex and starting vortex (as shown in Fig. 2.19). The starting vortices are far behind, and the attached vortices are superimposed on the airfoil and move at a constant speed with the airfoil, which has an important impact on the aerodynamic force of the airfoil (as shown in Fig. 2.20).

The earliest wings were made to imitate kites, with a piece of cloth on the skeleton, basically a flat plate. In practice, it is found that the bending plate is better than the flat plate, and can be used in a larger range of angle of attack. In 1903, the Wright brothers developed a thin airfoil with positive curvature.

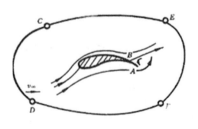

Fig. 2.17 Unbalanced airfoil around the boundary layer during start-up

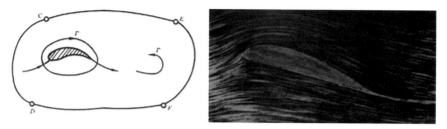

Fig. 2.18 The equilibrium state of the airfoil stationary around the boundary layer

Fig. 2.19 Attached vortex and starting vortex around an airfoil

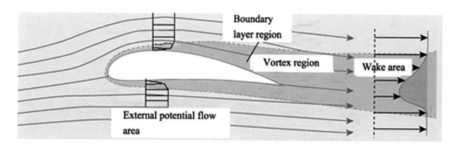

Fig. 2.20 Steady airfoil flow (equilibrium boundary layer structure)

After the wing theory of Joukowsky came out, it was clear that the low-speed airfoil should be composed of a round head, upper and lower wing surfaces (as shown in Fig. 2.21). The dome head can adapt to a larger range of angle of attack, the curved wing surface can avoid the separation of airflow around the wing surface, and further explain the mechanism of the attachment vortex generated by the asymmetric flow around the airfoil (as shown in Fig. 2.22). The lift–drag ratio of the flow around the flat is significantly different from that of the flow around the airfoil (as shown in Fig. 2.23).

In 1909, Joukowsky, using the conformal transformation method of complex function, studied the steady flow around an ideal fluid airfoil and proposed the famous Joukowsky airfoil theory (as shown in Fig. 2.24). In the 1920s, the theory of thin airfoil was put forward using the principle of singularity superposition of velocity potential function and the hypothesis of small

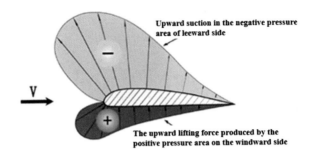

Fig. 2.21 Airfoil pressure distribution and its contribution to lift

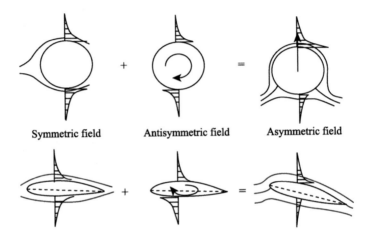

Fig. 2.22 Superposition principle

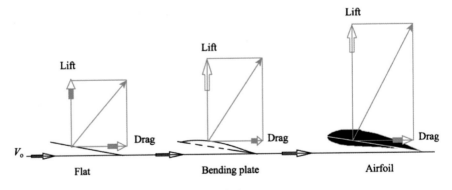

Fig. 2.23 Lift–drag ratio of plate and airfoil

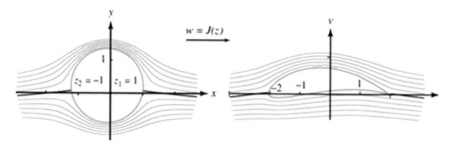

Fig. 2.24 Joukowsky airfoil

disturbance. It is concluded that for the ideal incompressible flow around a thin airfoil at a small angle of attack, the disturbance velocity potential, the boundary conditions of the material surface, and the pressure coefficient can be superposed linearly. The lift and moment acting on the thin airfoil can be regarded as the sum of the effects of curvature, thickness, and angle of attack. Therefore, the flow around a thin airfoil can be superposed by three simple flows.

During World War I, the belligerent countries have found some airfoils with better performance in practice, for example, Joukowsky airfoil, Gottingen airfoil in Germany, RAF airfoil in Britain (Royal Air Force, later changed to RAE airfoil), Clark-y in the United States; after the 1930s, NACA airfoil (National Advisory Committee for aeronautics, later NASA, National Aeronautics and Space Administration) in the U.S.A. and цАГИ airfoil (central air fluid Research Institute in Russia). In the late 1930s, the National Aeronautical Advisory Council of the United States (NACA, now NASA) made a systematic study on the performance of airfoil (as shown in Fig. 2.25)

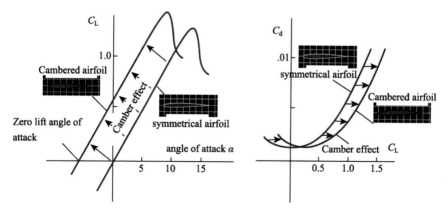

Fig. 2.25 Lift coefficient and drag coefficient of airfoil

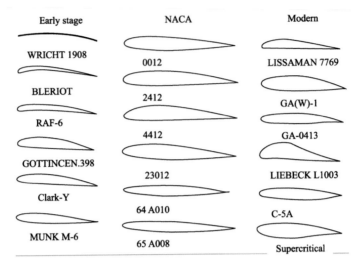

Fig. 2.26 Evolution of airfoil

and proposed four-digit airfoil family and five-digit airfoil family of NACA, as shown in Fig. 2.26. After a systematic study of the airfoil, they found that: (1) if the airfoil is not too thick, the thickness and curvature effects of the airfoil can be considered separately; (2) if a good airfoil obtained from experience in various countries is straightened, that is to say, it is changed into a symmetrical airfoil and converted into the same relative thickness, its thickness distribution almost coincides with each other. Therefore, it is proposed that the thickness distribution of NACA airfoil family is the best one of the time (as shown in Fig. 2.27).

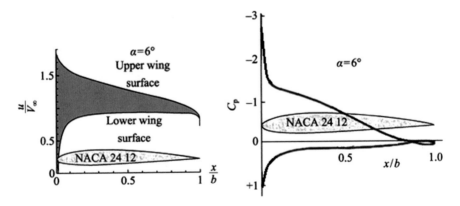

Fig. 2.27 Velocity and pressure distribution on the upper and lower surfaces of NACA2412 airfoil

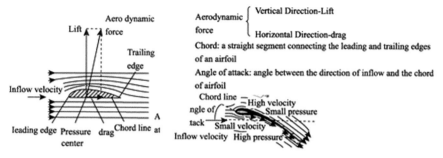

Fig. 2.28 Flow around airfoil

When the airfoil is flowing around (as shown in Fig. 2.28), why is the pressure on the lower wing greater than the pressure on the upper wing for the same incoming flow? It is a long struggle to understand the mechanics of flight, and it is a classic problem of aerodynamics.

In fact, there are different opinions about how the lift of airfoil is generated so far, and there are some mistakes in some opinions, which are reported in detail on NASA's official website (https://www.grc.nasa.gov/www/k-12/air plane/short.html).

1. Based on Bernoulli's view, "long path" or "equal time" is the most widely used. According to this theory, the airflow is divided into two parts at the leading edge of the airfoil, and finally converges at the trailing edge of the airfoil at the same time. However, due to the asymmetry of the shape of the upper and lower surfaces of the airfoil, the airflow moves along the upper surface of the airfoil for a long distance, and the velocity is fast. According to Bernoulli's theorem, the air pressure with fast speed is small, so that the pressure of the lower wing surface is greater than that of the upper wing surface, resulting in lift. Because this theory mainly depends on Bernoulli's law, it is called Bernoulli's effect later. The key point of this theory is that "the asymmetry of the upper and lower surfaces of the wing" is the source of lift. However, for the purpose of reducing the shock wave strength, the length of the lower surface is actually longer than that of the upper surface of the supercritical airfoil, which is widely used in modern aircraft. Therefore, the interpretation of lift is questionable. At the same time, this theory cannot explain the reason of the plane's inverted flight.

2. Based on the principle of force and reaction, Newton put forward the theory of skipping stone. This theory holds that the lift comes from the reaction force of air on the lower wing surface of the airfoil, just like the water floating, when the stone glides quickly across the water surface, it

will expel the water body to obtain the reverse force to leave the water body again, and the aircraft pushes the air downward continuously during the flight, so as to obtain the lift according to the reaction force. It is theoretically believed that the lift is mainly generated by the lower surface of the airfoil, and the contribution of the upper surface of the airfoil can be ignored. It is concluded that if the lower surface of the airfoil is unchanged, the shape of the upper surface will not change the lift, which is obviously wrong. A typical example is the spoiler. When the spoiler on the upper surface of the wing is opened, the lower surface does not change. It can be said that the shape of the upper surface does not change much, but it has a great impact on the aerodynamic force.

For the real airflow around the airfoil, there is a viscous boundary layer near the airfoil wall, which leads to the flow velocity along the airfoil. There are different sizes in the vertical direction of the airfoil. The closer to the object, the greater the difference.

3. Flow tube theory. When the air flows through the upper and lower surfaces, due to the bulge of the upper surface, the spacing between the upper streamlines becomes narrow, while the lower part is relatively flat, and the spacing between the streamlines becomes wider. According to the continuity theorem of fluid, when the fluid continuously flows through a pipe of different thicknesses, since the fluid in any part of the pipe cannot be interrupted or stacked, the mass of the fluid flowing into any section and the mass of the fluid flowing out of another section are equal at the same time, resulting in a greater flow velocity on the upper surface than on the lower surface, then according to Bernoulli's theorem, the lift is generated. The doubtful point is that this theory can only be established in a two-dimensional environment. A large amount of airflow around the real wing is affected, and the contraction deformation of the flow tube is not obvious.

4. Downwash airflow theory. The flow direction around the airfoil tends to deviate downward, and at the same time, a reaction force is generated to provide lift. This part of the lift does exist, which is called "impact lift", but it accounts for a relatively small proportion of the whole airfoil lift. For the "supercritical airfoil" used in large aircraft, the loading effect after an airfoil depends on the downwash generated by the downward bending of the trailing edge of the wing to provide lift.

All the above theories or viewpoints are obtained under the condition of inviscid flow. The author thinks that it is difficult to explain the whole region of airfoil flow with any theory or viewpoint. In fact, they are only

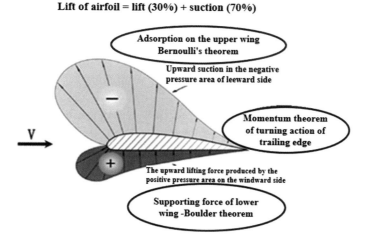

Fig. 2.29 Theoretical adaptability based on the ideal flow around an airfoil

suitable for the flow characteristics of different local regions of airfoil flow, as shown in Fig. 2.29.

A large number of wind tunnel tests show that the aerodynamic force acting on the airfoil is related to the shape, size, and attitude of the airfoil. In particular, the shape of the airfoil has an effect on the ratio of lift to drag. With a good airfoil shape, the ratio of lift to drag produced is large and the aerodynamic efficiency is high. That is to say, the resistance required to lift a Newtonian gravity is small and the efficiency is good. In aerodynamics, the optimization of the airfoil is specially discussed. In principle, it is to seek the shape with the largest ratio of lift to drag (as shown in Fig. 2.29). In today's aircraft wing design, it is an eternal aerodynamic optimization problem to seek the shape with a large ratio of lift to drag. For the airfoil of large civil aircraft, the designed ratio of lift to drag can reach between 80 and 110.

2.3 Development and Influence Mechanism of Boundary Layer Near Airfoil Surface

In the unbounded flow field, when an infinite cylinder rotating at a constant speed reaches equilibrium, a resultant force perpendicular to the flow direction acts on the cylinder, which is called lift force. If the rotating cylinder is regarded as the vortex core, the flow field in the vortex core is the vortex field with equal vorticity, and the flow field outside the vortex core is the flow field induced by the cylinder, which is the irrotational field. Such a flow model is a typical Rankine vortex model. At this time, the circulation of the rotating cylinder acts on the fluid through the cylinder edge interface contacting with the fluid, thus inducing the flow field outside the cylinder. The flow is intuitive and easy to understand, which is also verified by experiments. But for the low-speed ideal flow around an airfoil, the airfoil does not rotate, so how is the circulation generated? This is related to the physical mechanism of the formation process of the flow around the airfoil. One hundred and fourteen years ago, the theoretical explanation (or physical explanation) of the concept of starting vortex and attached vortex based on the ideal fluid motion is a classical content recognized by aerodynamics. Although there have been some objections over the years, it is generally accepted. I remember that at the beginning of contact, the author felt that it was a little abstract and difficult to understand. With the viscous flow around the airfoil, how does the attachment vortex exist near the airfoil surface? What is the function of the point at the back of the tip?

1. **Boundary layer characteristics and aerodynamic forces of steady viscous flow around airfoils**

   As shown in Fig. 2.30, in the steady viscous flow around an airfoil, a boundary layer flow around the near-wall region of the upper and lower airfoils will be formed, which is in a stable equilibrium state. At this time, the flow around the airfoil will form the boundary layer flow near the wall and the potential flow outside the boundary layer. The closed red line (including the boundary layer flow on the upper and lower wing surfaces) surrounding the airfoil is taken as clockwise (as shown in Fig. 2.30). The

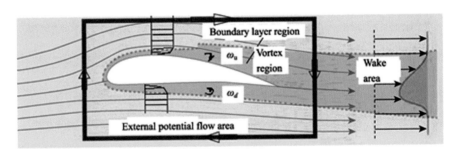

Fig. 2.30 Velocity loops around a steady airfoil

Stokes integral formula is as follows:

$$\Gamma = \oint_C \vec{V} \bullet d\vec{s} = \iint_A 2(\omega_u - \omega_d)d\sigma$$

The velocity circulation of the flow around the airfoil is positive clockwise, and the vorticity in the boundary layer near the wall of the upper wing is $2\omega_u$. The vorticity $2\omega_d$ in the boundary layer near the wall of the lower wing rotates counterclockwise, which is a negative contribution. Therefore, it can be further written as

$$\Gamma = \oint_C \vec{V} \bullet d\vec{s} = \int_0^b \left[\int_0^{\delta_u(x)} 2\omega_u dy - \int_0^{\delta_d(x)} 2\omega_d dy \right] dx = \int_0^b \gamma(x)dx$$

$$\gamma(x) = \int_0^{\delta_u(x)} 2\omega_u dy - \int_0^{\delta_d(x)} 2\omega_d dy = \gamma_u - \gamma_d$$

$$\gamma_u = \int_0^{\delta_u(x)} 2\omega_u dy, \gamma_d = \int_0^{\delta_d(x)} 2\omega_d dy$$

where $\gamma(x)$ is the surface vortex intensity along the chord, γ_u is the value of the upper wing (positive contribution), and γ_d is the value of the lower wing (negative contribution).

At the trailing edge of the airfoil, according to the Kutta and Joukowsky conditions, if the airflow is to leave the trailing edge smoothly, there is $\gamma(b) = 0.0$, so we can get

$$\gamma_u = \gamma_d, \int_0^{\delta_u(b)} 2\omega_u dy = \int_0^{\delta_d(b)} 2\omega_d dy$$

As an approximation, as shown in Fig. 2.31, by definition of

$$2\omega_u \approx \frac{V_u}{\delta_u(b)}, 2\omega_d \approx \frac{V_d}{\delta_d(b)}$$

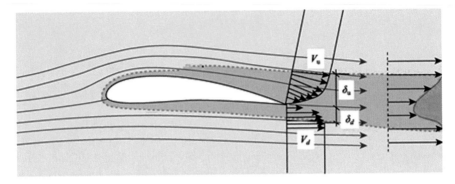

Fig. 2.31 Trailing edge conditions around steady airfoil

we get

$$\int_0^{\delta_u(b)} \frac{V_u}{\delta_u(b)} dy \approx \int_0^{\delta_d(b)} \frac{V_d}{\delta_d(b)} dy, V_u \approx V_d$$

This is the condition of Kutta and Joukowsky at the trailing edge, namely, the condition of leaving trailing edge smoothly is that the velocity of upper and lower wing surfaces leaving trailing edge is approximately equal. If you do the following

$$\gamma_0 = \int_0^{\delta_u(b)} 2\omega_u dy$$

$$\Gamma = \oint_C \vec{V} \cdot d\vec{s} = \int_0^b [\gamma_u - \gamma_d] dx = \int_0^b [\gamma_u - \gamma_0] dx$$

$$+ \int_0^b [\gamma_0 - \gamma_d] dx = \Gamma_u + \Gamma_d$$

$$L = \rho V_\infty \Gamma_u + \rho V_\infty \Gamma_d$$

the contribution of the upper and lower wings to the lift can be obtained. As shown in Fig. 2.32, at low angle of attack, the calculation formula of lift coefficient based on the pressure coefficient distribution acting on the

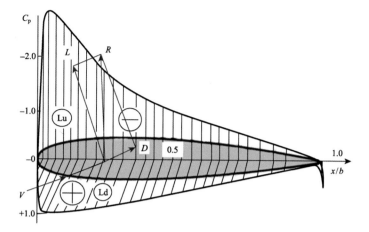

Fig. 2.32 Pressure coefficient and aerodynamic force along a chord line

airfoil is

$$C_L = \int_0^1 (C_{pd} - C_{pu}) d\xi = \int_0^1 C_{pd} dx + \int_0^1 (-C_{pu}) d\xi$$

$$L = \rho V_\infty \Gamma_u + \rho V_\infty \Gamma_d = \frac{1}{2} \rho V_\infty^2 b C_L$$

$$\Gamma_d = \frac{1}{2} V_\infty b \int_0^1 C_{pd} d\xi, \ \Gamma_u = \frac{1}{2} V_\infty b \int_0^1 (-C_{pu}) d\xi,$$

where C_{pu} and C_{pd} are the pressure coefficients acting on the airfoil surface, respectively. ξ is x/b.

Obviously, at any position away from the leading edge, the value of γ (x) depends on the difference of vorticity integral value in the local upper and lower wing boundary layer. According to the velocity distribution characteristics in the upper and lower wing boundary layer, the distribution of γ (x) along the chord line should be the change curve that gradually reduces from the leading edge to the rear edge, as shown in Fig. 2.33. It can be seen that the attached vorticity based on the concept of ideal fluid flow actually refers to the difference of vorticity integral value in the viscous boundary layer near the wall of the upper and lower wing surfaces in the steady flow around the airfoil. If the ideal fluid flow model is used instead of the boundary layer flow, it should be considered as the shape of the ideal fluid flow around the boundary curve of the airfoil surface

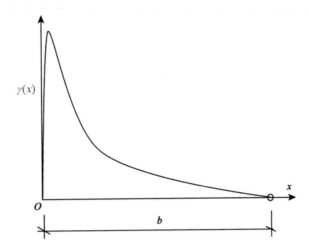

Fig. 2.33 Surface vortex intensity distribution along chord line

superimposed with the displacement thickness of the boundary layer. At the same time, the circulation value generated by the viscous boundary layer is added to the boundary, which is the attached vortex. This shows that the attached vortex is added to the external potential flow through the boundary layer of the wing surface.

For the flow in the near-wall boundary layer, the motion of the viscous fluid is always accompanied by the generation, diffusion, and dissipation of vorticity. When the Reynolds number of the incoming flow is large, the vortex flow in the boundary layer near the wall conforms to Prandtl's boundary layer approximation. Under the condition of no-slip boundary, it is equivalent to making the object surface a vortex surface source with certain intensity distribution. The relationship between the vorticity on the object surface Ω_b (clockwise is positive) and the wall shear stress τ_b is

$$\Omega_b = 2\omega_b = \left(\frac{\partial u}{\partial y} - \frac{\partial v}{\partial x}\right)_b = \left(\frac{\partial u}{\partial y}\right)_b = \frac{\tau_b}{\mu}$$

where u and v are the flow velocity components in the boundary layer. It can be seen that the vorticity on the airfoil is related to the wall shear stress, which indicates that the vorticity in the boundary layer is the largest on the airfoil and the vorticity away from the material surface decreases, which is caused by the viscous diffusion and dissipation of vorticity. For incompressible viscous flow around airfoil, the diffusion equation of the vorticity Ω is

$$\frac{\partial \Omega}{\partial t} + u\frac{\partial \Omega}{\partial x} + v\frac{\partial \Omega}{\partial y} = v\left(\frac{\partial^2 \Omega}{\partial x^2} + \frac{\partial^2 \Omega}{\partial y^2}\right)$$

In the steady boundary layer flow, the above equation can be simplified as

$$u\frac{\partial \Omega}{\partial x} + v\frac{\partial \Omega}{\partial y} = v\frac{\partial^2 \Omega}{\partial y^2}$$

In the boundary layer, on the one hand, the vorticity moves along the mainstream direction, and then gradually decreases. On the other hand, the vorticity diffuses along the vertical direction, and the vertical diffusion speed and attenuation speed depend on the viscosity coefficient of the fluid. The migration speed of vorticity depends on the flow speed, so the vorticity generated on the surface of the object will not spread to the whole field, but only in the boundary layer. The distance magnitude of vorticity diffusion in the normal direction of the wall is $\sqrt{vt}$, and the distance of vorticity migration along the flow direction is $V_\infty t$. for the airfoil with chord length b, the time required for vorticity migration from the leading edge to the trailing edge is b/V_∞, so the thickness of the boundary layer is

$$\delta \propto \sqrt{v\frac{b}{V_\infty}} = \sqrt{v\frac{b^2}{V_\infty b}} = b\sqrt{\frac{v}{V_\infty b}}$$

$$\frac{\delta}{b} \propto \frac{1}{\sqrt{Re}} \left(Re = \frac{V_\infty b}{v}\right)$$

2. **Evolution mechanism of boundary layer during starting airfoil**

For the unsteady flow around the airfoil during starting, it belongs to the formation and development process of the viscous boundary layer near the wall of the airfoil. The physical mechanism is complex, which involves the transformation of inviscid flow and viscous flow, the movement of separation points in the upper trailing edge of the airfoil, and the evolution and development process of separation region and separation vortex. It is obvious that the formation and development of the unsteady boundary layer in the starting process of the airfoil will eventually reach the steady equilibrium flow around the airfoil. The development of controlling this

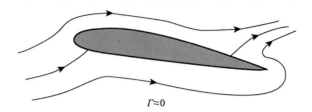

$\Gamma \approx 0$

Fig. 2.34 Early potential flow stage (accelerated)

process is an incompressible two-dimensional unsteady laminar boundary layer differential equation

$$\frac{\partial u}{\partial x} + \frac{\partial v}{\partial y} = 0$$

$$\frac{\partial u}{\partial t} + u\frac{\partial u}{\partial x} + v\frac{\partial u}{\partial y} = \frac{\partial V_e}{\partial t} + V_e\frac{\partial V_e}{\partial x} + v\frac{\partial^2 u}{\partial y^2}$$

where V_e is the outflow velocity of the boundary layer. In the initial stage of the airfoil starting from static condition, as shown in Fig. 2.34, the boundary layer has not been formed, the viscous shear force is large, the migration inertia force is small, and the unsteady inertia force of the outflow field is the main one. The above equation can be simplified as

$$\frac{\partial u}{\partial x} + \frac{\partial v}{\partial y} = 0$$

$$\frac{\partial u}{\partial t} - v\frac{\partial^2 u}{\partial y^2} = \frac{\partial V_e}{\partial t}$$

At the later stage of the start-up process, the boundary layer is basically formed and nearly stable. At this time, the unsteady inertial force is in the secondary position, and the boundary layer equation can be simplified as

$$\frac{\partial u}{\partial x} + \frac{\partial v}{\partial y} = 0$$

$$\frac{\partial u}{\partial t} - v\frac{\partial^2 u}{\partial y^2} = V_e\frac{\partial V_e}{\partial x} - u\frac{\partial u}{\partial x} - v\frac{\partial u}{\partial y}$$

Now, according to the development of boundary layer, the evolution of separation and vortex shedding, combined with the physical mechanism of viscous flow, the starting process of unsteady airfoil flow can be divided into the following stages.

(1) Initial potential flow stage

When the airfoil starts (as shown in Fig. 2.34), there is almost no boundary layer flow around the airfoil. The flow around the airfoil is dominated by the ideal flow. The rear stagnation point is located in the trailing edge area of the upper wing surface. The airflow around the lower wing surface reaches the rear stagnation point. The trailing edge point does not coincide with the rear stagnation point. There is no separation in the trailing edge area of the airfoil. At this time, the attached vorticity is almost zero, and the lift is close to zero. In this case, the velocity of the lower wing is slightly higher than that of the upper wing.

(2) Separation bubble stage

Due to the influence of centrifugal inertia force around the rear edge point, the reverse pressure gradient around the flow from the rear edge point to the rear stagnation point increases continuously, resulting in the separation of the rear edge area and the formation of the separation bubble (as shown in Fig. 2.35). At the same time, the rear stagnation point of the upper wing surface moves to the end of the separation bubble, and the separation point and the rear edge point do not coincide. At this time, the viscous flow in the trailing edge region of the airfoil begins to form, but the overall flow is still dominated by the ideal flow around the airfoil, and the attached vorticity is almost zero (the two counter-rotating vortices in the separation bubble cancel each other), and the lift is close to zero.

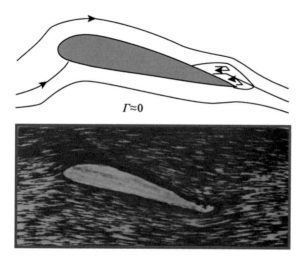

Fig. 2.35 Separation bubble stage (accelerated)

(3) Periodic shedding stage of trailing edge vortex

With the increase in airfoil velocity, the energy of vortex motion in the separation bubble increases continuously, which makes the kinetic energy accumulation in the bubble unable to self consume, so the separation bubble opens and forms vortex shedding (as shown in Fig. 2.36). At the same time, the centrifugal inertia force around the trailing edge is increased and the negative pressure near the trailing edge is increased due to the increase in air velocity, which causes the separation point of the trailing edge area of the upper wing surface to move to the trailing edge (flow from high pressure to low pressure). At this time, the boundary layer of the flow around the upper wing begins to form, which is relatively slower than the lower wing, thus increasing the speed of the outflow area of the upper wing boundary layer, and the boundary layer flow around the near area of the airfoil surface begins to play a role. The attached vorticity is not zero, and the lift begins to be greater than zero. With the increase in airfoil velocity, the vortices generated from the trailing edge are constantly shed out, and they are dumped downstream with the flow, and the flow velocity around the trailing edge is increasing. With the increase in centrifugal inertia force at the trailing edge, the negative pressure near the trailing edge increases continuously, which leads to the separation point of the trailing edge of the upper wing moving further to the rear edge. At this time, the flow boundary layer around the upper wing develops continuously, and the flow boundary layer around the lower wing begins to form and develop. The velocity difference in the outflow region of the upper and lower wing boundary layer further increases, the attached vorticity around the airfoil continues to increase, and the lift also increases.

(4) Stable equilibrium stage of boundary layer

The periodic shedding of the trailing edge separation vortices is repeated until the airfoil reaches a uniform speed and no longer accelerates. At this time, the separation point of the upper wing surface

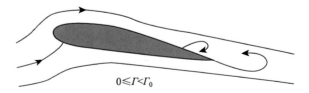

$$0 \leq \Gamma < \Gamma_0$$

Fig. 2.36 Period of falling edge separation vortex periodically falling off (acceleration)

moves to the trailing edge point, the vortex shedding stops, the airflow around the trailing edge point leaves smoothly, the boundary layer of the upper and lower wings forms a stable state, the velocity difference in the outflow area of the boundary layer reaches the maximum, the attached vorticity around the airfoil reaches the maximum, the lift reaches the maximum, and the airfoil flow around the airfoil completes the starting process (as shown in Fig. 2.37).

It should be noted that if the airfoil accelerates or decelerates again, the stable boundary layer changes from one stable boundary layer to another. The vortex shedding at the trailing edge will continue until a new stable boundary layer is formed (as shown in Fig. 2.38). After

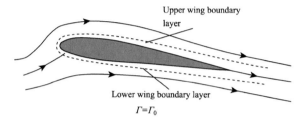

Fig. 2.37 Stable stage of boundary layer (uniform speed)

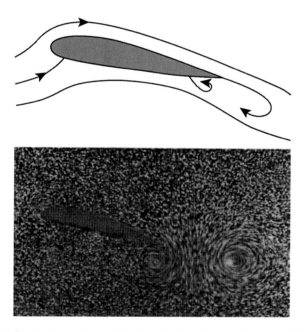

Fig. 2.38 Periodic shedding stage (deceleration) of rear edge separation vortex

reaching a new equilibrium state, the flow around the trailing edge will return smoothly leaving the trailing edge. It is just that the direction of the vortex shedding is different between the accelerating airfoil and the decelerating airfoil.

2.4   Low-Speed Flow Around Wing

The main components of the aircraft are the fuselage and wings (as shown in Fig. 2.39). The wings of the aircraft are the aerodynamic components that generate lift. In order to achieve a good aerodynamic effect, the wings are generally made of three-dimensional thin slender warped body structure, which is arranged on both sides of the fuselage (can be located on the upper, lower, and middle of the fuselage). There are various shapes of wings. The layout and shape of wings can be optimized according to different flight speeds and missions (as shown in Figs. 2.40 and 2.41). The earliest shape of the wing is a flat plate, such as the shape of a Chinese kite (with a hub to spread the cloth). The lift–drag ratio of the flat plate wing is the smallest, generally 2 to 3. Then there is the bend, which has a lift–drag ratio of 5.

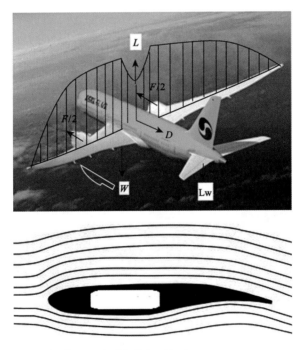

Fig. 2.39 Flow around wing and airfoil (section)

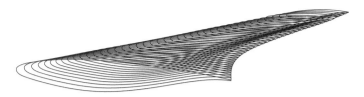

Fig. 2.40 B-spline parametric modeling of supercritical wing

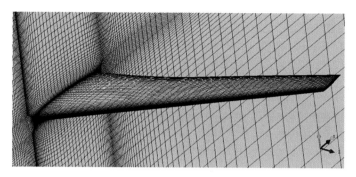

Fig. 2.41 Grid near critical wing

The later designed wing shape can produce lift–drag ratio of more than 20, for example, the lift–drag ratio of a pure wing of a large passenger plane can reach about 30. Due to the influence of wingtip, the lift–drag ratio of a three-dimensional wing is smaller than that of a two-dimensional airfoil. Because the fuselage mainly produces drag, the lift–drag ratio of the whole aircraft will be smaller when the fuselage and other drag components are added. For example, Boeing 747, when cruising, the lift–drag ratio of the aircraft is about 17 to 18, which is equivalent to lifting 1 kg of gravity only needs to overcome 55 g of drag.

In 1918, the German scientist Prandtl studied the aerodynamic problems of airfoils with chamber and put forward the thin wing theory. On this basis, the theory of lift line of a straight wing with a large aspect ratio is put forward. This work makes people realize the importance of the wingtip effect to the overall performance of the wing for a finite wingspan wing, and points out the essential relationship between wingtip vortex and induced drag, which has not been paid attention to for a long time. On the trailing edge of a straight wing with a large aspect ratio, a row of silk threads are evenly pasted along its aspect, and a small cotton ball is tied at the end of the silk threads. Then the wing is placed in a low-speed wind tunnel for the wind blowing test. The results show that for a finite span wing, due to the tip effect, the flow with high pressure on the lower surface of the wing will turn from the tip of the wing to the upper surface at a positive angle of attack, which makes the streamline of the upper surface deviate to the symmetry plane and the streamline of the lower surface deviate to the tip of the wing, and the skew increases gradually from the symmetry plane to the tip of the wing. When the airflow leaves the trailing edge of the wing, the airflow in the upper and lower wing surfaces, which is reversed along the span, will shear the air behind the wing, resulting in a free vortex surface (as shown in Fig. 2.42). D. Küchemann, the German aerodynamicist, once said, "vortices are the tendons of fluid motion." This is a famous saying in fluid mechanics, which profoundly summarizes the role of vortex in fluid motion. Professor Lu Shijia from Beijing University of Aeronautics and Astronautics once further pointed out that "the essence of fluid is vortex, because the fluid cannot withstand rubbing, and once rubbing, it will rub out vortex." This sentence not only shows the essential difference between fluid and solid but also points out the reason for vortex in fluid motion. Here, "rubbing" refers to the shearing effect on fluid movement. Therefore, the free vortex surface is generated by the spanwise flow of the upper and lower wings leaving the trailing edge.

Due to the interaction of the vortices, the free vortex surface will be rolled into two opposite vortices drawn from the wingtip at a place far away from

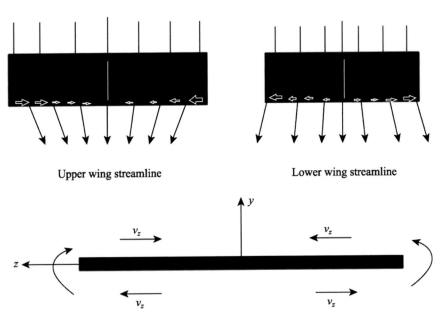

Fig. 2.42 Formation mechanism of free vortex surface at trailing edge of finite span wing

the trailing edge (about 1 time of the span). The axis of the vortices is approximately parallel to the direction of the incoming flow (as shown in Figs. 2.43, 2.44, 2.45 and 2.46). Figures 2.47 and 2.48 show the flow field characteristics and free vortex structure of bats in flight. Figure 2.49 shows the downwash of the aircraft wake and Fig. 2.50 shows the dissipation of the aircraft wake. Figure 2.51 shows the wake evolution of a large airliner during landing.

In order to lift the aircraft, when the gravity of the aircraft is fixed, the lift of the aircraft must reach the gravity. And because the lift is proportional to the square of the speed, it needs enough flight speed to make the wing produce the required lift. If the speed is not enough, the lift generated by it will not be enough to overcome the gravity of the aircraft. Therefore, before taking off, the aircraft needs to run a distance on the runway that is to make

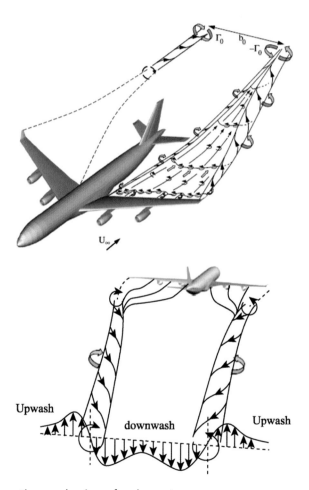

Fig. 2.43 Formation mechanism of wake vortex

Fig. 2.44 Aircraft wake

Fig. 2.45 Aircraft wake

Fig. 2.46 Numerical simulation of wing steady flow (streamline around wing)

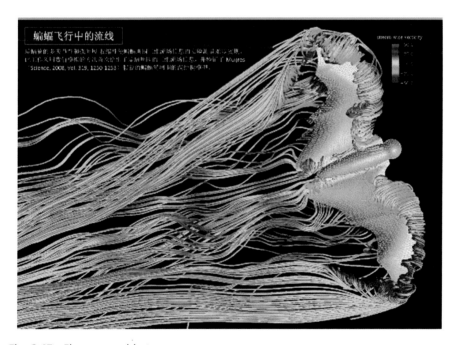

Fig. 2.47 Flow around bats

the aircraft accelerate to a sufficient speed and generate enough lift to make the aircraft rise. At this time, the pilot can fly off the ground and fly high as long as he pulls the pilot's pole. If the speed is not enough, the plane cannot get up even if the stick is pulled. The speed of the aircraft is displayed on the

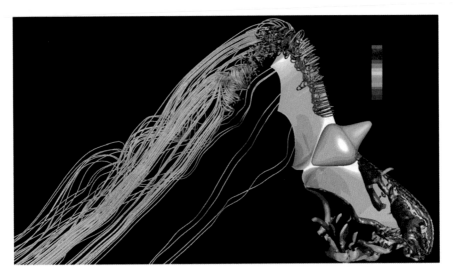

Fig. 2.48 Flow around bat

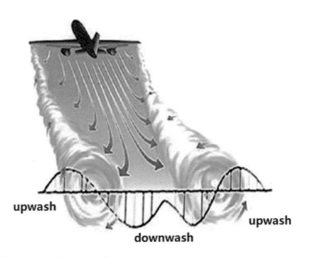

Fig. 2.49 Wake vortex downwash

speed dial at any time. When the ground speed is reached, the pilot can pull the rod to take off.

For tactical missiles and supersonic fighters, in order to reduce the wave drag of flight performance, a small aspect ratio wing (aspect ratio less than 3) is usually used, and a sharp edge, non-bending, torsional, symmetric, and large swept delta thin wing is usually used for the wing (as shown in Fig. 2.52). For the flow around a swept delta wing with a small aspect ratio

Fig. 2.50 Tail vortex dissipation

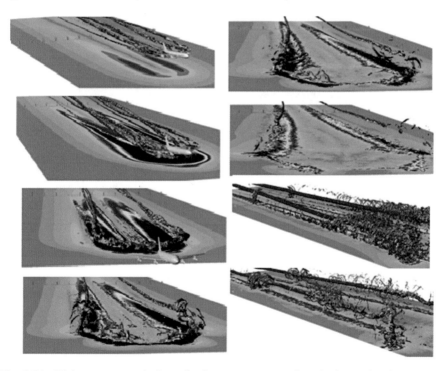

Fig. 2.51 Wake vortex evolution of a large passenger aircraft during landing

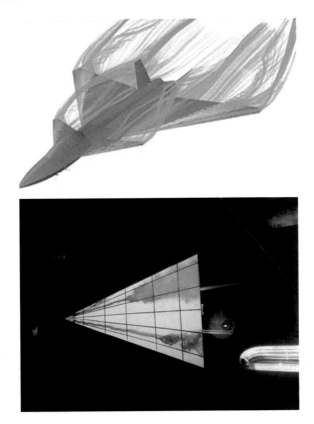

Fig. 2.52 Leading-edge vortex

and a large swept back, at a small angle of attack (3–40), because the high-pressure airflow on the lower wing surface bypasses the side edge and flows to the upper surface, there will be separation on the side edge, and a detached vortex will be formed on the upper wing surface (also known as the leading-edge vortex, as shown in Figs. 2.53, 2.54 and 2.55). The emergence of these detached vortices will produce greater negative pressure on the upper wing surface, resulting in a greater lift, which is often called vortex lift (as shown in Fig. 2.56). The lift characteristic curve of the wing with a small aspect ratio is nonlinear. In 1966, the United States aerodynamic scientist Polhamus E.C. (long-term work in NASA) proposed the leading-edge suction analogy method. The basic idea of this method is to divide the total lift of airfoil with detached vortex into the sum of potential flow lift and vortex lift. The normal force generated by the vortex on the wing surface is equal to the suction generated by bypassing the circular leading edge, and the direction is turned up 90°. Physically speaking, this analogy actually assumes that when the airflow is separated at the leading edge and attached to the upper surface

Fig. 2.53 Leading-edge vortex of wing at high angle of attack

of the wing, the force required to maintain the flow balance around the separation vortex is equal to the suction generated by the leading edge maintaining the flow around the body in the potential flow. According to the comparison of the leading-edge suction, the normal force increment caused by the leading-edge separation vortex is equal to the leading-edge suction. The vortex lift is equal to the projection of the normal force increment in the direction perpendicular to the incoming flow.

2.5   Basic Theory of Compressible Flow

With the appearance and rapid development of turbojet engine, the flight speed of aircraft is increasing rapidly. It is found that the air density cannot be regarded as a constant when the speed of flight in the air exceeds 100 m/s. But with the change in velocity, the influence of density on the flow cannot be ignored. According to the state equation of ideal gas, the key to solve this problem is to couple the fluid motion equation with the thermodynamic equation, establish the relationship between the motion parameters and the thermodynamic parameters, and correctly obtain the solution of the high-speed aerodynamics problem. In 1887–1896, the Austrian scientist Ernst Mach (1836–1916, as shown in Fig. 2.57) pointed out that the propagation characteristics of the disturbance caused by the projectile are different in different flows smaller than or larger than the speed of sound. In high-speed flow, the ratio of flow velocity to local sound velocity is an important dimensionless parameter. In 1929, the German aerodynamics scientist Akerlett first connected this dimensionless parameter with the name of Mach, and used the Mach number to describe the influence of fluid motion on compressibility.

Fig. 2.54 Development of wing leading-edge vortex

1. *One-dimensional steady compressible flow and its equations*

The problem of incompressible flow is more complicated than that of compressible flow. In the incompressible flow, the change in density and temperature caused by velocity difference is small, so the density and temperature can be regarded as constant approximately, so the parameters of gas flow are only velocity and pressure. However, in the compressible flow, the larger velocity difference causes a greater change in density and temperature, which has a significant impact on the flow. At this time, the parameters of airflow are velocity, pressure, density, and temperature.

Fig. 2.55 Development of duck-wing vortex

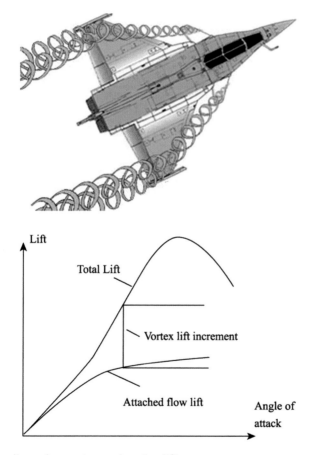

Fig. 2.56 Leading-edge vortex and vortex lift

Fig. 2.57 Ernst Mach (1836–1916, Austrian physicist)

In one-dimensional steady compressible flow, because the flow parameters increase from two to four, it needs four basic equations to solve. The basic control equations needed are as follows: state equation, continuous equation, momentum equation, and energy equation of ideal flow.

The equation of state for a complete gas is

$$p = \rho RT$$

where p is the pressure, ρ is the density, T is the temperature, and R is the gas characteristic constant (= 287.053 N.m/(kg.K)).

The continuity equation is

$$\rho AV = const., \quad \frac{d\rho}{\rho} + \frac{dV}{V} + \frac{dA}{A} = 0$$

where A is the pipe area. The momentum equation is

$$\frac{dp}{\rho} = -VdV$$

For the moving system of unit mass gas, the energy equation can be obtained from the first law of thermodynamics

$$dq = de + pd\left(\frac{1}{\rho}\right) + \frac{1}{\rho}dp + VdV$$

where dq is the heat transferred from the outside to the system, de is the internal energy increment of the system, $pd\left(\frac{1}{\rho}\right)$ is the expansion work of the system to the outside, $\left(\frac{1}{\rho}\right)dp$ is the pressure difference work of the moving system, and VdV is the kinetic energy increment of the system, which can also be written as

$$dq = d(e + \frac{p}{\rho} + \frac{V^2}{2})$$

In the case of adiabatic flow, the energy equation becomes

$$e + \frac{p}{\rho} + \frac{V^2}{2} = C$$

The above formula shows that the sum of internal energy, pressure energy, and kinetic energy per unit mass of gas is constant along the same streamline in a one-dimensional steady flow. In a compressible gas, the enthalpy is defined as

$$h = e + \frac{p}{\rho}, h = C_p T$$

where h is the sum of internal energy and pressure energy of unit mass gas. C_p is the specific heat coefficient of constant pressure ($= 1004.7$ N.m/(kg.K)). The energy equation of a one-dimensional steady adiabatic flow is

$$C_p T + \frac{V^2}{2} = C$$

The internal energy of unit mass gas can be expressed as

$$e = C_v T, C_v = 717.6 \, \text{N.m/(kg.K)}$$

where C_v is the constant volume specific heat coefficient of unit mass gas. Introducing specific heat ratio $\gamma = C_p / C_v$, the energy equation can be written as

$$\frac{\gamma RT}{\gamma - 1} + \frac{V^2}{2} = C, \frac{a^2}{\gamma - 1} + \frac{V^2}{2} = C, \frac{\gamma}{\gamma - 1}\frac{p}{\rho} + \frac{V^2}{2} = C$$

For the adiabatic steady flow of ideal fluid, it is also isentropic flow. The above energy equation can also be obtained by the integration of the Euler equation along the streamline. Using isentropic relation, the integral along streamline of the Euler equation is obtained as

$$\frac{V^2}{2} + \int \frac{dp}{\rho} = C$$

Using isentropic relation $p = C\rho^\gamma$, we get

$$\frac{V^2}{2} + \frac{\gamma}{\gamma - 1}\frac{p}{\rho} = C$$

In thermodynamics, adiabatic process and isentropic process are two different things. For the adiabatic flow of ideal fluid, it must be isentropic. In the case of viscous fluid, when there is friction between the flow layers, although it is adiabatic, the friction makes the mechanical energy converted into heat energy, which makes the entropy of the airflow increase, and the adiabatic entropy must be unequal. In adiabatic flow, the effect of viscous friction does not change the sum of kinetic energy and enthalpy, but part of kinetic energy is converted into enthalpy. (The above energy equation is applicable to adiabatic flow and adiabatic isentropic flow.) For one-dimensional steady adiabatic flow, the relationship between flow parameters and streamline integration can be determined, and the reference point is often needed to determine the constant. The reference point used is either the stagnation point (or the virtual stagnation point on the streamline) or the critical point.

Stagnation point refers to the point with zero flow velocity or kinetic energy on the same streamline, which can exist in the flow field or be a virtual reference value. According to the energy equation of one-dimensional adiabatic flow, the enthalpy of the fluid at the stagnation point reaches the maximum, which is called the total enthalpy $h0$, the corresponding temperature is called the total temperature T_0, the pressure is the total pressure p_0, and the density is the total density ρ_0. Relatively speaking, the velocity is not equal to zero, such as static pressure, static temperature, and static density.

$$\begin{cases} \frac{T_0}{T} = 1 + \frac{\gamma-1}{2}Ma^2 \\ \frac{p_0}{p} = (\frac{T_0}{T})^{\frac{\gamma}{\gamma-1}} = (1 + \frac{\gamma-1}{2}Ma^2)^{\frac{\gamma}{\gamma-1}} \\ \frac{\rho_0}{\rho} = (\frac{p_0}{p})^{\frac{1}{\gamma}} = (1 + \frac{\gamma-1}{2}Ma^2)^{\frac{1}{\gamma-1}} \end{cases}$$

where Ma is the local Mach number of the gas ($=V/a$, the $a = \sqrt{\gamma RT}$).

Based on the one-dimensional steady compressible flow equation, the quantitative relationship between the motion parameters and the static parameters of each point can be established.

In 1889, Swedish engineer Karl Gustaf Patrik De Laval (1845–1913, inventor of single-stage impact turbine, as shown in Fig. 2.58) successfully obtained the supersonic flow through the first contraction and then expansion pipe, manufactured the impact steam turbine, and proposed the famous pull valve nozzle. According to the continuous equation of one-dimensional steady flow

$$\frac{\mathrm{d}\rho}{\rho} + \frac{\mathrm{d}V}{V} + \frac{\mathrm{d}A}{A} = 0$$

Substituting the speed of sound $a^2 = \frac{\mathrm{d}p}{\mathrm{d}\rho}$ into the momentum equation, there are

$$\frac{\mathrm{d}\rho}{\rho} = -Ma^2\frac{\mathrm{d}V}{V}$$

Fig. 2.58 Karl Gustaf Patrik de Laval (1845–1913, Swedish engineer)

Substituting into the continuous equation to get

$$(Ma^2 - 1)\frac{\mathrm{d}V}{V} = \frac{\mathrm{d}A}{A}$$

It can be seen from the above formula:

(1) for subsonic (including low velocity) flow, if the pipe section shrinks, the flow velocity increases, and the area expansion flow velocity decreases;
(2) for supersonic (including low-speed) flow, if the cross section of the pipe shrinks, the flow rate will decrease, and the flow rate of area expansion will increase;
(3) the change rule of velocity and cross-sectional area of supersonic section is opposite to subsonic speed, because the contribution of density change to the continuous equation is different at supersonic speed and subsonic speed. The change in density in subsonic is slower than that in velocity, while that in supersonic is faster than that in velocity.

It can be seen that for a one-dimensional isentropic pipe flow, if the airflow is to be continuously accelerated from subsonic speed to supersonic speed along the pipe axis, that is, to keep $\mathrm{d}V > 0$ all the time, the pipe should be contracted first and then expanded, with the minimum section in the middle, that is, the throat. This shape of the pipe is called a Laval pipe or nozzle, and the contraction type and Laval nozzle are shown in Fig. 2.59. Figure 2.60 shows the supersonic jet flow in the tail nozzle of the engine.

2. **Broadcast interface and Mach wave of disturbance wave**
   When the object is moving in the still air, the influence range and the influence way of different moving speeds on the air are different. The so-called disturbance refers to the change in the speed, density, pressure,

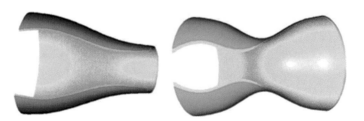

Fig. 2.59 Retractable and Laval Nozzles

Fig. 2.60 Supersonic jet flow of the engine nozzle

and other parameters of the airflow. For subsonic flow field and supersonic flow field, the propagation range of disturbance is different. In a uniform flow field, it is assumed that the gas is stationary and the disturbance source is moving. The small disturbance emitted by the disturbance source propagates around at the speed of sound, and its influence area has the following four situations (as shown in Fig. 2.61).

(1) Static disturbance source ($Ma = 0$)

From a certain moment, the disturbance wave surface emitted in the first n seconds is a concentric sphere with the disturbance source O as the center and na as the radius. As long as the time is long enough, any point in the space will be affected by the disturbance source, that is, the influence area of the disturbance source is the whole flow field.

(2) Subsonic disturbance source ($Ma < 1$, $V < \alpha$)

In the first N seconds, the spherical wave with a radius of na emitted by the disturbance source should move down from O to the On point along the direction of the incoming flow, OOn = nV. Because of nV < na, the disturbance can still cover the whole flow field. The small disturbance of subsonic flow field can be found in the whole flow field, and the turbulence of the airflow has been felt before the arrival of the disturbance source, so the flow direction and airflow parameters will be changed gradually to meet the requirements of the disturbance source.

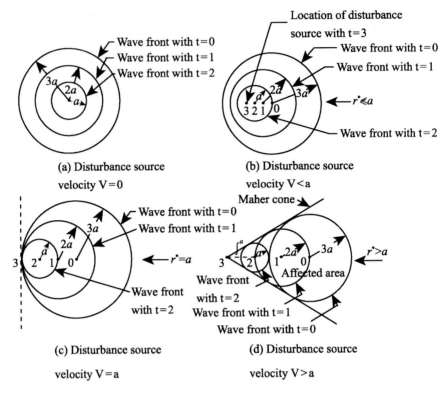

Fig. 2.61 The range of influence of disturbance sources with different speeds

(3) Sound velocity disturbance source ($Ma = 1$, v = a)

In the flow field where the disturbance source moves at the speed of sound, the small disturbance will not be transmitted to the upstream of the disturbance source, and the airflow does not feel any disturbance before the disturbance source arrives, so the existence of the disturbance source is not known.

(4) Supersonic disturbance source ($Ma > 1$, V > a)

If the disturbance source moves at supersonic speed, the influence range of a small disturbance will be smaller. When a small disturbance occurs to the supersonic flow, it will propagate around at the speed of sound. The envelope surface of the disturbed spherical wave is called the disturbed interface, which is also called the Mach wave array surface, which is called the Mach wave for short. For the point disturbance source, it is also called the Mach cone because of the conical shape of the array (the shape of the Mach wave array with different shapes is different, for example, the influence area of the thin wedge-shaped object is wedge-shaped, for the slender cone-shaped object, the Mach cone is cone-shaped). In the upstream of the Mach wave, the flow is not affected, and in the downstream of the Mach wave, the flow is disturbed. Let the half vertex angle of the Mach cone be called the Mach angle μ. According to the geometric relationship, the normal velocity of the airflow perpendicular to the Mach line is sound speed a, which has

$$\sin(\mu) = \frac{V_n}{V} = \frac{a}{V}, \mu = \arcsin\left(\frac{a}{V}\right) = \arcsin\left(\frac{1}{Ma}\right)$$

As shown in Fig. 2.62, the control body shown in Fig. 2.63 is taken around the CL line on the Mach wave, which can be seen from the continuous equation

$$m = \rho V_n = (\rho + d\rho)(V_n + dV_n)$$

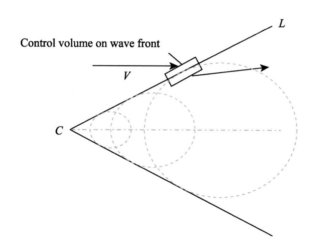

Fig. 2.62 Control body on Mach wave front

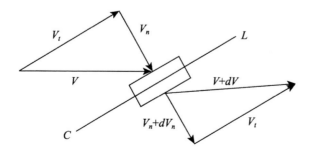

Fig. 2.63 Speed decomposition of over-control body

From the tangent momentum equation

$$V_t m - V_t' m = 0$$

From the normal momentum equation

$$\rho V_n d V_n = -dp$$

Solving jointly, we get

$$V_n = a, \ V_t = V_t'$$

It is shown that the normal velocity component in front of the Mach wave is sound velocity, and the tangential velocity through the Mach wave is constant.

3. **Expansion wave**

In compressible flow, the change in density cannot be ignored. The process of pressure and density increasing is called compression process, and the process of pressure and density decreasing is called expansion process. In high subsonic flow, although there are compression and expansion processes, there is no disturbance boundary, and the disturbance can spread to the whole flow field. In supersonic flow, both compression and expansion processes have disturbed boundary, which is similar to the Mach wave. This disturbed boundary is called wave front. As shown in Fig. 2.64, for supersonic flow around the expansion angle of the wall outside fold dδ. At the O-point, the wall is turned outward at a small angle dδ, which enlarges the flow area. Then the O-point is a small disturbance source, and the propagation range of the disturbance is downstream

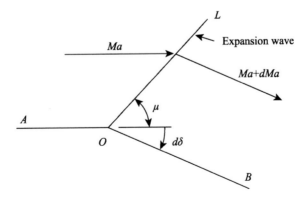

Fig. 2.64 Supersonic flow bypasses the expansion angle of microelements

of the Mach wave oL emitted by the O-point. The result of the disturbance is that the airflow is also deflected by an angle of the same size of dδ. For supersonic flow, the wall fold will increase the cross-sectional area of the passage, which will increase the flow speed, while the pressure and density will decrease, and the flow will expand. At this time, the role of the Mach wave line oL is to accelerate and reduce the pressure of supersonic airflow, and the airflow will undergo adiabatic accelerating expansion process, so the Mach wave oL is called expansion wave. For a limited expansion angle, when the supersonic flow is bypassed, each expansion wave will be connected to form a continuous expansion band (as shown in Fig. 2.65). In 1908, Prandtl and his student Theodor Meyer proposed the expansion wave theory, which became the theoretical basis of the design of supersonic wind tunnel.

4. **Shock Wave**

When the aircraft is flying at supersonic speed, the disturbance cannot be transmitted to the front of the aircraft. As a result, the gas in front of the aircraft is suddenly compressed, forming a concentrated strong disturbance (composed of numerous micro compression waves). At this time, an interface of the compression process appears, which is called a shock wave. Shock wave is a strong fault wave formed by the superposition of weak compression wave, which has a strong nonlinear effect. After the shock wave, the pressure, density, and temperature of the gas will suddenly rise, and the flow rate will suddenly drop. The jump in pressure produces an audible bang. For example, when an aircraft is flying at supersonic speed in a lower airspace, people on the ground can hear the sound, which is called a sonic boom. The shock wave can be photographed by optical instrument

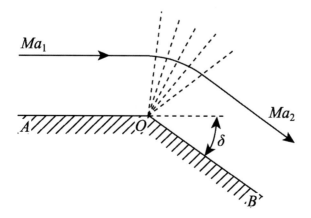

Fig. 2.65 Supersonic flow bypasses a finite expansion angle

according to the sudden change in shock gas density. The shock wave of ideal gas has no thickness and is a mathematical discontinuity. The actual gas has viscosity and heat transfer, which makes the shock wave continuous, but the process is still very rapid. Therefore, the actual shock wave has thickness, but the value is very small. Only in a certain multiple of the free path of gas molecules, the larger the relative supersonic Mach number of the wave front, the smaller the thickness value. Figure 2.66 shows the shock cloud of the supersonic fighter, and Fig. 2.67 shows the shock disk at the exit of the nozzle.

(1) Positive shock

The angle between the shock wave (wave front) and the direction of the incoming flow is called the shock angle. When the shock angle is 90°, it is called a positive shock. Its wave front is perpendicular to the airflow direction. If the relative coordinate system is used to establish the relationship between the flow parameters before and after the shock wave, the problem is relatively simple. The advantage of using relative coordinates is that the flow is steady relative to the wave front, and the basic equations of steady flow can be directly applied. As shown in Fig. 2.68, take the control surface shown by the dotted line before and after the shock wave. When the shock wave does not move, the static airflow flows to the shock wave at the velocity V_1, and the airflow velocity after the shock wave is V_2 (less than a_2).

From the continuous equation

$$\rho_1 V_1 = \rho_2 V_2$$

Fig. 2.66 Shock cloud generated when a fighter breaks through the sound barrier

Fig. 2.67 Shock formed by fighter jet nozzle

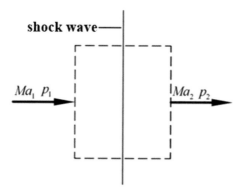

Fig. 2.68 Control body in relative coordinates

The momentum equation is applied to the dotted control surface

$$-\rho_1 V_1^2 + \rho_2 V_2^2 = p_1 - p_2$$

From the energy equation of adiabatic flow

$$V_1^2 + \frac{2}{\gamma - 1} a_1^2 = V_2^2 + \frac{2}{\gamma - 1} a_2^2 = \frac{\gamma + 1}{\gamma - 1} a_*^2$$

In 1908, Prandtl derived the relation of supersonic shock wave as

$$\lambda_1\lambda_2 = 1, \lambda_1 = \frac{V_1}{a^*}, \lambda_2 = \frac{V_2}{a^*}, a^* = \sqrt{\frac{2}{\gamma + 1}} a_0$$

where a^* is the critical section velocity. The above formula is the famous Prandtl shock wave formula, which represents the relationship between the velocity coefficient before and after the wave. It is shown that the velocity coefficient λ_2 after a positive shock wave is exactly the reciprocal of the velocity coefficient λ_1 before the shock wave. Because the wave front must be supersonic flow, $\lambda_1 > 1$, the velocity coefficient after the wave $\lambda_2 < 1$, that is to say, the supersonic flow must be subsonic after passing through the positive shock wave. Other physical relations before and after a positive shock can also be derived. If the density ratio is

$$\frac{\rho_2}{\rho_1} = \frac{V_1}{V_2} = \frac{\lambda_1}{\lambda_2} = \lambda_1^2 = \frac{\frac{\gamma+1}{2} Ma_1^2}{1 + \frac{\gamma-1}{2} Ma_1^2}$$

the static temperature relation is

$$\frac{T_2}{T_1} = \frac{1}{\lambda_1^2} \frac{1 - \frac{\gamma+1}{\gamma-1}\lambda_1^2}{\lambda_1^2 - \frac{\gamma+1}{\gamma-1}}$$

The relationship between static pressure strength ratio is

$$\frac{p_2}{p_1} = \frac{2\gamma}{\gamma+1} Ma_1^2 - \frac{\gamma-1}{\gamma+1} = \frac{1 - \frac{\gamma+1}{\gamma-1}\lambda_1^2}{\lambda_1^2 - \frac{\gamma+1}{\gamma-1}}$$

After shock wave, the total temperature is constant, the total pressure decreases, and the entropy increases.

(2) Oblique shock

For the flow around the body with different head shapes, the shock wave shape is different when it flies at hypersonic speed. For example, for an aircraft with a rhombus wing shape, if the top angle of the leading-edge wedge of the wing is small under a certain $Ma_1 > 1$, two simple oblique shock waves will be formed, the wave surface and the direction of motion are at a certain angle, and the shock wave is attached to the tip of the object. This kind of shock is different from the normal shock in form. Its wave front is obliquely intersected with the flow direction, which is called the oblique shock. In oblique shock wave, the angle β between the front of shock wave and the direction of incoming flow is called shock angle. Similarly, the airflow direction after the oblique shock wave is not perpendicular to the shock wave surface, nor parallel to the airflow direction before the wave, but parallel to the sharp split plane. The included angle δ is called the airflow angle, which means the angle of the airflow after the oblique shock wave (as shown in Fig. 2.69).

Compared with normal shock, oblique shock belongs to weak shock, and the smaller the wedge angle is, the weaker the shock is. If the half apex angle δ of the wedge is infinitesimal, it is obvious that the disturbance of the very thin wedge to the supersonic flow must be weak, the disturbance wave must be the Mach wave, and the disturbance angle must be the Mach angle. With the increase in wedge angle δ, the shock wave β increases. The larger δ is, the larger β is. Therefore, for a positive shock, as long as Ma_1 is determined, the increment of other parameters is determined; for an oblique shock, it is necessary to determine the shock slope angle β by Ma_1 and δ, and then other physical quantities can be solved according to the shock intensity determined by β. Figure 2.70 shows the relationship between oblique shock angle and wedge angle at different Mach numbers.

(3) Internal structure of shock wave

It is possible to treat the shock wave as a sudden jump surface (discontinuity) without thickness in dealing with the general flow problems without causing large errors. However, when considering the effect of viscosity, shock waves cannot be regarded as thickness free. In fact, the velocity gradient is infinite when the thickness of the shock wave is zero, and the viscosity has a great influence. Under the action of viscosity, the velocity cannot decelerate suddenly from V_1 to V_2 without thickness, that is to say, there must be a transition zone, and the thickness of this transition zone is the thickness of shock wave. It is found that the thickness of shock wave is a small quantity, which is the same order of magnitude as the average free path of molecule. In the sea atmosphere, the molecular free path is 70×10^{-6} mm. In the case of $Ma = 3$, the shock thickness calculated by continuum theory is 66×10^{-6} mm. Some people use the equation of continuous medium considering viscosity and heat transfer to analyze the change process of airflow parameters in the shock wave. It is found

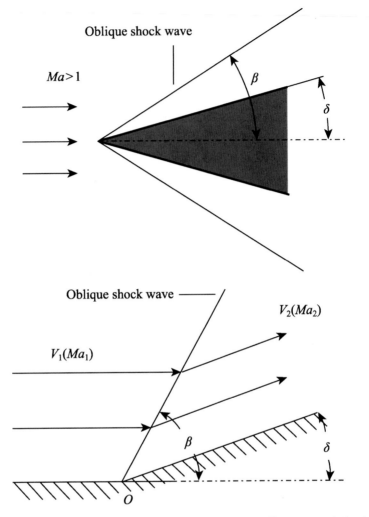

Fig. 2.69 Oblique shock waves passing by a supersonic flow around the leading edge

that the parameters such as velocity change continuously in the shock wave, as shown in Fig. 2.71. The change in velocity from V_1 at the front to V_2 at the back is a gradual process, so the shock thickness cannot be clearly defined. Generally, the tangent of the curve is made at the inflection point of the v-x curve, and the distance between the intersection of the curve and the horizontal line with V_1 and V_2 as constant values is the shock thickness.

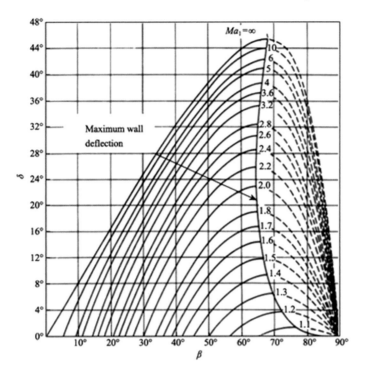

Fig. 2.70 Relationship between oblique shock angle and cleavage angle at different Mach numbers

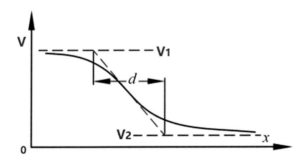

Fig. 2.71 The change curve of the shock velocity obtained from the continuum hypothesis

2.6 Solution of Compressible Flow

In the flow around the aircraft, whether the compressibility of air must be considered depends on whether the pressure change in the flow process can cause significant density change. When Ma number is less than 0.3, the density change is less than 5% and the change in pressure coefficient is less than 2%. Generally, this flow can be regarded as incompressible approximately; only when Ma number is greater than 0.3, the compressibility effect can be considered. Different compressibility has different flow characteristics and different influences on aerodynamics. For the flow around an aircraft, it is usually divided by the undisturbed Mach number Ma far ahead. When Ma is less than 0.3, it is similar to the incompressible flow, which is called low-speed flow; when Ma is between 0.3 and 0.8, it is subsonic flow, at this time, the influence of compressibility on aerodynamic characteristics can be corrected by compressibility on the results of low-speed flow. When Ma is between 0.8 and 1.2, it is a transonic flow. At this time, there will be a local supersonic or local subsonic region in the flow field, and generally shock waves will appear. In this range, the aerodynamic coefficient will change greatly with the increase in Ma. When Ma is between 1.2 and 5, it is supersonic flow. When Ma exceeds 5, the flow is hypersonic flow.

From the analysis of the propagation characteristics of the disturbance wave in the flow field, it can be seen that for the subsonic flow with Ma number less than 1, the disturbance propagates throughout the whole flow field. For the supersonic flow with Ma number greater than 1, the disturbance only propagates to the downstream Mach cone. These characteristics will affect the properties of control equations and the methods of solving them.

1. *Subsonic flow*

When the subsonic flow is an inviscid potential flow, the governing equation is a nonlinear second-order elliptic partial differential equation. The main approximate method to study this kind of flow is the theory of small perturbation linearization. Prandtl in 1922 and H. Glauert (British aerodynamic scientist) in 1928 established, respectively, the compressibility correction rule for the subsonic flow, which is called Prandtl and Glauert rule. According to this law, the influence of compressibility on aerodynamic characteristics can be obtained by modifying the compressibility of the results of low-speed flow, and it is not necessary to solve the compressible flow equation. In 1939, von Karman and Qian Xuesen further amended the Prandtl–Glauert rule and put forward the famous Karman Qian rule. This law is better to establish the correction relation of air compressibility to the surface pressure in subsonic flow, and the range of application is obviously larger than that of Prandtl–Glauert's rule, especially the correction of the pressure coefficient on the leeward side of the airfoil is more reasonable.

2. *Supersonic flow*

When the supersonic flow is an inviscid potential flow, the governing equation is nonlinear second-order hyperbolic partial differential equation. Similarly, based on the theory of small perturbation linearization, the second-order hyperbolic partial differential equation is established and solved by the characteristic method. In supersonic flow, the effects of compression wave, expansion wave, and shock wave on the flow are studied. The shock wave of an ideal gas has no thickness and is a discontinuity in mathematical sense. Rankine in 1870 (a British scientist) and Hugoniot in 1887(a French scientist) independently derived the relationship between flow parameters before and after shock wave (called the Rankine Hugoniot relationship) using the continuous equation, momentum equation, and energy equation. Later, Prandtl established the relationship between velocity coefficients before and after normal shock wave. For the small disturbance problem of thin wing, Ackeret put forward the linear theory of two-dimensional airfoil in 1925, and then correspondingly appeared the linear theory of three-dimensional wing. For two-dimensional and three-dimensional steady supersonic flows, the boundary between disturbed and undisturbed areas is the Mach wave. If the supersonic flow is accelerated by a series of Mach waves, it is called expansion wave. Prandtl and his student T. Meyer (1907–1908) established the expansion wave relation. Figure 2.72 shows the oblique shock wave and the positive shock wave around the head of the body. Figure 2.73 shows the shock system and the Mach disk of supersonic aircraft.

Prandtl Glauert condensation clouds (as shown in Fig. 2.74) are formed by the condensation of water vapor in the air into clouds when the air stream passes through the shock wave and compresses the surrounding air. After the condensation of water vapor into tiny water droplets, it is like a cloud. But it is not always accompanied by acoustic explosion, and it is not necessarily the shock wave when the sound barrier is broken.

3. *Transonic flow*

For the transonic inviscid potential flow, part of the supersonic flow area (as shown in Fig. 2.75 with the emergence of shock wave) will be around the flow field. The flow change is complex, and the flow control equation is a second-order nonlinear mixed partial differential equation, which is difficult to solve theoretically. Especially when the speed of flight or flow is close to the speed of sound, the aerodynamic performance of the aircraft changes rapidly, the resistance increases abruptly, and the lift drops abruptly. The maneuverability and stability of the aircraft are extremely deteriorated, which is the famous sound barrier in the aviation history.

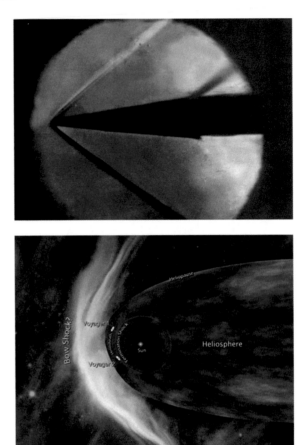

Fig. 2.72 Oblique shock and forward shock (bow shock)

The appearance of a high thrust engine rushed over the sound barrier, but it did not solve the complex transonic flow problem well. Until the 1960s, due to the requirements of transonic cruise flight, maneuvering flight, and the development of high-efficiency jet engine, the research of transonic flow has been paid more attention and developed greatly.

2.7   Hypersonic Aerodynamics

The term hypersonic speed was put forward by Qian Xuesen in 1946. Hypersonic aerodynamics is a newly developed subject, which mainly studies the law of hypersonic airflow and the interaction between air and hypersonic vehicle. In modern times, with the promotion of the aerospace industry, the

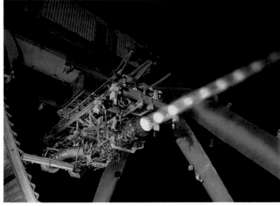

Fig. 2.73 Supersonic aircraft shock system (360doc.com) Mach disk

theory, calculation, and experimental technology of hypersonic aerodynamics have developed rapidly. Hypersonic flow is generally defined as a flow with a velocity more than five times that of sound, i.e., $Ma_\infty > 5$. In fact, this definition is not absolute. Whether the flow is hypersonic or not depends on the specific shape of the aircraft. For the flow around a blunt body, the characteristics of hypersonic flow begin when the Ma_∞ number is greater than 3; for the flow around a slender body, the hypersonic flow begins when the Ma_∞ number is greater than 10. The main problems of hypersonic aerodynamics are aerodynamic force (lift, drag, moment, pressure distribution, etc.), aerodynamic heat (heat flow calculation, heat release measures, etc.), and aerodynamic physics (photoelectric characteristics of flow field). The main means of research are theoretical analysis, numerical calculation, and wind tunnel test. The main parameters of hypersonic flow simulation include the free flow Mach number, Reynolds number, total flow enthalpy, density

Fig. 2.74 Prandtl-Glauert Condensate Cloud

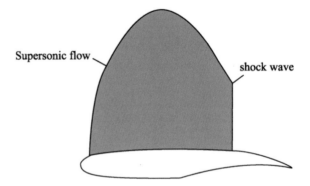

Supersonic flow

shock wave

Fig. 2.75 Transonic airfoil flow

ratio before and after shock wave, test gas, wall temperature to total temperature ratio, and thermal and chemical properties of the flow field. Common ground simulation equipment includes shock tube, arc heating wind tunnel, hypersonic wind tunnel, and free ballistic target.

Hypersonic flow characteristics are closely related to the shape of the aircraft (as shown in Fig. 2.76), and its physical phenomena are very complex. For space vehicles, the high-altitude non-equilibrium thermochemical phenomena, viscous interaction, and rarefied effect must be considered. For long-range ballistic missile (as shown in Fig. 2.77), they are subject to severe aerodynamic heating rate and high dynamic pressure, but the time is very short. Ablation thermal protection system can be used to resist severe turbulence heating rate, so it is necessary to study the thermal interaction between flow field and exothermic layer erosion reduction. As shown in Fig. 2.78, a hypersonic vehicle equipped with an inspiratory propulsion system must work at a lower altitude to meet the requirements of engine performance. At this time, high dynamic pressure and high Reynolds number will cause huge aerodynamic loads. Boundary layer transition and severe surface heating will become important issues in the development of such vehicles.

In summary, the main characteristics of hypersonic flow are as follows:

(1) Small density ratio and thin shock layer

In the flow around a hypersonic vehicle, the flow area between shock wave and object is called shock layer, which is a feature of hypersonic flow. The larger the free flow Mach number Ma, the stronger the shock wave, the greater the compressibility of the gas after the shock wave,

Fig. 2.76 The X-51A hypersonic aircraft being developed in the United States (cruising speed is up to 5.1 Mach)

Fig. 2.77 Long-range ballistic missile

Fig. 2.78 US pre-researched SR-72 hypersonic vehicle (Mach 6)

and the smaller the density ratio before and after the shock wave. For example, for the high Mach flow with complete gas, the density ratio before and after the positive shock wave is about 1/6, while the density ratio of the real gas in the reentry flight of Apollo spacecraft is about 1/20. According to the conservation of mass, the shock wave is close to the surface. In this case, with the help of the boundary layer theory, the order of magnitude analysis of the physical quantities of convective animals can be carried out and the approximate theory of thin shock layer can be established.

(2) Flow field controlled by strong viscous effect

For hypersonic flows, the order of magnitude relations among the laminar boundary layer δ, the incoming Mach number Ma_∞, and the incoming Reynolds number Rex (the characteristic scale of the incoming velocity and the length of the surface) are

$$\frac{\delta}{x} \propto \frac{Ma_\infty^2}{\sqrt{Rex}}$$

Under the condition of hypersonic flow at high altitude, the thickness δ of the laminar boundary layer becomes very large, which changes the shape of the flow around the aircraft surface and seriously affects the flow of the outflow field. Especially because of the thin shock layer, the thickness of the boundary layer cannot be omitted compared with the shock layer, and even the whole shock layer is affected by viscosity. At this time, the viscosity effect affects the whole flow field, and Prandtl's boundary layer theory fails.

(3) High entropy layer effect of flow around blunt head

In hypersonic flow around a blunt body, the convective heat transfer at the stagnation point of the head is inversely proportional to the square root of the curvature radius of the head, so the passivation of the head is beneficial to reduce the heat load. Because the shock around the blunt head has a high bending behavior, the streamline passing through different positions of the curve shock has experienced a different entropy increase, so the gas layer with a strong entropy gradient will cover the surface of the object to form a high entropy layer, and extend to a considerable distance downstream of the head. Because of the different entropy values of streamlines entering into the outer edge of the boundary layer, the characteristics of the outer edge of the boundary layer are affected by the high entropy layer, resulting in vortex interaction.

(4) High temperature effect of flow around blunt head

When the hypersonic flow decelerates through shock compression and viscous blockade, the kinetic energy of part of the flow changes into the internal energy of the random movement of molecules, which makes the gas temperature increase. This temperature rise leads to the failure of the traditional complete gas hypothesis. For example, the temperature of Apollo spacecraft at 53 km, $T = 283$ K and Mach number of incoming flow is 32.5, and the stagnation point temperature of gas around blunt head is as high as 11600 K.

(5) Rarefied gas effect

When a hypersonic vehicle is flying at a very low atmospheric density, the rarefied gas flow around it will appear. When the air density is very low, the ratio of the average free path of the gas molecules to the characteristic scale of the aircraft is not small, or even of the same magnitude. At this time, the movement of the gas medium is no longer continuous (the airflow and the surface do not meet the condition of no slip); the gas temperature on the surface is different from that on the wall, and there is temperature jump. Therefore, it is necessary to study this kind of flow by different methods than continuous flow, which is usually treated by molecular motion theory, which represents discrete characteristics.

2.8   Principle of Aeroacoustics

1. *Classic and Aeroacoustics*

The sound wave is a tiny pressure fluctuation phenomenon that propagates through the medium and can be sensed by human or animal auditory organs (as shown in Fig. 2.79). The sound wave is produced by the solid vibration in contact with the fluid medium, or by the vibration force directly acting on the fluid, the violent motion of the fluid itself (such as jet flow), the thermal effect of the vibration, etc. The space of sound wave propagation is called sound field. Sound wave can be understood as the propagation of disturbance wave whose medium deviates from the equilibrium state. This transmission process is only the transmission of energy, but not the transmission of quality. If the disturbance is

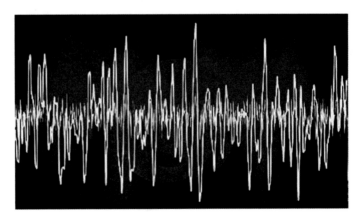

Fig. 2.79 Sound wave propagation

small, the transmission of sound wave satisfies the classical wave equation, which is a linear wave. If the disturbance is large, the linear wave equation is not satisfied, and the wave dispersion and shock wave will appear.

The experiment shows that the propagation speed of sound wave is constant in the same medium and normal temperature and pressure. For example, the propagation speed of sound wave in air is 340 m/s; in water is 1345 m/s; and in solid is 4000–8000 m/s. If dissipation is not taken into account, the waveform of sound wave will remain unchanged in the process of propagation. The sound wave heard by the human ear is called an audible sound wave, which is between 20 Hz and 20 kHz. The wave less than 20 Hz is called infrasound wave, and the wave greater than 20 kHz is called ultrasonic wave.

The sound pressure level is defined as

$$L_p = 10 \lg \frac{p^2}{p_0^2} = 20 \lg \frac{p}{p_0} \quad (\text{unit}: \ \text{dB})$$

Among them, $p_0 = 2 \times 10^{-5}$ Pa, which is the reference sound pressure. The range of sound pressure that the human ear can bear is as follows: the lowest sound pressure $p_0 = 2 \times 10^{-5}$ Pa, the highest sound pressure $p_m = 20$ Pa, which are called hearing threshold and pain threshold, respectively, with a difference of one million times, between 20 and 120 dB.

Aeroacoustics is a new interdisciplinary subject based on aerodynamics and classical acoustics. It studies the mechanism and control of the sound production of gas itself and the interaction between gas and solid boundary. It has successfully applied the classical acoustics, especially the physical concepts, basic rules, and skills of dealing with problems similar to the acoustics of moving media. The noise radiated by the disturbance caused by the gas flow or the interaction between the object and the gas is called aerodynamic noise. The main excitation mechanism of aerodynamic noise is the propagation of internal stress and pressure disturbance in the medium caused by the relative movement of solid and gas and the irregular movement of flow itself. The sound source of aerodynamic noise is generated by moving vortex, or free gas movement, or the interaction between solid and gas. In addition, the heat source can also generate aerodynamic noise. The sound sources of aerodynamic noise can be divided into monopole (pulsating volume excitation), dipole (oscillating force excitation), and quadrupole (turbulence excitation) according to their different sound generation mechanism. Monopole sound source is the sound source formed when the mass or heat inflow into the medium is not uniform, and it can also be generated by moving objects. Monopole sound source generated by moving objects is also called thickness noise, which is a kind of surface sound source. It is caused by the mass change in the object surface caused by the movement or vibration of the object. The typical monopole sources are the pulse jet of high-speed airflow periodically discharged through the nozzle, the flute of stable airflow periodically modulated, and the thickness noise of propeller, rotor, and other moving parts which make the air displace periodically. Monopole source can also be considered as a point source of pulsating mass. If the center of a balloon is placed on this point source, we can observe that the balloon expands or contracts as the gas enters or exits. This motion is always purely radial, and the surrounding gas is compressed or expanded to adapt to its motion, which is the formation of a spherically symmetric sound field. Setting a mathematical spherical boundary around the sound source, we can observe the outflow and inflow of the accumulated net flow through the boundary, which is called monopole (point sound source). The amplitude and phase of the sound field are the same at each point on the spherical surface, and the monopole sound source direction is shown in Fig. 2.80. For a moving monopole sound source, the sound radiation power W is

$$W \propto \frac{\rho^2 V_0^4 D^2}{\rho_0 a_0}$$

where ρ is the fluid density of the moving sound source, V_0 is the fluid velocity of the moving sound source, and D is the regional characteristic

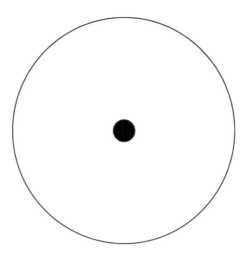

Fig. 2.80 Monopole sound source direction diagram

scale of the moving sound source. ρ_0 is the density of the environment medium, and a_0 is the sound velocity of the environment medium.

Dipole sound source is a kind of aerodynamic sound source formed by unsteady aerodynamic force between fluid and object when there are obstacles in the fluid. Dipole noise, also known as load noise, belongs to surface acoustic source, which is caused by the stress change in the surface of the object acting on the adjacent fluid. The dipole can be described by two monopoles with equal size and opposite phase. The common dipole is a small ball with constant shape and volume. Any object with low-frequency vibration is a dipole. For example, aircraft, propeller, helicopter, and turbine flying in the turbulent atmosphere will produce dipole noise. When integrating along the whole spherical boundary, the net flow of the fluid always shows zero, because the flow-in equals the flow-out. However, since the inflow and outflow flow are in the same direction and their momentum is additive, there is a net momentum in the system. According to Newton's second law, a force related to a dipole can be found. The second way to describe a dipole is to think of it as a ball driven by an oscillating force. Under these two descriptions, the fluid motion on the observation boundary is equivalent. There is radial flow along the change in momentum or the axial direction of the force. It can be inferred that the compressible motion or acoustic motion is the largest. There is no radial movement in the direction 90° from the force axis. Therefore, the sound field has a maximum occupation direction, and the direction perpendicular to the axis is zero. Each lobe of the sound field has a 180° difference, just as the phase difference between the outflow and inflow of the fluid at the sound source. As shown in Fig. 2.81, the orientation of the dipole is in the shape of "8". The sound radiation power W of the dipole is

$$W \propto \frac{\rho^2 V_0^6 D^2}{\rho_0 a_0^3}$$

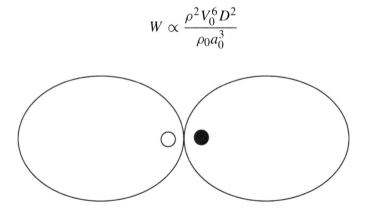

Fig. 2.81 Direction diagram of dipole sound source

The quadrupole source is a body source, which is closely related to the nonlinear flow in the control plane. If there is no mass or heat injection into the medium, and there is no obstacle, only the stress related to viscosity can radiate sound waves (turbulent stress), then this is the quadrupole sound source (stress field sound source). If there is no object in the airflow, the force can only be generated in the same size and opposite direction, which is equivalent to a pair of dipoles with the same size and opposite phase. The main sound source in the high subsonic turbulent jet is the quadrupole sound source. There is neither net mass flow nor net force in the integration along the spherical boundary around the quadrupole source. The orientation diagram of the quadrupole is shown in Fig. 2.82, which is in the form of "four lobes". The sound radiation power W of the quadrupole is

$$W \propto \frac{\rho^2 V_0^8 D^2}{\rho_0 a_0^5}$$

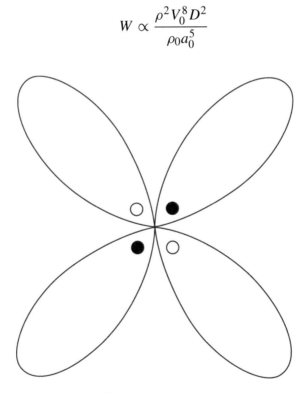

Fig. 2.82 Direction diagram of dipole sound source

In most cases, the sound power of quadrupole source is so small that it can be ignored.

2. *Development of aeroacoustics theory*

Since the theory of sound was published by the British scientist Rayleigh in the middle of the nineteenth century, the research of classical acoustics has reached a climax at the end of the nineteenth century. Its basic theory has been quite mature at the beginning of the twentieth century, but after entering the twentieth century, its development is less. With the end of World War II, jet propulsion technology began to enter the aviation industry. The aerodynamic noise generated by powerful jet flow has become a serious obstacle to the promotion of jet engines. This technical difficulty makes people realize the urgency of solving the aerodynamic noise problem. Since then, the aerodynamic sound has developed as a new discipline. In 1952, the British fluid mechanics Michael James Lighthill (1924–1998, as shown in Fig. 2.83) published his famous Lighthill equation and sound simulation theory in the Journal of the Royal Society of England. Today, people generally regard this work as a sign of the birth of aeroacoustics. The Lighthill equation has become the most basic equation for the study of aeroacoustics. Different from the linear wave equation that assumes that the medium is static and uniform and the amplitude of the harmonic quantity is small, the Lighthill equation is derived directly from the Navier–Stokes equation (N-S equation for short), without any

Fig. 2.83 Michael James Lighthill (1924–1998, British Fluid Mechanics)

simplification and assumption. It expresses the left side of the equation as the classical acoustic wave equation and moves all the terms deviating from the wave equation to the right side of the equation as the source term. So people can first use the experimental or computational methods (DNS, Les, or even turbulence model theory) to obtain the expression of these source terms, and then regard the sound field as the sound generated by the sound source propagating in the static medium, and then use the mature classical acoustic methods to calculate the sound field. This method of treating the flow field and sound field separately is the famous aeroacoustics analogy. Theoretically, it is possible to solve the sound field parameters directly from the N-S equation by numerical method, but the pressure disturbance of sound wave is a very small quantity. Even at the current level of numerical calculation, it is still a challenging work to obtain the convergence solution from the equation and ensure that the results have appropriate accuracy, which was even more impossible 50 years ago. Lighthill proposed a famous 8-power law using the equation, that is, the radiated sound power of the turbulent jet aerodynamic noise is directly proportional to the 8-power of the flow characteristic velocity. His theory of acoustic simulation has been widely used in practical problems, which ushered in a new era of aeroacoustics.

The theory of Lighthill acoustic simulation is established for the aerodynamic noise of airflow in unbounded space. It is applicable to the case that the solid boundary does not play a major role, such as jet aerodynamic noise. However, for many practical situations, such as the noise of stationary objects and moving objects in turbulence, the influence of solid boundary cannot be ignored. In 1955, Curle first used Kirchhoff's method to generalize Lighthill's theory to consider the influence of the static solid boundary and obtained that the effect of the solid boundary is equivalent to the distribution of the dipole sound source on the whole solid boundary, and the intensity of the dipole sound source at each point is equal to the force of the solid surface acting on the fluid at that point. Therefore, the sound field at this time is composed of a dipole on the solid surface and a quadrupole (Lighthill stress tensor) source outside the solid wall. The problem of Aeolian sound and aerodynamic noise induced by cylinder vortex separation in turbulent flow is solved successfully by Curle's theory. In 1969, Ffowcs Williams and Hawkings extended Curle's achievement to consider the influence of moving solid wall on sound, that is, the sound generation of moving objects in fluid, using the generalized function method. The equation named after them, FW-H equation, was obtained. This equation shows that the sound field produced by the inter-action between the moving object and the fluid is composed of quadrupole source, dipole source, and monopole source. The FW-H equation is the most effective tool to solve the aerodynamic noise radiation problem of fans, propellers, and compressor rotors.

At the same time, Lighthill's theory has achieved great success, but has also led to some new problems. Because the sound field and the flow field are essentially unified, their governing equations are all N-S equations. Theoretically, the flow field solution and the sound field solution can be obtained directly from the equations. However, the Lighthill sound simu-lation theory divides the solution of the flow field and the sound field into two steps, which cannot explain such basic problems as how the sound field and the flow field interact, how the sound energy is generated and transferred in the fluid, etc. Therefore, after Lighthill, the development of aeroacoustics continues in accordance with Lighthill 's thought on the one hand, and on the other hand, the connotation and research scope of aeroacoustics continue to expand. In 1964, Powell Alan put forward the theory of vortex sound, which shows that the generation of sound wave is closely related to the interaction between vortices, potential flow, and vortices in the fluid, and the formation and transformation of sound wave energy are also completed through these nonlinear interactions.

3. *Calculation method of aerodynamic noise*
   Aerodynamic noise is mainly related to aerodynamics and acoustics, so the calculation of aerodynamic noise can be divided into two parts: the generation of aerodynamic noise in the flow field and the propagation in the sound field. The two parts can be calculated at the same time or separately. Farassat (1975), the aeroacoustics expert, summed up the calculation methods of aerodynamic noise as follows: pure theoretical method; semiempirical method; pure numerical method; CFD combined with "acoustic analogy". It is specified below.

   (1) Pure theoretical method
       The pure theory method is to obtain the analytical results of the flow field and sound field using the mathematical theory tools directly. This method is usually suitable for the basic research of the relevant simplified model, which is an important basis for the development of other methods, and also a standard tool to verify the correctness of other methods. The model solved by this method is as simple as possible on the premise of catching some physical phenomena before obtaining the analytical solution. For example, Ffowcs Williams and Hall (1970) solved the aerodynamic noise at the trailing edge of a

wing using a simple quadrupole motion model. The results of this kind of method are very valuable, because it can help researchers to obtain some important qualitative knowledge, such as the influence of law of speed and other parameters on aerodynamic noise intensity, similarity criterion, aerodynamic noise spectrum characteristics, directivity law, and noise reduction guiding principle.

(2) Semiempirical method

Because the turbulence problem has not been solved completely, there are some difficulties in numerical simulation and quantitative analysis, so the semiempirical method based on experimental database and theoretical analysis is more often used. This method has the advantages of intuition and stability, and is an important means to study aerodynamic noise. The aircraft noise prediction program of the NASA Langley experimental center relies heavily on semiempirical methods. In recent years, through microphone array, a large number of high-quality models and acoustic data of full-scale aircraft have been measured, which improves the accuracy of semiempirical method.

(3) Pure numerical method

In essence, sound wave is a part of the flow field, so the generation and propagation process of the sound wave can be obtained by directly solving the N-S equation in theory (as shown in Fig. 2.84). Based on this idea, the pure numerical method unifies the flow field and the sound field and calculates the turbulent flow and the sound wave propagation through the complete numerical method. This method is the same as the hybrid method in the calculation of aerodynamic noise, and the difference is the propagation process of aerodynamic

Fig. 2.84 Wing body fusion engine noise radiation (NASA)

noise. The pure numerical method uses the numerical method to calculate the propagation, so the calculation model allows obstacles between the sound source and the observation point, while the mixed method does not. However, due to the magnitude difference between the sound field pressure and the flow field pressure, this method is very demanding for calculation. On the one hand, it needs a grid with dense near-field and far-field, on the other hand, it needs a low dissipation and low dispersion format. Therefore, the main difficulty of this method is the large amounts of calculation of far-field aerodynamic noise.

(4) Method of combining CFD with "acoustic analogy"

The combination of CFD and "acoustic analogy" is also known as a hybrid method, which is the most commonly used method to solve aerodynamic noise. In this method, the unsteady RANS, LES/DES, and other methods are used to calculate the sound source, or the RANS is added to the disturbance factor to simulate the sound source, and then the LEE (linearized Euler equation) method or FW-H, Kirchhoff, and other integration methods are used to calculate the propagation or radiation of the sound wave. The hybrid method is usually divided into two parts. First, the source information (flow field characteristics) is calculated, and then the sound propagation or far-field radiation is calculated according to the needs. At present, the main aerodynamic noise calculation method is the hybrid method because of the influence of grid, the accuracy of discrete format, the amount of calculation, and the time of calculation.

2.9   Stall Characteristics of Low-Speed Airfoil and Wing

1. *Stall characteristics of airfoil*

As the angle of attack of airfoil increases, the lift coefficient of airfoil will appear maximum and then decrease. This is the result of the separation of the flow around the airfoil. Stall characteristics of airfoil refer to the aerodynamic performance near the maximum lift coefficient. The phenomenon of airfoil separation is closely related to the flow and pressure distribution on the leeward side of the airfoil.

At a certain angle of attack, when the low-speed airflow bypasses the airfoil, it can be seen from the pressure distribution and velocity change in the upper wing surface that the airflow on the upper wing surfacestarts to accelerate and decompress rapidly to the maximum speed point (forward pressure gradient area) after passing the front stagnation point, and then starts to decelerate and pressurize to the rear edge point (reverse pressure gradient area) of the airfoil. The lower wing is slowly accelerated to the trailing edge of the lower wing as shown in Fig. 2.85.

However, with the increase in the angle of attack, the forward stagnation point moves backward, and the suction peak of the airflow around the near area of the front edge increases, which makes it difficult for the airflow behind the peak point to flow backward against the reverse pressure gradient, and the airflow slows down seriously. This not only causes the boundary layer to thicken and become turbulent, but also when the angle of attack is high to a certain extent, the adverse pressure gradient causes the flow to be unable to continue to slow down against the adverse pressure, resulting in separation and the main flow leaving the airfoil. At this time, the airflow is divided into two parts: the flow inside the separation zone and the flow outside the separation zone.

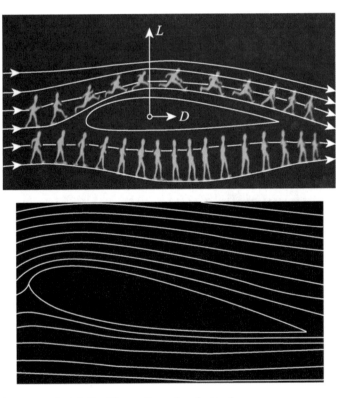

Fig. 2.85 Low-speed airfoil with small angle of attack

In the separation boundary (called free boundary), the static pressure of the two must be equal everywhere. After separation, the mainstream will no longer slow down and increase pressure. The air flows in the separation zone, because the main flow continuously takes away the mass through the viscous effect on the free boundary, the central part will be filled by the airflow continuously from the back, forming the backflow of the central part as shown in Fig. 2.86.

According to a large number of experiments, the separation of flow around a low-speed airfoil at a large Reynolds number can be divided into two parts according to the thickness:(1) trailing edge separation (turbulence separation); (2) leading-edge separation (leading-edge short bubble separation); (3) thin wing separation (leading-edge long bubble separation) as shown in Fig. 2.87.

(1) Trailing edge separation (turbulence separation)

The relative thickness of the corresponding airfoil is more than 12%. The negative pressure of the airfoil head is not very large. The separation starts from the trailing edge area of the upper airfoil surface. With the increase in the angle of attack, the separation point gradually develops to the leading edge, and the slope of the lift line deviates

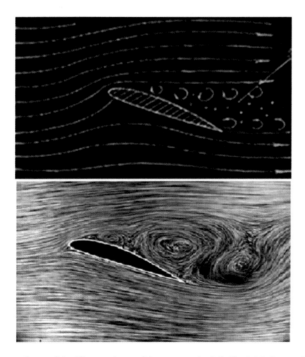

Fig. 2.86 Separation of trailing edge of low-speed airfoil at high angle of attack

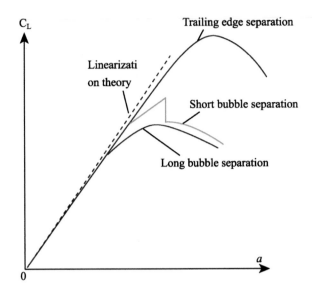

Fig. 2.87 Lift coefficient curve of airfoil with different thicknesses under high angle of attack

from the straight line at first (but the increase in the leading-edge suction is still greater than the decrease in the trailing edge separation). When the angle of attack reaches a certain value, when the separation point develops to a certain position on the upper wing surface (about half of the wing surface), the lift coefficient reaches the maximum (the increase in leading-edge suction is balanced with the decrease in trailing edge separation), and then the lift coefficient decreases (the increase in leading-edge suction is smaller than the decrease in trailing edge separation). The development of trailing edge separation is relatively slow, the change in flow spectrum is continuous without sudden jump, and the lift curve of stall area also changes slowly, with good stall characteristics as shown in Fig. 2.88.

For the airfoil with medium thickness (6–9%), the radius of leading edge is small, and the negative pressure is large when the air flows around the leading edge, which results in a large reverse pressure gradient. Even at a small angle of attack, the flow separation takes place near the leading edge, and the separated boundary layer turns into turbulence, which obtains energy from the outflow, and then attaches to the wing surface, forming separated bubbles. At first, this kind of short bubble is very short, only 0.5–1% of the chord length. When the angle of attack reaches the stall angle, the short bubble suddenly opens and the airflow can no longer be attached, resulting

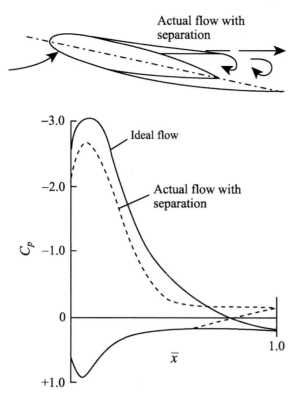

Fig. 2.88 Separation characteristics of trailing edge of low-speed airfoil at high angle of attack leading-edge short bubble separation

in the sudden complete separation of the upper wing surface and the sudden change in lift and moment. As the angle of attack decreases, the lift coefficient curve cannot return to the original path, as shown in Fig. 2.89.

For a thin airfoil (4–6% thickness), the leading-edge radius is smaller, and the negative pressure is larger when the air flows around the leading edge, which results in a large reverse pressure gradient. Even at a small angle of attack, the flow separation is caused near the leading edge. After separation, the boundary layer turns into turbulence, which obtains energy from the outflow, flows for a long distance, and then attaches to the wing surface, forming a long separation bubble. At first, the bubble is not long, only 2–3% of the chord length. However, with the increase in the angle of attack, the reattachment point moves to the downstream continuously. When the stall angle of attack is reached, the bubble no longer attaches, the upper wing surface is completely separated, and the lift reaches the

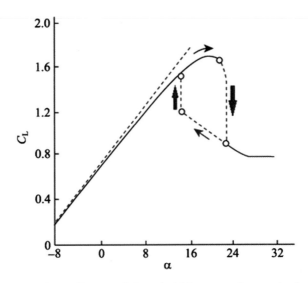

Fig. 2.89 Curve of lift coefficient of short bubble separation at the leading edge of a low-speed airfoil long bubble separation at leading edge (thin airfoil separation)

maximum value. The angle of attack continues to increase and the lift decreases.

In addition to the above three types of separation, there may also be a mixed separation form, such as when the airflow bypasses the airfoil, the separation occurs at the leading edge and the trailing edge at the same time.

2. *Stall characteristics of high aspect ratio straight wing*

At a small angle of attack, the lift coefficient C_L of the wing has a linear relationship with the angle of attack α. However, when α continues to increase to a certain extent, $C_L \sim \alpha$ curve begins to deviate from the linear relationship. At this time, the boundary layer near the trailing edge of the airfoil began to be partially separated, but it did not extend to the whole airfoil. Therefore, C_L will still increase when it continues to increase. Then, as the separation area gradually expands, it almost covers the whole wing surface. When C_L rises to a maximum value C_{Lmax}, if α increases again, C_L will decline. Now the wing stalls.

There are many factors that affect the stall characteristics of wings, such as airfoil, the Reynolds number, the Mach number, and plane shape of wings. The following only discusses the low-speed flow stall characteristics of non torsional elliptical, rectangular, and trapezoidal wings.

(1) Stall characteristics of elliptical wing

According to the theory of lift line, for an elliptical wing, the downwash velocity is constant along the span (the downwash angle is constant), so the effective angle of attack of each wing section along the span is constant. Therefore, if the same airfoil is used to design an elliptical wing, with the increase in α, the whole wingspan will separate from each wing section at the same time, reaching C_{Lmax} (maximum lift coefficient of the airfoil), and stall will occur at the same time, with good stall characteristics, as shown in Fig. 2.90.

(2) Stall characteristics of rectangular wing

The downwash speed of the rectangular wing increases from the root to the tip (also the downwash angle), so the effective angle of attack of the root section is larger than that of the tip section, and the lift coefficient of the corresponding section is larger than that of the tip. Therefore, the separation first occurs in the root part of the wing, then the separation area gradually extends to the wing end, and the stall is gradual, as shown in Fig. 2.91.

(3) Stall characteristics of trapezoidal wing

In the case of trapezoidal straight wing, the downwash speed decreases from the root to the tip. Therefore, the effective angle of attack of the airfoil section increases toward the wingtip, and the trend is more obvious with the increase in the root tip ratio. Therefore, the separation first occurs near the wingtip, which not only reduces the

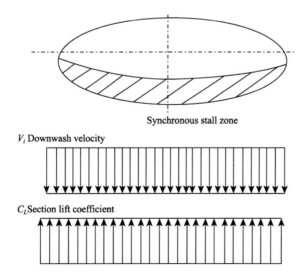

Synchronous stall zone

V_i Downwash velocity

C_L Section lift coefficient

Fig. 2.90 Stall characteristics of an elliptical wing

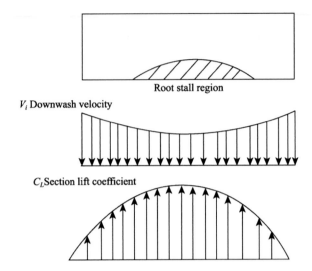

Fig. 2.91 Stall characteristics of rectangular wing

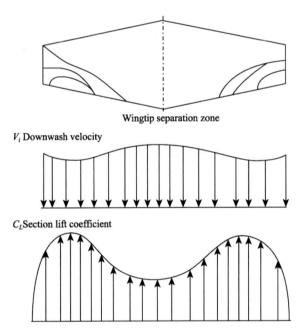

Fig. 2.92 Stall characteristics of trapezoidal wing

maximum lift coefficient of the wing, but also reduces the efficiency of the aileron and other control surfaces, as shown in Fig. 2.92. The stall characteristics of swept wing are similar to those of trapezoidal wing.

It can be seen that the elliptical wing not only has a good lift–drag characteristics at medium and small angles of attack, but also has good stall characteristics at high angles of attack. The lift–drag characteristics of the rectangular wing are not as good as those of the elliptical wing at medium and small angles of attack, and the C_{Lmax} at high angles of attack is also small. However, the separation of the wing root first will not cause the deterioration of the aileron characteristics, which can give the pilot a warning of impending stall. Because the lift–drag characteristics of trapezoidal wing are similar to those of elliptical wing at medium and small angle of attack, the structure weight of trapezoidal wing is also lighter, so it is widely used. However, the separation first occurs near the wingtip, causing the wingtip to stall first. So in terms of stall characteristics, trapezoidal straight wing is the worst of the three wings. Especially, the reduction in aileron efficiency caused by the first separation of wingtip may lead to serious flight safety problems, which is a serious and even inadmissible disadvantage in terms of aerodynamics. But as has been pointed out before, the plane shape of trapezoidal wing is the closest to the optimal plane shape, so trapezoidal wing is often used, but measures should be taken to improve its stall characteristics. Common methods are as follows:

(1) geometric torsion, such as external wash torsion, is adopted to reduce the angle of attack in the wingtip area to avoid the wingtip reaching the stall state too early. The value of torsion angle is $\varphi = -2° \sim -40°$.
(2) the airfoil with a high stall angle of attack is adopted near the wingtip.
(3) the leading-edge slat is used in the outer part of the wing, which makes the airflow with high pressure flow from the lower wing surface to the upper surface through the leading-edge slot, accelerating the airflow on the upper wing surface, thus delaying the separation of the boundary layer of the outer part of the wing.

2.10 Interaction Between Shock Wave and Boundary Layer in Supersonic Flow

1. *Overview*

When the flight speed of the aircraft exceeds the critical speed or supersonic speed, the supersonic flow will appear when the airflow bypasses the aircraft, which will form shock wave outside the boundary layer around the body. The interaction between shock wave and boundary layer almost exists in the transonic or supersonic flow. It involves the stability of compressible flow, transition, separation, shock oscillation, and turbulence pulsation, as well as the correlation among vortex, wave, and flow. Especially in recent years, with the development of transonic and supersonic vehicle research, the mechanism of compressible boundary layer transition and separation caused by shock wave and boundary layer interference has been paid more and more attention, because they directly affect the resistance, surface thermal protection, and flight performance of the vehicle.

As we all know, in 1904, Professor Ludwig Prandtl (1875–1953), the famous German hydrologist, first studied the flow problem affected by viscosity in the thin layer near the wall at low speed, put forward the famous boundary layer theory, and explained the mechanism of resistance generation and heat exchange around the flow object physically, which made it a widely used and studied theory. The study on the interaction between shock waves and compressible boundary layer flow was first carried out by Liepmann H.W. in 1946 and Jakob Ackeret (1898–1981), the Swiss aerodynamics scientist, in 1947. After that, it developed slowly. But with the advent of supersonic vehicles, people begin to pay more attention to this problem. Especially in the past ten years, driven by the development of supersonic vehicles, transonic transporters, and reusable space–time hypersonic vehicles, with the rapid development of computational and experimental hydrodynamics, the research of supersonic compressible boundary layer flow and shock wave interference has been pushed to a new climax. Based on different Reynolds numbers, there are laminar and turbulent flow states in the boundary layer, which have different effects on the wall friction resistance and heat transfer performance. If there is an interaction between shock wave and compressible boundary layer, the flow in the boundary layer will be more complex, and complex flow problems such as laminar flow, transition, turbulence, separation, and reattachment may occur (as shown in Fig. 2.93), which will seriously affect lift, resistance, and surface thermal protection of the aircraft.

2. *The interaction between normal shock wave and laminar boundary layer*

It has been known that the presence of a positive shock in the supersonic flow will reduce the Mach number of the main flow to the subsonic value, which is accompanied by a rapid increase in pressure, density, and

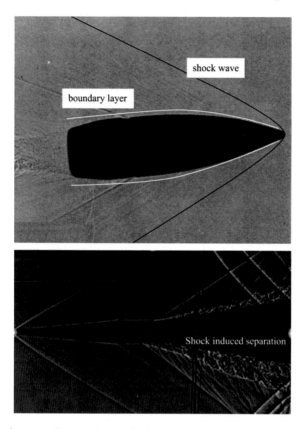

Fig. 2.93 Shock waves interacting with the boundary layer

temperature along the flow direction. If the wall boundary layer is encountered, the effect of shock wave is similar to imposing a sudden inverse pressure gradient on the boundary layer, which will seriously change the flow characteristics of the boundary layer. If the shock wave is strong, the interference between the shock wave and the boundary layer will cause the separation of the boundary layer behind the wave, at the same time, it will cause strong unsteady flow and strong heat conduction in the local area, which will seriously affect the performance of the aircraft. Therefore, the study of the interaction mechanism between shock wave and boundary layer is of great significance to the prediction of transition position, separation position, shock oscillation control, separation control, and other complex problems.

For the flow around the supersonic object, the flow outside the boundary layer is supersonic flow, but the fluid velocity in the boundary layer is rapidly reduced to zero velocity on the wall due to the influence of viscosity, so there are subsonic and supersonic regions in the layer, and the shock pressurization outside the boundary layer will propagate upstream to the front of the shock through the subsonic region in the layer. This reverse pressure gradient from the downstream to the upstream obviously changes the boundary layer state in the upstream of the shock wave, resulting in the thickness of the boundary layer in the disturbed area, changes in the distribution of velocity, temperature, pressure, and density in the layer, the reduction in friction resistance, and the change in the local shock structure near the wall.

In the transonic flow around airfoil, when the Mach number of the incoming flow is greater than the critical Mach number, the supersonic flow region will appear when the flow is around the upper wing surface. Obviously, the supersonic region is connected with the downstream subsonic flow by almost positive shock form. The shock acceleration and deceleration behavior cannot be carried forward in the supersonic region outside the boundary layer. However, when encountering the subsonic boundary layer near the wall, the shock acceleration will propagate upstream along the upstream of the boundary layer, resulting in the thickening of the boundary layer. In serious cases, it will cause the separation of the boundary layer (the separation bubble or the end of the wave) full separation, related to shock intensity and boundary layer characteristics), shock oscillation, and other complex phenomena, as shown in Fig. 2.94.

The interference characteristics of positive shock and boundary layer are closely related to shock intensity and boundary layer characteristics. For

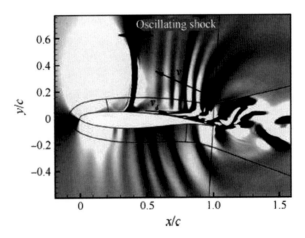

Fig. 2.94 Interaction between transonic airfoil shock waves and boundary layer

example, in the strong interference region, the definition and magnitude of the boundary layer are not applicable, because the flow velocity gradient and the normal velocity gradient have the same magnitude. The interference flow in the laminar or turbulent region is also obviously different. For example, when the boundary conditions and shock intensity are the same, the upstream propagation distance of the reverse pressure in the laminar boundary layer is far from the turbulent boundary layer, and the laminar boundary layer has weak resistance to flow separation. The interference characteristics of positive shock and laminar boundary layer are related to the Mach number, Reynolds number, and shock intensity outside the boundary layer. According to the different shock intensity, three kinds of interference may occur.

(1) Weak interference

In the transonic flow around the airfoil, the shock wave is relatively weak in the upper wing area of the airfoil, which interferes with the laminar boundary layer. The weak shock pressure increases the front boundary layer slowly, without transition and separation. The thickened boundary layer in front of the wave causes the airflow to deflect inward, and a series of weak compression waves are reflected on the boundary layer of the wave front, and merge with the main shock wave to form the so-called λ-shaped wave, resulting in the included angle between the main wave front and the boundary layer being less than 90 degrees, as shown in Fig. 2.95. After the main shock wave, the

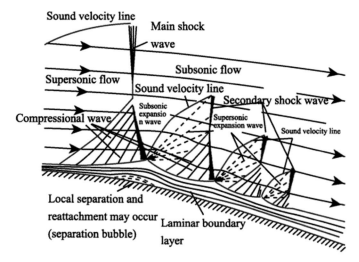

Fig. 2.95 Interference between weak shock waves on a curved surface and a laminar boundary layer (laminar boundary layer)

boundary layer is attached to the wall again, the thickness is thinner, the outer boundary is inclined to the wall, the airflow turns outward, and the subsonic region becomes the supersonic region again, thus the fan-shaped supersonic expansion wave system appears, the pressure drops, and then the secondary shock appears again, which causes the shock wave and the laminar boundary layer to interfere again, the interference characteristics are basically similar, but the intensity drops obviously. If the condition is suitable, it may be repeated several times to form a series of λ-shaped wave systems.

(2) Medium interference

If the velocity of the incoming flow increases, the Mach number and Reynolds number of the wave front will increase. Although the wave front is a laminar boundary layer, under the strong shock wave (the increase in the reverse pressure gradient), the boundary layer after the wave will be separated, and soon turn into turbulence, and then attach to the wall. The results are as follows: (1) a separation bubble appears on the wall; (2) a series of reflected compressions appear in front of the main shock wave, forming a λ wave system with the main shock wave; (3) after the main shock wave, the boundary layer is attached to the wall again, the outer boundary is inclined to the wall, the airflow turns outward, forming a fan-shaped supersonic expansion wave system, the pressure drops, and the secondary shock appears; 4) when the boundary layer becomes flat, a series of compressional wave systems appear again. Due to the large Reynolds number and strong shock wave, the transition is generally in turbulent boundary layer, as shown in Fig. 2.96.

(3) Strong interference

With the increase in the Mach number of the incoming flow, the shock intensity also increases. The interference between the shock and the boundary layer is enough to cause the separation of the laminar boundary layer. Thus, the main direction outside the boundary layer turns obviously, and a stable oblique shock wave appears before the main shock wave, which forms an obvious λ shock wave above the boundary layer. Because the boundary layer can no longer be attached, the secondary shock wave no longer exists, as shown in Fig. 2.97. In this case, the flow around the airfoil will suddenly separate, the lift will decrease, the drag will suddenly increase, and shock-induced stall will appear.

3. ***The interaction between oblique shock wave and boundary layer in supersonic flow***

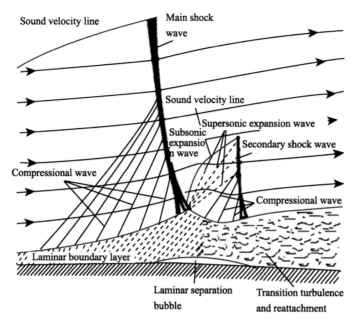

Fig. 2.96 Interference (transition) between a medium shock wave and a laminar boundary layer on a curved surface

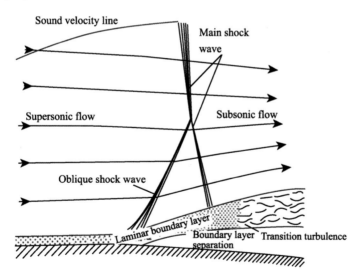

Fig. 2.97 Interference (separation) of a strong shock wave on a curved surface and a laminar boundary layer

One of the main differences between supersonic flow and subsonic flow is that the change in pressure gradient along the flow direction is the inverse sign with the change in cross-sectional area along the flow direction. For example, in the contraction pipe, the supersonic flow is compressed, the pressure increases, and the velocity decreases; in the expansion pipe, the supersonic flow is expanded, the airflow accelerates, and the pressure decreases. For the subsonic flow, the flow characteristics in the contraction and expansion pipes are just opposite to that in the supersonic flow. Therefore, for the supersonic flow around the expansion angle, there are a series of expansion waves outside the boundary layer, the pressure gradient is barotropic, and the boundary layer around the angle will not separate. In this way, the interference between the ultrasonic expansion wave and the boundary layer is weak, and the interaction is small, as shown in Fig. 2.98. If the flow around the boundary layer is compressed, the oblique shock appears. The interference of the oblique shock wave with the boundary layer will cause the boundary layer to thicken in front of the wave and may cause the boundary layer to partially separate at the corner behind the wave, forming a series of compression waves outside the boundary layer, as shown in Fig. 2.99. If the oblique shock wave is very strong and the laminar boundary layer is thin, the shock wave will penetrate into the depth, and the high-pressure countercurrent area behind the wave will be smaller, forming a dense reflection wave system in front of the incident point, and quickly merging into the first reflection wave. Under the action of the reverse pressure gradient, the separated bubbles of the boundary layer are formed. The expansion wave appears downstream of the incident wave, and then the second reflection wave appears. After the reflection wave, the laminar boundary layer turns into turbulent boundary layer, as shown in Fig. 2.100.

In addition, the interaction between oblique shock wave and turbulent boundary layer is different from that of laminar boundary layer.

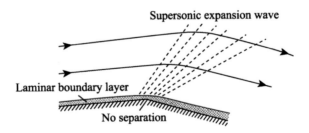

Fig. 2.98 Supersonic expansion angle flow around the expansion wave system and boundary layer interference

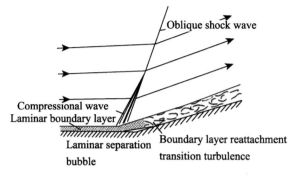

Fig. 2.99 Supersonic compression angle oblique shock around boundary layer interference (transition turbulence)

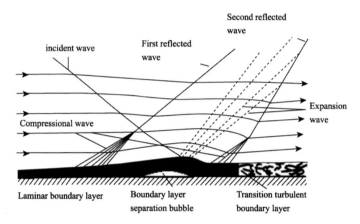

Fig. 2.100 Supersonic oblique shock and laminar boundary layer interference (separation bubble, transition turbulence)

Under the same inflow conditions, the time-average velocity distribution of the turbulent boundary layer is full, and the subsonic velocity area in the boundary layer is thinner than that in the laminar boundary layer, which leads to the deep penetration of the shock wave and the small front passage of the high-pressure countercurrent after the wave, so the distance of the pressure countercurrent is smaller than that of the laminar boundary layer. The momentum exchange caused by wall turbulence can restrain the action of shock wave's back pressure gradient and make the boundary layer difficult to separate. If the boundary layer is not separated after the interaction (as shown in Fig. 2.101), a λ wave system will appear, forming a dense reflection wave system before the incident point, and quickly merging into a reflection wave close to the ideal flow. If the incident wave is strong, local

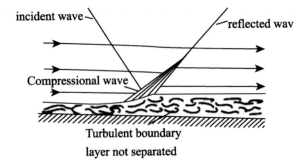

Fig. 2.101 Interference of oblique shock wave and turbulent boundary layer (turbulent boundary layer is not separated)

separation and reattachment of the turbulent boundary layer will be generated, forming a separation bubble, as shown in Fig. 2.102. At this time, the boundary layer bulges greatly, and a large range of compressed wave system appears in front of the incident point. After converging to form a reflected wave, it passes through the incident wave, forming a λ-shaped shock wave.

4. *Interference between head shock and boundary layer*

   The interference between bow shock wave and boundary layer is shown in Fig. 2.103. Influenced by the rapid thickening of the boundary layer around the head, the head of the flow around the outside of the boundary layer becomes blunt, resulting in an in vitro shock wave (bow shock wave), and a small area of subsonic velocity area is formed after the bow shock wave. The expansion wave from the head boundary layer intersects with the shock wave, which weakens and bends the shock wave. In the same way, the effect of bow shock will also change the boundary layer. If the boundary layer is thin, the interaction mainly occurs at the head. For the

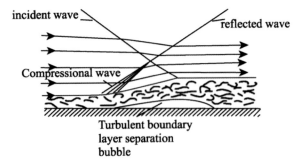

Fig. 2.102 Interference between oblique shock and turbulent boundary layer (turbulent boundary layer separation bubble)

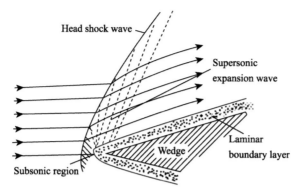

Fig. 2.103 Wedge-shaped head shock and boundary layer interference

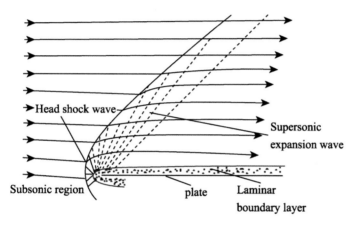

Fig. 2.104 The interference of the shock on the flat head and the boundary layer

supersonic flow around a flat plate, there will also be a weak separated shock wave at the head, but it will soon weaken to the Mach wave, as shown in Fig. 2.104.

2.11 The Leading Role of Aerodynamics in the Development of Modern Aircraft

As the foundation and forward-looking discipline of Aeronautics and Astronautics, aerodynamics has always played a leading role in the development of various aircraft. Therefore, its development level plays a decisive role in the advanced nature of aircraft. For example, in the development process of fighter (as shown in Fig. 2.105), in the 1950s, the emergence of jet engines

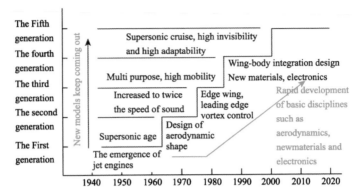

Fig. 2.105 The important role of aerodynamics in the development of modern fighter aircraft

developed the first generation of supersonic fighter (MiG-15, F-86, etc.); in the 1960s, the breakthrough of the problem of large swept wing and area law aerodynamics led to the development of the second generation fighter, which increased the aircraft speed to twice the sound speed (MiG-21, F-4, French Mirage III, etc.); in the 1980s, the third generation fighters (Su-27, F-15, etc.) were developed by the breakthrough of new aerodynamic technologies such as nonlinear lift technology and wing layout, which significantly improved weapon performance, airborne equipment,and maneuverability; after the 1990s, with the breakthrough of wing body integration design, new materials, electronics, and other new technologies, the fourth-generation fighter, represented by the U.S. F-22, has been developed, with stealth, supersonic cruise, over the horizon combat capability, high mobility, and agility. It can be seen that in the development of fighter aircraft, in addition to the application of propulsion technology, electronic technology, new material technology and stealth technology, and other high-tech achievements, the aircraft presents a more serious challenge to aerodynamics. How to expand the application range of angle of attack as much as possible, how to improve the agility of the aircraft, how to achieve the minimum detectability while meeting the flight performance, how to play the high efficiency of the propulsion system, etc. are all the problems that need to be solved in aerodynamics. For this reason, NASA has listed aerodynamics as one of the key technologies in future research strategy in recent years. The goal of aerodynamics is to develop new concepts, put forward physical understanding and theory, test, and CFD calculation verification, and finally ensure the effective design and safe operation of the aircraft.

In addition, from the perspective of the development trend of civil aircraft, high-performance power devices and excellent aerodynamic characteristics are the guarantee for civil aircraft to obtain excellent cruise performance, takeoff and landing performance, and economy. Since comet in 1952, the development of large civil aircraft is almost inseparable from the progress of aerodynamics. The introduction and breakthrough of supercritical wing and accretion device design, winglet, flow control, deformable wing, and other technologies greatly promote the development of civil aircraft (as shown in Fig. 2.106). In order to solve all kinds of challenging problems faced by aerodynamics, it is generally believed that it is very important to strengthen the basic research work of disciplines, and it is necessary to continuously explore the mechanism and law of various complex flow phenomena such as (1) turbulence structure; (2) laminar transition process; (3) leading-edge vortex and its breakup; (4) the cause and control of asymmetric vortex under high angle of attack; (5) effective control of laminar flow and turbulent drag reduction; shock wave and boundary layer interference, and these will be the main research directions of this subject.

At the same time, another problem that attracts people's attention is aeroacoustics. Because aviation noise pollutes the environment, damages human health, affects the comfort of the airliner, and also affects the concealment of the military aircraft, in the development of modern aircraft, the requirement of noise index is higher and higher, which makes the aeroacoustics (aeroacoustics) develop rapidly. Thousands of aircraft have been grounded around the world since ICAO set ground noise standards for aircraft and helicopters. Reducing aircraft noise has become a necessary condition for obtaining an airworthiness certificate and entering the world aviation market.

Driven by the rapid development of aerodynamics, especially the application of numerical simulation technology, higher requirements are put forward for aerodynamic wind tunnel experiments. On the one hand, numerical

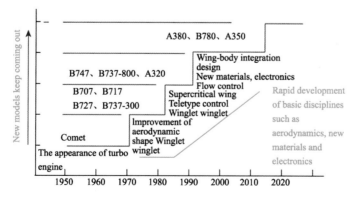

Fig. 2.106 The important role of aerodynamics in the development of large passenger aircraft

calculation can replace some experimental items, can significantly improve the aerodynamic design speed of the aircraft, can provide necessary aerodynamic data more quickly and economically than the experiment, and make the wind tunnel experiment more faced with the cutting-edge topics and complex flow mechanism problems in aerodynamics. On the other hand, no matter the development of science or the development of aircraft, numerical simulation can never replace wind tunnel experiment. This is because various theories of aerodynamics are formed and developed on the basis of a large number of experimental studies, and many major breakthroughs are inseparable from experiments. Numerical calculation can only solve the flow phenomenon that the flow mechanism has been clearly studied by experiments, but for the flow problem with complex mechanism and the verification of numerical calculation results, experiments are the inevitable way. In addition, the performance of modern aircraft is more and more advanced, and the aerodynamic problems are very complex. Therefore, the flow field index and technical level of wind tunnel experiment are required to be higher and higher, and the experimental items and hours are also increasing. For example, in the 1950s, the wind tunnel test hours needed to develop a large aircraft amounted to 10,000 h. The aerodynamic problems involved were mainly adhesion flow and its reduction in drag. The development means were theoretical analysis and wind tunnel test. In the 1970s, the number of wind tunnel experiments for the development of supersonic airliner reached 40,000 h. The aerodynamic problems involved were mainly shock wave, supersonic flow, and shock resistance reduction. The development means were theoretical analysis, numerical calculation, and wind tunnel experiments. In the 1980s, the number of wind tunnel experiments for the development of the space shuttle reached 100,000 h. The main aerodynamic problems involved were low-speed supersonic flow, high angle of attack flow around, and thermal barrier problems. The development means were theoretical analysis, numerical calculation, and wind tunnel experiments. This shows that wind tunnel test is still an important guarantee for the development of modern aerodynamics and aircraft development.

3

Hydrodynamics

3.1 Development of Hydrodynamics

Liquid dynamics is the study of the law of the liquid at rest and motion. Because the main research medium is water, it is also called hydrodynamics. Human beings began to study the laws of the static and motion of liquids very early.

In the early period of human beings, because of flood control and canal digging, the flow law of water was summarized. For example, the book on Mohism written by Mozi (about 478–392 BC) and his disciples discusses this aspect. Archimedes' calculation of buoyancy in ancient Greece is an important achievement of mechanics. Dujiangyan project in China is located in the west of Chengdu City, Sichuan Province, and on the Minjiang River in the west of Chengdu Plain. It was built in the last year of King Zhao of the Qin Dynasty (about 256–251 BC). It is a large-scale water conservancy project (as shown in Fig. 3.1) organized and built by Li Bing and his son, the governor of Shujun County, on the basis of the excavation of Bieling by his predecessors. It was composed of fish mouth, flying sand weir, precious-bottle-neck, etc. For more than two thousand years, it has been playing a role in flood control and irrigation, making Chengdu plain a "land of abundance" with thousands of miles of water and drought. So far, the irrigation area has reached more than 30 counties and cities, covering an area of nearly ten million mu. It is the longest, still the only one in use in the world. The grand water conservancy project, characterized by the diversion without dam, embodies the hard work, courage, and wisdom of the ancient Chinese working people.

© Science Press 2021
P. Liu, *A General Theory of Fluid Mechanics*,
https://doi.org/10.1007/978-981-33-6660-2_3

Fig. 3.1 Dujiangyan project

Fig. 3.2 Lord Kelvin (1824–1907, British physicist)

From the fifteenth to the seventeenth century, Leonardo da Vinci, Galileo, Torricelli, Pascal, and Newton studied the problems of water static pressure, atmospheric pressure, water shear stress, and orifice outflow by experimental methods. In 1643, Torricelli proved the basic law of constant orifice outflow and proposed that the orifice outflow velocity is proportional to the square root of the head value.

Fig. 3.3 Horace Lamb (1849–1934, British mathematician and mechanic)

After the eighteenth century, liquid dynamics developed rapidly. On the basis of Euler's equations of ideal fluid motion, the study of the law of fluid flow can be roughly divided into two categories. One is to use mathematical methods to carry out strict derivation and obtain some results that have guiding significance for practical problems and the theoretical basis of liquid dynamics (microelement flow theory). The other is to deal with the integral equation of motion of one-dimensional flow or to summarize and analyze the experimental results, so that the results can be used to solve engineering problems, that is, the one-dimensional flow theory of hydraulics (including one-dimensional pipe flow, also known as the total flow analysis method). The main scientists who have made important contributions to the former research are Navier, Saint Venant, Stokes, and Reynolds. Lagrange established the velocity potential and stream function and studied the wave theory with Cauchy and Gerstner. Helmholtz studied the vortex motion. The British physicist Lord Kelvin (1824–1907, as shown in Fig. 3.2) and the British physicist Rayleigh (1842–1919) studied the stability of wave and fluid motion. Joukowsky studied the lifting force and water hammer of airfoil. In 1945, the British mathematician and mechanics, Lamb (as shown in Fig. 3.3) summarized the above research results in his book hydrodynamics.

Fig. 3.4 Giovanni Battista Venturi (1746–1822, Italian physicist and hydraulician)

Henri Pitot (1695–1771) who was the French mathematician and water engineer invented a pitot tube to measure the velocity of flow. Giovanni Battista Venturi, an Italian physicist and hydrologist, designed a venturi tube to measure the flow through the pipeline (1746–1822, as shown in Fig. 3.4). D'Alembert (1717–1783), the famous French physicist, mathematician, and astronomer, established a towing pool to carry out the resistance experiment of diving objects. In 1852, the French F. Reech proposed the similarity criterion under the action of gravity by observing wave motion and ship model test. The British scholar William Froude (1810–1879, as shown in Fig. 3.5) gave the method of calculating ship friction resistance. Poiseuille (1799–1869), the French physiologist, discussed the flow of blood and gave the calculation formula of capillary resistance to flow and flow distribution. The French engineer Henry Darcy (1803–1858, as shown in Fig. 3.6) carried out a seepage experiment and obtained the movement rule of liquid through porous media. The French hydrologist A. Chezy (1718–1798) established the theory of steady uniform flow in open channel. According to the flow resistance of river and pipeline and the roughness of the boundary, the velocity and discharge were calculated from the empirical formula. Manning (1889), the Irish engineer, introduced the roughness coefficient to characterize the influence of the wall and established the empirical relationship between the chezy coefficient and the hydraulic radius of the river cross-section. Later,

Fig. 3.5 William Froude (1810–1879, British scholar)

on this basis, people developed the theory of steady nonuniform gradual flow in open channel and proposed the calculation method of water surface profile. Jean-Charles de Borda (1733–1799, as shown in Fig. 3.7) the French mathematician and physicist, put forward the calculation formula of local energy loss caused by the sudden expansion of pipeline based on experimental research. Saint Venant (1797–1886, as shown in Fig. 1.31) the French hydrologist, established Saint Venant equations (differential equations of conservation of mass and momentum) to describe the unsteady one-dimensional gradual flow in open channels. Later, people gave equations of the unsteady two-dimensional gradual flow based on the average of water depth. Saint Venant equations belong to the first-order hyperbolic quasilinear partial differential equations. The unknown function can be obtained by solving the equations with initial and boundary conditions. In practice, approximate calculation methods are often used, such as characteristic method, direct difference method, transient method, and finite element method. With rapid development of numerical solution, the theory of unsteady flow in open channel is widely used in flood control, irrigation, shipping, power generation, coastal reclamation, and environmental protection. Its research objects include natural rivers, artificial channels, river networks, reservoirs, lakes, tidal estuaries, harbors, and urban sewer systems.

Fig. 3.6   H Darcy (1803—1858, French engineer)

3.2   Liquid Motion

3.2.1   Ideal Liquid Motion

A liquid that ignores viscosity is called an ideal liquid. According to Prandtl's boundary layer theory, in the region outside the boundary layer, the viscous force cannot be considered, so the motion law of ideal liquid can still be applied under specific conditions. Before Prandtl, there had been a lot of research in this field. The compressibility of liquid is very small; only in cases such as water hammer, sound wave in water, shock wave propagation, etc., the compressibility of liquid should be considered.

3.2.2   Viscous Liquid Motion

Some liquids, such as lubricating oil, are very viscous and must be taken into account when analyzing their flow states (see Stokes flow). In addition, the viscosity of the liquid must be considered in the analysis of the friction resistance of the ship, the interference between the boundary layer and the wave, and the wake of the ship and the submerged body. The viscosity of

Fig. 3.7   Jean-Charles de Borda(1733–1799, French physicist)

the liquid must be considered when studying the friction resistance of the channel or pipe wall.

3.2.3 Cavitation and Cavitation Erosion

When the liquid flows through an area with low enough pressure, it will vaporize and form a cavity inside the liquid or at the liquid-solid interface (as shown in Fig. 3.8). Water often contains bubbles (called gas cores) with diameters ranging from tens to hundreds of microns. Cavitation occurs only when there are gas cores. The collapse of the cavity produces impact, which causes erosion and destruction of sidewall materials.

3.2.4 Multiphase Flow

A liquid flow with solid particles, bubbles or both is called multiphase flow (as shown in Fig. 3.9). The most common flow is the sediment laden flow (sediment movement), followed by the aerated flow and cavitating liquid flow (cavitation flow theory). The gas nuclear energy affects the propagation of

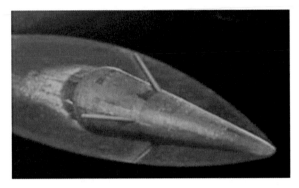

Fig. 3.8 Underwater movement of a supercavity torpedo (www.xatrm.com)

Fig. 3.9 Sediment movement

sound waves. When the volume ratio of the gas core and water in the water is greater than 10^{-3}, the sound velocity in the water will be less than that in the air (the sound velocity in the water is about five times of that in the air).

3.2.5 Non-newtonian Fluid Flow

The shear stress of some liquids (such as water with high sand content) is not linearly related to the shear deformation rate. These liquids are non-Newtonian fluids. The water added to the polymer is also a non-Newtonian fluid, which has a much lower resistance to the objects moving in it (non-Newtonian fluid mechanics, as shown in Fig. 3.10).

Fig. 3.10   Milk Movement

3.2.6   Non Pressure Flow (Open Flow)

The partial boundary of liquid flow can be the interface of liquid and air, and the pressure along the boundary is close to constant (usually atmospheric pressure). Channel, channel and ocean flow belong to this type, which is called non pressure flow. The free surface flow has a wide range, including open channel flow (constant and unsteady flow, uniform flow, gradual flow, rapid change flow, etc.), river unsteady flow, etc.

3.2.7   Pressure Flow

The flow around the liquid which is constrained by the solid side wall is called pressure flow (as shown in Fig. 3.11), and also called as full pipe flow. The flow between the rotating blades of hydraulic machinery and ship propeller is also the pressure flow. In the early stages, in order to calculate the flow distribution of water supply system, we began to study the characteristics of pipe flow. Pressure pipes are often connected with hydraulic machinery, so there are problems of elastic vibration and water hammer. Two or more layers of the liquid with different densities can form a stratified flow. The density difference can be caused by different liquids (such as water and oil), or by different salt content, sediment content, or temperature. In the research of oil exploitation, seawater immersion, submarine navigation, reservoir sediment discharge, and cooling water of power station, the stratified flow is very important (see pressure flow, density flow, rotating fluid, and stratified fluid flow).

Fig. 3.11 Long distance oil pipeline

3.2.8   Flow Induced Vibration (Hydroelastic Problem)

When the liquid flows through the solid side wall, it can cause the vibration of the side wall under some conditions, which in turn, changes the flow characteristics. The theory of the interaction between liquid, water, and solid is called hydroelastic theory.

3.3   One-Dimensional Flow Theory and Mechanical Energy Loss

3.3.1   Theory of One-Dimensional Flow

In hydraulics, if the flow coordinates s and time t are taken as independent variables, the hydraulic elements such as velocity and pressure can be expressed as (s, t) functions, thus the so-called one-dimensional flow theory is proposed. If a microelement area dA perpendicular to the streamline is taken in the flow field, a microelement flow tube can be formed through the perimeter of the microelement surface as the streamline, as shown in Fig. 3.12. A beam of liquid filled with a microchannel is called a beam. According to the principle that streamline cannot intersect, the flow in the microelement flow tube will not flow outward through the flow tube wall, and the flow outside the flow tube will not flow inward through the flow tube wall. When the liquid flow is constant, the shape and position of the micro flow tube do not change with time. However, in the case of unsteady flow,

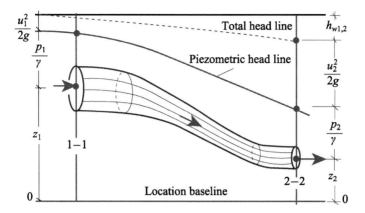

Fig. 3.12 Micro-tubestream beam and energy equation

the shape and position of microelement flow tube generally change with time, unless the position of the flow tube is fixed. For the constant incompressible flow, if the cross sections 1-1 and 2-2 (perpendicular to the streamline) are taken along the microelement flow tube, the continuity equation and energy equation of microelement flow can be obtained according to the conservation law of mass and energy. Namely

$$z_1 + \frac{p_1}{\gamma} + \frac{u_1^2}{2g} = z_2 + \frac{p_2}{\gamma} + \frac{u_2^2}{2g} + h_{w1-2}$$

$$u_1 dA_1 = u_2 dA_2$$

where z_1, p_1, and u_1 are the location, pressure, and velocity of the cross-section 1-1; z_2, p_2, and u_2 are the location, pressure, and velocity of the cross-section 2-2; h_{w1-2} is the mechanical energy loss of the cross-section 1-2. The above continuous equation shows that the flow rate of incompressible liquid keeps constant along the same microelement flow beam. The energy equation shows that along the same microelement beam, the total head of 1-1 cross-section element flow is equal to the total head of 2-2 cross-section element flow plus the mechanical energy loss of 1-2 cross-section element flow.

Any actual liquid flow with a boundary is called a total flow. Obviously, the total flow can be regarded as liquid flow consisting of numerous multi-micron beams. The section that is perpendicular to the streamline is called the cross section. Obviously for the total flow, if the streamline is a parallel straight line, the cross section is flat (as shown in Fig. 3.13), otherwise it is a curved surface. The volume of water passing through the section per unit

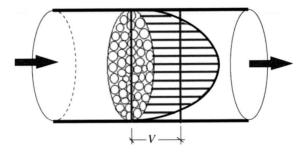

Fig. 3.13 Section-velocity distribution and average velocity of the cross-section

time is called the discharge (volume discharge) and its value is

$$Q = \iint_A u\,dA = V A$$

where A is the total flow cross-section and V is the average velocity of the section.

$$V = \frac{Q}{A} = \frac{\iint_A u\,dA}{A}$$

If each physical quantity is expressed by the average value of the cross section of the water flow, the total flow at this time can also be regarded as one-dimensional flow. The theory, thus obtained is called the total flow theory, and this method is also called the total flow analysis method. Its biggest feature is that it ignores the change in the physical quantity perpendicular to the cross section and uses the average value of the cross section.

By integrating the energy equation of the stream of microelements on the cross section of the total flow, the energy equation expressed by the average physical quantity of the cross section can be obtained as follows:

$$z_1 + \frac{p_1}{\gamma} + \frac{\alpha_1 V_1^2}{2g} = z_2 + \frac{p_2}{\gamma} + \frac{\alpha_2 V_2^2}{2g} + h_{w1-2}$$

This is the total flow energy equation. The equation shows that the average total head of 1-1 cross section is equal to the average total head of 2-2 cross section plus the mechanical energy loss of the total flow between 1-2 cross sections, where α_1 is the kinetic energy correction coefficient of section 1-1, and α_2 is the kinetic energy correction coefficient of section 2-2.

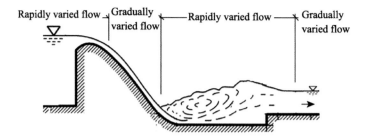

Fig. 3.14 Flow through the dam

The kinetic energy correction coefficient is related to the velocity distribution on the cross section. The velocity distribution is 1 for uniform and greater than 1 for nonuniform. In the cross section of gradual flow, $\alpha \approx 1.05–1.1$. The application conditions of the total flow energy equation are as follows:

(1) the total flow must be incompressible and constant;
(2) the mass force acting on the liquid flow is only gravity;
(3) the two selected cross sections must be uniform flow or gradually varied flow;
(4) between the two selected cross sections, the total discharge remains unchanged, and there is no flow output or input;
(5) there is no energy input or output for the total flow between the two selected cross sections.

In the total flow analysis, the case of uniform flow (the streamline keeps parallel straight-line flow, and the flow cross-section is plane) is rare, but most cases belong to the nonuniform flow (the flow cross-section is a curved surface). The nonuniform flow can be divided into gradually varied flow and rapidly varied flow according to the degree of non parallel and curved streamline. When the streamline of the total flow is almost a parallel straight line, it is called gradually varied flow, otherwise it is called rapidly varied flow, as shown in Figs. 3.14 and 3.15.

3.3.2 Mechanical Energy Loss

In the actual liquid flow, due to the viscosity of the liquid flow and the blocking effect of the boundary wall, the liquid flow resistance will be generated in the flow process. The liquid flow must overcome the flow resistance and do work, so that a part of the mechanical energy consumed will be

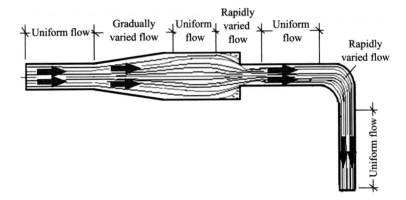

Fig. 3.15 Pipe flow

converted into heat energy loss (which can no longer be used by the mechanical energy). In hydraulics, the mechanical energy consumed per unit weight per unit time is called head loss, which is expressed in h_w. The head loss is closely related to the physical properties, internal structure, and boundary characteristics of the liquid. From the surface, the head loss of liquid flow can be divided into the head loss along the way and the local head loss. If the flow channel or flow pipe is smooth, the head loss is mainly caused by the work done by the liquid flow to overcome the boundary friction resistance, as shown in Fig. 3.16, represented by h_f, which is called the head loss along the way. If the liquid flow is partially separated, the head loss will mainly

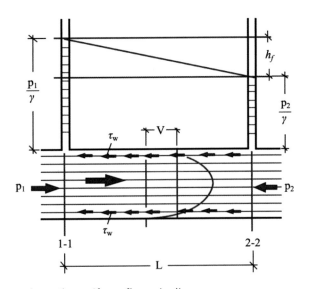

Fig. 3.16 Losses along the uniform flow pipeline

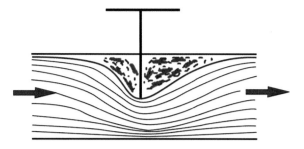

Fig. 3.17 Partially separated flow of liquid flow around a flat gate

appear in the liquid flow. This part of loss is called the local head loss, which is represented by h_j, as shown in Fig. 3.17. It can be seen that the total head loss of liquid flow can be expressed as

$$h_w = h_f + h_j$$

For a pipe of horizontal equal diameter (as shown in Fig. 3.16), if the pipe area is A, the boundary perimeter is P, and the distance between section 1-1 and section 2-2 is L, then it can be known from the continuous equation that the flow rate is unchanged along the way, $V = Q/A$, obtained from the energy equation

$$\frac{p_1}{\gamma} + \frac{\alpha V^2}{2g} = \frac{p_2}{\gamma} + \frac{\alpha V^2}{2g} + h_f$$

$$h_f = \frac{p_1}{\gamma} - \frac{p_2}{\gamma}$$

For the constant pipe flow, establish the equation of the total flow between sections 1-2, and get

$$(p_1 - p_2)A = \tau_w LP$$

Solve the above two formulas, there are

$$\tau_w = \gamma \frac{h_f}{L} \frac{A}{P} = \gamma R J$$

In the formula, τ_w is the frictional shear stress on the pipe wall surface, $R = A/P$ is the hydraulic radius of the pipe section (for a circular pipe, $P = \pi d$, $R = d/4$), and $J = h_f/L$ is the hydraulic gradient. Let the wall frictional shear stress be proportional to the square of the velocity, take

$$\tau_w = \frac{1}{2}\rho V^2 C_f = \frac{1}{2}\rho V^2 (\frac{\lambda}{4})$$

Among them, C_f is the wall friction resistance coefficient, and λ is the resistance coefficient along the pipeline. Take the former

$$h_f = \lambda \frac{L}{4R}\frac{V^2}{2g} = \lambda \frac{L}{d}\frac{V^2}{2g}, \; \lambda = f(Re, \frac{\Delta}{d})$$

In the formula, λ is a dimensionless coefficient, which is related to Reynolds number Re of liquid flow and relative roughness of pipe wall. This formula was first proposed by the German scientist Julius Weisbach (1806–1871, as shown in Fig. 3.18) in 1850. The French scientist Darcy (as shown in Fig. 3.6) verified it with the experimental method in 1858, so it is called the Darcy–Weisbach formula, also known as the general formula of head loss along the way. Darcy–Weisbach formula is suitable for laminar and turbulent flows in smooth and rough pipes of any cross-section shape, which has important engineering application values. Generally, for industrial pipelines, it can be obtained by the corresponding Reynolds number Re and the relative roughness Δ/d Moody diagram (as shown in Fig. 1.75). In industry, a simpler empirical formula can also be used. As early as 1769, the French engineer A. Chezy (1718–1798) proposed a formula for calculating the loss of uniform flow head by summarizing the measured data of uniform flow in

Fig. 3.18 Julius Weisbach (1806–1871, German scientist)

open channels, namely.

$$V = C\sqrt{RJ}$$

where C is The Chezy coefficient. Substituting $J = h_f/L, R = A/P$ into the above formula, we can get

$$h_f = \frac{8g}{C^2}\frac{L}{4R}\frac{V^2}{2g}, \quad \lambda = \frac{8g}{C^2}, \quad C = \sqrt{\frac{8g}{\lambda}}$$

Chezy coefficient C is a dimensionally empirical coefficient in $m^{1/2}/s$. Later, in 1889, Robert Manning gave a simpler formula for coefficient C. Namely

$$C = \frac{1}{n}R^{1/6}$$

where n is the roughness coefficient of the pipeline, which is called roughness for short. In practical application, the value of n can be found in the table.

For the local loss h_j, which involves the size and degree of the separation zone of the liquid flow in the pipeline, it is generally determined by the experiment according to the specific situation, and an empirical formula is usually used

$$h_j = \xi\frac{V^2}{2g}$$

where ξ is the coefficient of local head loss, which is determined by experiments, mainly determined by geometry, Reynolds number of overflow, etc. The French physicist J.C. Borda (1733–1799, as shown in Fig. 3.7) used the total flow momentum and energy equation to get the formula of local loss caused by the sudden expansion of the pipeline, as shown in Fig. 3.19, which is short for the formula of Balda, namely

$$h_j = \frac{(V_1 - V_2)^2}{2g} = \left(1 - \frac{V_2}{V_1}\right)^2\frac{V_1^2}{2g} = \left(1 - \frac{A_1}{A_2}\right)^2\frac{V_1^2}{2g} = \xi\frac{V_1^2}{2g}$$

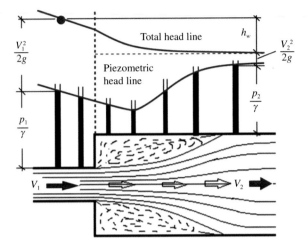

Fig. 3.19 Partially separated flow of sudden expansion pipeline

The mechanical energy utilization efficiency of a sudden expansion pipeline is defined as

$$\eta = \frac{\frac{V_1^2}{2g} - h_j}{\frac{V_1^2}{2g}} = 1 - \xi$$

Figure 3.20 shows the relationship between the local head loss coefficient and mechanical energy utilization efficiency of the sudden expansion pipe as a function of area ratio.

Taken together, for equal-section pipes with several different flow components, the total head loss can be expressed as

$$h_w = \left(\lambda \frac{L}{d} + \sum \xi\right) \frac{V_1^2}{2g}$$

3.4   Steady Flow Along a Pressure Pipeline

3.4.1   Simple Pipe Flow

Pressure pipe flow refers to the flow without free surface on the cross section of the pipe. In engineering practice, it is commonly used in water diversion projects (including water diversion, water supply, water discharge, and other

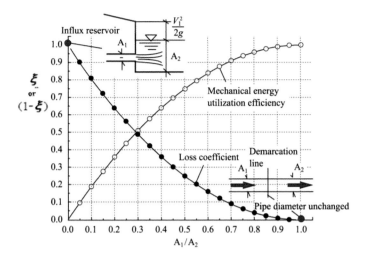

Fig. 3.20 Relationship between local head loss coefficient, mechanical energy utilization efficiency, and area ratio of A_1/A_2

projects) to supply water for agriculture and urban life and industry. For the case of single pipe free outlet flow, as shown in Fig. 3.21, one end of the pipe is connected to the pool and the other end is a free outlet in the atmosphere. If the horizontal plane passing through the centerline of the pipeline outlet is taken as the datum plane, the section 1-1 is located at a certain distance from the inlet in the pool, and the section 2-2 is located at the outlet of the pipeline, the total flow energy equation between 1-1 and 2-2 is established

$$H + \frac{\alpha_1 V_0^2}{2g} = H_0 = \frac{\alpha_2 V^2}{2g} + h_{w1-2} = \frac{\alpha_2 V^2}{2g} + \left(\lambda \frac{L}{d} + \sum \xi\right)\frac{V^2}{2g}$$

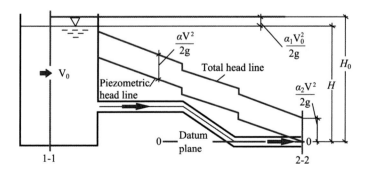

Fig. 3.21 Free flow from a simple pressure pipe

where V0 is the approaching velocity of the pool, V is the average velocity at the pipeline section 2-2, H is the height difference between the centerline of the pipeline outlet and the water surface of the pool, which is called the effective head of the pipeline and H_0 is the total head of the pipeline including the approaching velocity head. Take $\alpha_2 \approx 1.0$, and the velocity at the pipeline section 2-2 obtained from the above formula is

$$V = \mu\sqrt{2gH_0}$$

$$\mu = \frac{1}{\sqrt{1 + \lambda\frac{L}{d} + \sum\xi}}$$

where μ is the velocity coefficient or flow coefficient of the pipeline. The discharge through the pipeline is

$$Q = VA = \mu A\sqrt{2gH_0}$$

3.4.2   Water Pump System

The water pump system drives the water pump runner to rotate to work on the liquid flow, so as to make the liquid flow from the low place to the high place, as shown in Fig. 3.22. When the liquid flows through the water pump, certain energy is obtained from the water pump, which is input to the flow in the pressure pipe, and then flows into the water tower. In hydraulic calculation, suction pipe flow and pressure pipe flow can be calculated, respectively.

For the flow section of the suction pipe, the diameter of the suction pipe and the installation elevation of the water pump are mainly determined. The diameter of the suction pipe can be determined according to flow discharge and economic velocity. If flow Q is known, the allowable flow velocity V in the suction pipe is given, so the diameter d of the suction pipe is

$$d = \sqrt{\frac{4Q}{\pi V}}$$

For the installation elevation z_s of the water pump, it is mainly determined by the maximum allowable vacuum degree h_v and the head loss of the suction pipe. As shown in Fig. 3.22, the energy equation between section 1-1 and section 2-2 is established with the water level of the water storage well as the

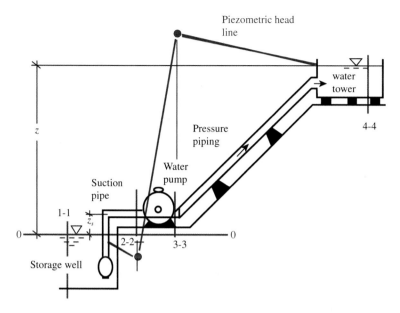

Fig. 3.22 Water pump system

datum

$$\frac{p_a}{\gamma} = z_s + \frac{p_2}{\gamma} + \frac{\alpha_2 V_2^2}{2g} + h_{w1-2}$$

$$z_s = \frac{p_a - p_2}{\gamma} - \left(\alpha_2 + \lambda\frac{L}{d} + \sum\xi\right)\frac{V_2^2}{2g}$$

Because the vacuum degree at the front section 2-2 of the pump cannot be greater than the allowable vacuum value h_v of the pump, the maximum installation height z_s of the pump is

$$z_s \leq h_v - \left(\alpha_2 + \lambda\frac{L}{d} + \sum\xi\right)\frac{V_2^2}{2g}$$

The hydraulic calculation of the pressure pipeline mainly determines the lift and installed capacity of the water pump. The diameter of the pressure pipe is determined by the economic speed. Now, the energy equation between section 1-1 and section 4-4 is established

$$H_P = z + h_{w1-4} = z + h_{w1-2} + h_{w3-4}$$

where z is the height difference between the water level of the water tower and the water level of the water storage well. h_{w1-4} is the head loss of liquid flow from the water storage well to section 4-4 of the water tower. H_P is the head of the water pump, which represents the mechanical energy absorbed by the liquid per unit time and weight, also known as the total head of the water pump. For total discharge Q, the total mechanical power obtained from the pump is

$$N_P = \frac{\gamma Q H_P}{\eta}$$

where η is the total efficiency of the water pump.

3.4.3   Water Turbine System

The working principle of the water turbine system is different from that of the water pump. It introduces the water of the reservoir into the water turbine through the pressure pipeline, and drives the water turbine to rotate, thereby driving the generator to generate electricity. In other words, the hydraulic turbine converts the mechanical energy of the liquid flow into rotational kinetic energy and outputs it to the generator set. It extracts energy from the liquid flow, as shown in Fig. 3.23. Using the water level of the tail channel as the reference surface, the energy equation between section 1-1

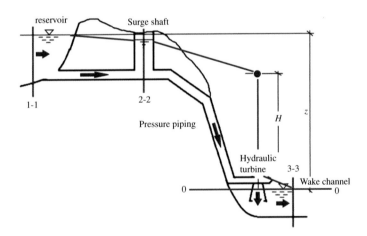

Fig. 3.23 Hydraulic turbine system

of the upstream reservoir and section 3-3 of the tail channel is established.

$$z + \frac{\alpha_1 V_1^2}{2g} = H + \frac{\alpha_3 V_3^2}{2g} + h_{w1-2} + h_{w2-3}$$

If we ignore the kinetic energy at sections 1-1 and 3-3, then

$$H = z - (h_{w1-2} + h_{w2-3})$$

where z is the height difference between the reservoir water level and the tail water level, and is called the reservoir head. h_{w1-2} is the head loss of the diversion tunnel from the reservoir to the surge well, and h_{w2-3} is the head loss of the flow from the surge well to the pressure channel of the tailrace channel. H is the hydraulic head of the turbine, which represents the mechanical energy of the liquid per unit time and unit weight extracted by the hydraulic turbine from the liquid flow. For the total discharge Q, the total mechanical power (unit power) extracted by the turbine is

$$P_w = \frac{\gamma Q H}{\eta}$$

where η is the total efficiency of the turbine.

3.5 Steady Flow in Open Channel

3.5.1 Overview

Open channel is a kind of channel which is constructed manually or formed naturally. When the liquid flows through these channels, it will form a free surface contacting with the atmosphere. The pressure of each particle on the surface is atmospheric pressure. This kind of flow with no pressure and free surface is called open channel flow or no pressure flow. The flows in the water conveyance channel (as shown in Figs. 3.24 and 3.25), aqueduct (as shown in Fig. 3.26), culvert, and natural river channel are open channel flows. When the hydraulic elements of open channel flow do not change with time, it is called open channel steady flow; otherwise, it is open channel unsteady flow. In the steady flow of open channel, if the streamline is a group of parallel straight lines, the water depth, distribution of velocity, and the section average velocity are constant along the flow direction, it is called the steady uniform flow of open channel or the steady nonuniform flow. In the nonuniform flow

Fig. 3.24 Main Canal of south-to-north water diversion project

Fig. 3.25 Hongqi Canal of Water Diversion Project) in Linzhou City, Henan Province

of open channel, it can be divided into the steady gradually varied flow and the steady rapidly varied flow according to the curvature degree and angle of streamline.

Fig. 3.26 Changgangpo Aqueduct in Guangdong

3.5.2 Steady Uniform Flow in Open Channel

According to the definition, the steady uniform flow in open channel refers to the flow with a parallel straight streamline, and the pressure on the free surface is atmospheric pressure, as shown in Fig. 3.27. Obviously, in this case, the depth and velocity distribution of the channel remains unchanged. If a section of ABCD is taken for analysis in the uniform flow, the weight of water W, the friction force on channel boundary F_f, the hydrodynamic pressure at both ends of the flow section P_1 and P_2, respectively, and the inclination angle of the channel θ (the angle between the bottom line of the channel and the horizontal direction), the force balance is established along the flow

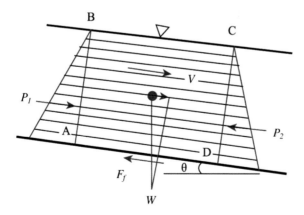

Fig. 3.27 Open channel uniform flow

direction

$$P_1 + W \sin \theta - P_2 - F_f = 0$$

Because of the steady uniform flow, $P_1 = P_2$ simplified as

$$W \sin \theta = F_f$$

It shows that in the steady uniform flow of the open channel, the frictional resistance between the wall surface and the component force of the water gravity in the flow direction is balanced.

The shape, size, and bottom slope of the open channel have an important effect on the flow regime. Common artificial open channel sections have rectangular, trapezoidal, and circular geometries, and natural river channels will show irregular shapes. For a rectangular section as shown in Fig. 3.28, let the water depth be h and the channel width be b, then the hydraulic radius R is

$$R = \frac{A}{P} = \frac{bh}{b + 2h}$$

The degree of a longitudinal inclination of the open channel bottom is called the bottom slope. The bottom slope is usually expressed as $i = \sin\theta$. $i > 0$ for down slope channel, $i < 0$ for reverse slope channel, $i = 0$ for flat slope channel. The uniform flow characteristics of the open channel are as follows:

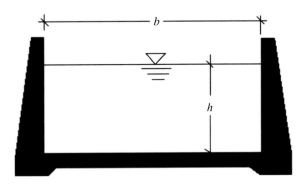

Fig. 3.28 Rectangular section of open channel

(1) the shape, size, and water depth of the cross-section remain unchanged along flow direction;

(2) the average velocity and velocity distribution of the cross-section remains unchanged along the flow direction and the corresponding kinetic energy correction coefficient and velocity head remain unchanged along the flow direction;

(3) the total headline, water surface line, and bottom slope line are parallel to each other.

The main task of hydraulic calculation of open channel is to determine the velocity and depth of water. According to the Chezy formula, the discharge of uniform flow is

$$Q = VA = AC\sqrt{Ri}, \quad Q = K\sqrt{i}$$

where $K = AC\sqrt{R}$ is called flow modulus (unit: m^3/s). It comprehensively reflects the influence of the shape, size, and roughness of the open channel section on the flow discharge. If Manning's formula is used, the above formula can be further written as

$$Q = AC\sqrt{Ri} = A\frac{1}{n}R^{1/6}\sqrt{Ri} = A\frac{R^{2/3}}{n}\sqrt{i} = \frac{A^{5/3}\sqrt{i}}{nP^{2/3}}$$

3.5.3 Steady Nonuniform Gradually Varied Flow

(1) Flow pattern of open channel

According to the definition, in the gradually varied flow of open channel, the streamline is a nearly parallel straight line, that is to say, the included angle between the streamline is very small and the curvature radius of the streamline is very large. Because the water surface line, the bottom slope line, and the total water headline are not parallel, the depth and velocity of the gradually varied flow in the open channel vary along the flow direction. In order to facilitate the analysis, the uniform water depth is often called the normal water depth, which is expressed by h_0. The influence of the free surface of open channel flow on water depth is the main, and the gravity wave propagation of free surface is the key to affect the shape of the water surface. Using the equation of one-dimensional flow continuity and energy, the propagation velocity of microgravity wave in the static water of rectangular channel with a depth of h is $V_w = \sqrt{gh}$. When the channel water moves, the relationship between its velocity and microwave velocity will determine the flow pattern in the channel. According to the Froude number (Fr), it can be divided into three flow patterns: subcritical flow, supercritical flow, and critical flow. For rectangular channels, Fr defined by the average velocity V of the section and the water depth h is

$$Fr = \frac{V}{\sqrt{gh}}$$

When $Fr < 1$, the water flow is a subcritical flow. In this case, the perturbation wave propagates throughout the flow field, that is, it can propagate up against the current. When $Fr > 1$, the water flow is a supercritical flow. In a supercritical flow, disturbance waves can only propagate downstream. When $Fr = 1$, the water flow is a critical flow. In that case, the disturbance wave can only propagate downstream. From the perspective of the force on the moving water body, the Fr number represents the ratio of the inertia force of the water to the gravity. Therefore, when Fr is less than 1, it means that the inertia force of the water is less than the gravity force. When Fr is greater than 1, it means that the inertial force of the water is greater than gravity and the inertial force plays a leading role in the water flow. If $Fr = 1$, the effect of inertial force on the water is equal to the effect of gravity and the water flow is in a critical flow.

(2) Section specific energy and critical water depth

In 1911, B.A. Bakhmeteff (1880–1951, as shown in Fig. 3.29), the Russian hydraulician, proposed the unit time unit weight water mechanical energy as an important physical quantity to study the flows of open channels, which is expressed by the lowest particle of any cross-section as a base particle, as shown in Fig. 3.30. For the case of a small slope, the expression of section specific energy is as follows:

$$E_s = h \cos \theta + \frac{\alpha V^2}{2g} \approx h + \frac{V^2}{2g}$$

In rectangular channel, if q (= Q/b) is used to represent unit width discharge, then $v = q/h$. In this way, E_s can also be expressed as

$$E_s = h + \frac{q^2}{2gh^2}$$

Fig. 3.29 Boris Alexandrovich Bakhmeteff, 1880–1951, Russian hydraulician)

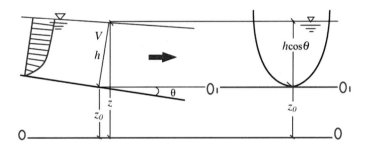

Fig. 3.30 Section specific energy at any cross section of open channel

In fact, the gravity potential energy at the bottom of the channel is deducted from the specific energy of the section. Under a given discharge, the trend of its variation with water depth is the theoretical basis for people to analyze water surface profile. Under the given discharge, draw the relationship curve between E_s and h (called specific energy curve), as shown in Fig. 3.31. It is found that the relationship between specific energy E_s and water depth is similar to the quadratic parabola. The straight line passing through the origin at an angle of 45° between the upper end of the curve and the horizontal axis is the asymptote, the lower end of the curve is the asymptote with the horizontal axis, and there is a minimum section (E_{smin}) specific energy particle

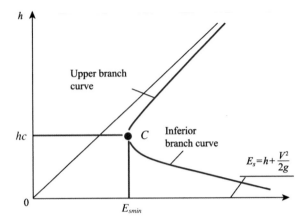

Fig. 3.31 Section specific energy curve

in the middle, called the critical particle C, The corresponding water depth is called critical water depth, which is expressed by h_C. The section specific energy above particle C increases with the increase in water depth, while the section specific energy below particle C decreases with the increase in water depth. The specific energy E_s along water depth h is

$$\frac{dE_s}{dh} = 1 - \frac{q^2}{gh^3} = 1 - \frac{V^2}{gh} = 1 - Fr^2$$

The above formula shows that the change in the specific energy of the cross-section of the open channel with the water depth is a function of the Froude number of the cross-section, as shown in Fig. 3.32. For $Fr < 1$, it is in the subcritical flow region, $\frac{dE_s}{dh} > 0$, and is located on the upper branch of the specific energy curve, and the specific energy of the section increases with the increase in water depth. For $Fr > 1$, it is in the supercritical flow region, $\frac{dE_s}{dh} < 0$, and is on the lower branch of the specific energy curve. The specific energy decreases with increasing water depth. For $Fr = 1$, it is in a critical state, $\frac{dE_s}{dh} = 0$, the specific energy at the section is the minimum, and the corresponding water depth is the critical water depth $h_c = 0$. The calculation formula is

$$\frac{dE_s}{dh} = 0, \quad h_c = \left(\frac{q^2}{g}\right)^{1/3}$$

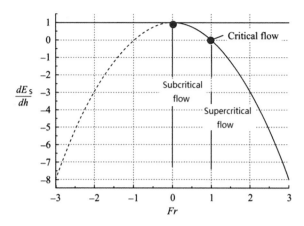

Fig. 3.32 Relationship between the change rate of section specific energy with water depth and *Fr* number

From the above formula, at the critical water depth, the average kinetic energy of the section is half of the critical water depth, and the minimum specific energy at the section is 1.5 times the critical water depth. That is

$$\frac{V_c^2}{2g}=\frac{h_c}{2}, \ E_{smin}= h_c + \frac{V_c^2}{2g} = \frac{3}{2}h_c$$

Critical water depth is used to determine flow regimes: when *Fr* < 1, *h* > *h_c*, it is a subcritical flow; when *Fr* > 1, *h* < *h_c*, it is a supercritical flow; when *h* = *h_c*, *Fr* = 1, it is a critical flow.

(3) Critical slope, mild slope, steep slope, and adverse slope

For prismatic open channel flow with a given discharge, section shape, and size (straight channel with constant section shape and size along the flow), when the flow is uniform, if the bottom slope of the open channel is changed, the corresponding normal depth of uniform flow h_0 will also be changed. However, when the bottom slope increases, the uniform water depth monotonically decreases, as shown in Fig. 3.33. If the uniform water depth is greater than the critical water depth, the corresponding bottom slope is a mild slope and the uniform flow is the subcritical flow; if the uniform water depth is less than the critical water depth, the corresponding bottom slope is a steep slope and the uniform flow is a supercritical flow; if the uniform water depth is equal to the critical water depth, the corresponding bottom slope is a critical slope and the uniform flow is a critical flow. For the critical

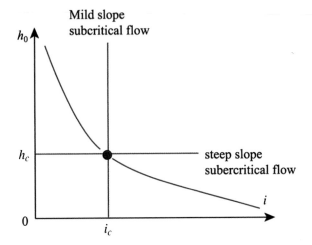

Fig. 3.33 Relation between uniform water depth and bottom slope

slope i_c, it can be obtained by solving the equations of steady uniform flow and critical water depth, namely

$$i_c = \frac{gh_c(b + 2h_c)}{C_c^2 bh_c}$$

The above formula shows that the critical bottom slope is related to the shape, size, discharge, and roughness of the channel, but not to the actual bottom slope of the channel. For the channel with bottom slope i, there may be three different bottom slopes when given different discharge, section shape, size, and roughness. That is to say, when $i < i_c$, the uniform water depth $h_0 > h_c$ and the slope is the mild slope; when $i > i_c$, the uniform water depth $h_0 < h_c$ and the slope is the steep slope; when $i = i_c$, the uniform water depth $h_0 = h_c$ and the slope is the critical slope.

3.5.4 Water Surface Curves for the Steady Gradually Varied Flow

Jean Baptiste Charles Joseph BéLanger (1790–1874, as shown in Fig. 3.34), the French hydraulician, studied the gradually varied flow surface curves of steady nonuniform flow in an open channel in 1828, established the differential equation of water depth change, and made a detailed analysis of water surface change on different slopes. For the case of bottom slope i, take any microelement (as shown in Fig. 3.35) in the gradually varied flow in a prism

Fig. 3.34 Jean Baptiste Charles Joseph BéLanger(1790–1874, French hydraulician)

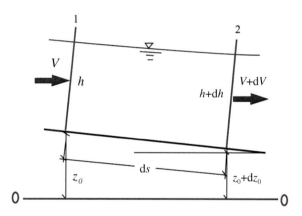

Fig. 3.35 Differential equation of steady gradually varied flow in open channel

rectangular channel, and obtain the differential equation of water depth change from the energy equation and continuous equation

$$i\mathrm{d}s = \mathrm{d}h \cos\theta + \mathrm{d}\left(\frac{V^2}{2g}\right) + \frac{V^2}{C^2 R}\mathrm{d}s$$

For the case of small bottom slope, i < 0.1, cos $\theta \approx 1.0$, from which we can get

$$\frac{dh}{ds} = \frac{i - \frac{V^2}{C^2 R}}{1 - \frac{V^2}{gh}} = \frac{i - J}{1 - Fr^2}$$

Using the above formula, the variation in water depth along the channel can be calculated under the condition of given discharge, section size, and bottom slope. On a mild slope channel, the critical depth is less than the uniform water depth, there are three areas. The area larger than the normal water depth line N-N and the critical water depth line is defined as zone 1, the area between the normal water depth line and the critical water depth line is zone 2, the area smaller than the normal water depth line and the critical water depth line is zone 3. M (M refers to the mild slope) represents the mild slope channel, so there are three types of water surface lines M1, M2, and M3. According to the above differential equation, M1 line is the backwater line at a subcritical flow, M2 line is the drawdown line at a subcritical flow, and M3 line is the backwater line at a supercritical flow, as shown in Fig. 3.36a. Similarly, there are three types of water surface profiles on a steep slope, namely S1, S2, and S3; on a critical slope, there are two types of water surface profiles, namely C1 and C3; on a horizontal slope, there are two types of water surface profiles, namely H2 and H3; on the adverse slope, there are two types of water surface profiles, namely A2 and A3. There are 5 bottom slopes and 12 water surface profiles in which there are 3 backwater lines at subcritical flows, 5 backwater lines at supercritical flows, 3 drawdown lines at subcritical flows, and 1 drawdown line at a supercritical flow, as shown in Fig. 3.36b–e. Figures 3.37 and 3.38, respectively, show that the connection and transition of water surfaces of open channel system composed of different bottom slopes and gates.

3.5.5 Rapidly Varied Flow in the Open Channel

In hydraulic projects, there are various over-flowing buildings (such as sluices and weirs in channels), which will inevitably destroy the gradually varied flow in the original channel and make the flow patterns and hydraulic elements of these over-flowing buildings' violent changes, and thus cannot meet the conditions of the gradually varied flow, and become a rapid change flow. Generally speaking, the abrupt flow is basically a complex flow in space. However, if the width of the over-flowing building is large, it can also be

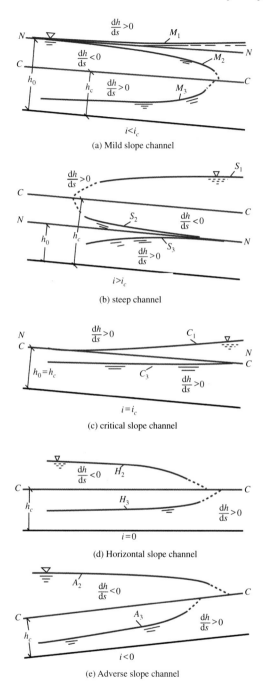

Fig. 3.36 Water surface profiles on open channels with different bottom slopes

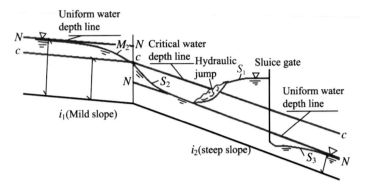

Fig. 3.37 Connection and transition of water surface profiles on open channel in mild slope, steep slope, and sluice system

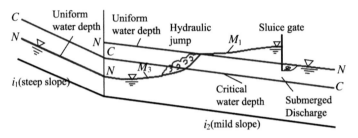

Fig. 3.38 Connection and transition of water surface profiles on open channel in steep slope, mild slope, and sluice system

approximated using one-element flow theory. This section mainly analyzes three typical rapid changes in weir, sluice, and hydraulic jump.

(1) Weir flow

Weir flow is a kind of hydraulic drop phenomenon that water flows over buildings under the action of gravity (the water surface line is a smooth down curve). When the flow is close to the top of the weir, the streamline shrinks, the flow velocity increases, and the free water surface drops sharply. The flow pattern is related to the relative ratio of the thickness δ of the overflow structure and the head H on the weir crest (the height difference between the upstream water level and the weir crest), as shown in Fig. 3.39. Generally, at the distance of $3H$ to $5H$ in front of the weir, a cross-section 0-0 is set in front of the weir. The flow velocity at this cross-section is the approaching flow velocity V_0, and the cross-section at the intersection of the horizontal line passing through the weir top and the water tongue is 1-1. The height

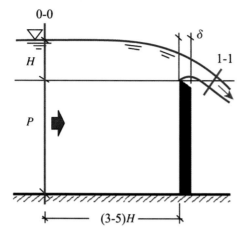

Fig. 3.39 Free flow over the sharp crested weir

difference P between the top of the weir and the bottom of the upstream river is called the weir height. According to the experimental data, the relationship between the flow pattern and δ/H is as follows:

(1) The sharp crested weir

As shown in Fig. 3.39, when $\delta/H < 0.67$, the water nappe curve over a sharp crested weir is not affected by the thickness of the weir. The lower edge of the nappe is a line from the crest of a weir and the water surface presents a monotonous falling curve. In practical applications, the weir crest is made with a sharp edge.

(2) The spillway

This kind of weir requires $0.67 < \delta/H < 2.5$. Due to the thickening of the weir, the lower edge of the water nappe is in surface contact with the weir crest and the water nappe is supported by the top surface over the weir. However, the jacking effect has little effect, and the water flow over the weir top shows a free-fall shape mainly under the action of gravity. Polyline weirs (as shown in Fig. 3.40) and curve weirs (as shown in Fig. 3.41) are commonly used in engineering.

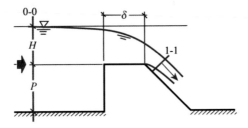

Fig. 3.40 Free flow over the polyline weir

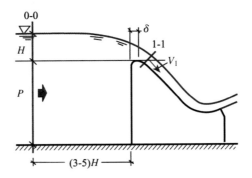

Fig. 3.41 Free flow over the spillway

(3) The Broad crested weir

The weir requires $2.5 < \delta/H < 10$. As the weir is thickened, the lower edge of the water nappe is in contact with the weir crest, and the water nappe is supported by the weir crest. The flow into the weir crest is restricted by the weir crest in the vertical direction, the cross-section of the water flow decreases, and the flow velocity increases continuously, resulting in the increase in the kinetic energy and the decrease in the potential energy of the water per unit weight. In addition, the local energy loss caused by the flow entering the weir crest makes the water surface at the inlet drop rapidly, and then the water surface is almost parallel to the weir crest under the action of the weir crest. The flow out of the weir will form a secondary drop, as shown in Fig. 3.42. The experiment shows that the head loss of broad crested weir is still mainly local head loss and the loss along the way is ignored.

The main purpose of studying weir flow is to determine the flow over the weir. Now we take the horizontal line passing through the weir crest as the datum plane (as shown in Fig. 3.39) and establish the energy equation between the weir front section 0-0 and the weir crest section 1-1. It is assumed that the 0-0 section is a gradual flow section, while the 1-1 section

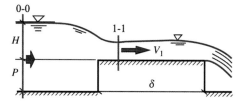

Fig. 3.42 Free flow over the broad crested weir

is located in the curved region of the streamline (which belongs to the rapid flow) and the piezometric head on 1-1 section is not constant. So the average piezometric head value on 1-1 section is used. The resulting energy equation is

$$H_0 = H + \frac{\alpha_0 V_0^2}{2g} = \overline{\left(z + \frac{p}{\gamma}\right)} + (\alpha_1 + \xi)\frac{V_1^2}{2g}$$

where H_0 is the total head of weir crest (including the head of approaching velocity); α_0 and α_1 are the kinetic energy correction coefficients at corresponding sections; ξ is the local head loss coefficient. Let us assume

$$\overline{\left(z + \frac{p}{\gamma}\right)} = \lambda H_0$$

where λ is the correction coefficient. From the energy equation, the velocity at 1-1 section can be shown to be

$$V_1 = \frac{\sqrt{(1-\lambda)}}{\sqrt{\alpha_1 + \xi}}\sqrt{2g H_0}$$

where the width of the weir crest is b. The thickness of weir crest water nappe is μH_0 in which μ is the vertical contraction coefficient reflecting the water nappe at the crest. In this way, the discharge Q over the weir crest is

$$Q = b\mu H_0 V_1 = \mu \frac{\sqrt{(1-\lambda)}}{\sqrt{\alpha_1 + \xi}} b\sqrt{2g} H_0^{3/2} = mb\sqrt{2g} H_0^{3/2}$$

$$m = \mu \frac{\sqrt{(1-\lambda)}}{\sqrt{\alpha_1 + \xi}}$$

where m is the discharge coefficient over the weir. The above formula is the standard weir flow formula, which shows that the discharge over the weir is proportional to the power of 1.5 of the total head of the weir crest. The discharge coefficient m is mainly determined by experiments, and the value is different for different weir flows.

(2) Gate outlet

The gate installed on the top of the channel or weir can be used to adjust the discharge and water level, so the main task of the gate outlet flow calculation is to determine the discharge through the gate. As shown in Fig. 3.43, H is the head in front of the gate and e is the opening of the gate. When the water flow is close to the gate hole, the streamline bends sharply under the restriction of the gate, and the flow out of the gate shrinks first and then expands under the inertia action, so a contraction section appears after the gate hole. Generally, the contraction section is about (0.5–1.0) times the gate hole opening e from the gate hole position. At the contraction section 1-1, water depth h_1 and flow velocity V_1 are set. When the flow is from the free gate hole, free hydraulic jump occurs in the downstream, and the discharge of the gate hole is not affected by the downstream water depth. Now, the energy equation between the upstream section 0-0 and the contraction section 1-1 of the gate shown in Fig. 3.43 is established

$$H_0 = H + \frac{\alpha_0 V_0^2}{2g} = h_1 + (\alpha_1 + \xi)\frac{V_1^2}{2g}$$

where H_0 is the total head of the sluice hole. We can get

$$V_1 = \frac{1}{\sqrt{\alpha_1 + \xi}}\sqrt{2g(H_0 - h_1)} = \varphi\sqrt{2g(H_0 - h_1)}$$

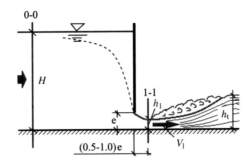

Fig. 3.43 Free flow out of plate gate hole

where φ is the velocity coefficient. The discharge through the gate is

$$Q = bh_1V_1 = \varphi bh_1\sqrt{2g(H_0 - h_1)}$$

For the water depth h_1 of the contraction section, it can be expressed by the product of gate opening and vertical shrinkage coefficient. That is, $h_1 = \varepsilon e$, and take $\mu_0 = \varepsilon\varphi$ as the discharge coefficient of sluice hole. Finally, the discharge formula of the free outflow of the gate hole is as follows:

$$Q = \mu_0 be\sqrt{2g(H_0 - \varepsilon e)}$$

The formula shows that the discharge of the sluice outlet is directly proportional to the power of 0.5 of the head in front of the sluice.

(3) Hydraulic jump

Hydraulic jump is a local hydraulic phenomenon that occurs when the water surface suddenly jumps up in the open channel flow from the supercritical flow to the subcritical flow. It is also called standing wave in wave dynamics. When the supercritical flow from the gate and dam is connected with the subcritical flow of the natural river, the phenomenon of hydraulic jump will appear, as shown in Figs. 3.44 and 3.45. As early as 1818, Giorgio Bidone, the Italian hydraulician, reported the experimental results of hydraulic jump in Paris, France; in 1828, J. BéLanger, the French hydraulician, further observed the phenomenon of hydraulic jump in steep channels; in 1860, J.A.Ch. Bresse, the French hydraulician, derived the conjugate depth equation of free hydraulic jump on horizontal rectangular channels; In 1865, the French engineer Darcy (as shown in Fig. 3.6) gave a further study for hydraulic jumps.

Fig. 3.44 Free hydraulic jump in a rectangular water tank

Fig. 3.45 Hydraulic jump behind the sluice

As shown in Fig. 3.46, the water flow in the hydraulic jump area can be divided into two parts: one is the surface roll caused by the supercritical flow rushing into the subcritical flow, which is full of air, called the surface roll area. The other part is the main flow area under the surface water roll. The velocity changes from fast to slow and the water depth changes from small to large. However, the main flow and the surface water roll are not completely separated, because the velocity gradient at the interface of the two is very large, and the turbulent mixing is very strong, and the mass, momentum and energy exchange between the two sections are continuously carried out. In the process of a sudden change in hydraulic jump, there are strong turbulent shear and mixing effects in the water flow. The water flow will undergo dramatic change and readjustment, thus consuming a lot of mechanical energy, some of which are as high as 60–70% of the inflow energy. Therefore, the flow velocity drops sharply, and the water flow will soon change into a subcritical flow. Because of the better energy dissipation effect of hydraulic jump, it is

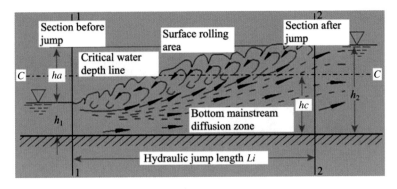

Fig. 3.46 Hydraulic jump

often used as an effective energy dissipation way to connect the downstream flow of the discharge structure.

Generally, the section at the beginning of surface water roll is called the section before jump, and the corresponding water depth is called the water depth before jump; the section at the end of surface water roll is called the section after jump, and the corresponding water depth is called the water depth after jump. However, the position of water roll on the surface is not stable. It swings back and forth along the direction of water flow, usually taking the average value in a period of time. The difference between the water depth after the jump and the water depth before the jump is called the jump height. The distance between the section before the jump and the section after the jump is called the jump length for short.

(1) Classification of hydraulic jump

The form of hydraulic jump is related to Froude number Fr_1 (Froude number, $Fr_1 = V_1/\sqrt{gh_1}$, V_1 is the average velocity, h_1 is the depth of water and g is the acceleration of gravity). It is found that the hydraulic jump can be classified as follows according to the Froude number Fr_1 of the section before the jump.

When, there will be a series of undulating waves on the surface of the hydraulic jump, the wave crest will decrease along the current, and finally disappear. This form of hydraulic jump is called wavy hydraulic jump. Because there is no rotation in the wave jump, the mixing effect is poor, the energy dissipation effect is not significant, and the wave energy will be attenuated after a long distance.

When $Fr_1 > 1.7$, the hydraulic jump becomes a typical hydraulic jump with surface water roll, which is called a complete hydraulic jump. In addition, the complete hydraulic jump can be subdivided according to the Froude number Fr_1 of the section before the jump. However, this classification is only different in the strength and weakness of the surface phenomenon of the hydraulic jump turbulence, and there is no essential difference, as shown in Fig. 3.47a.

When $1.7 \leq Fr_1 < 2.5$, it is called weak water jump. There are many small rolls on the water surface, the energy dissipation effect is not good, the energy dissipation efficiency is less than 20%, but the cross-section after the jump is relatively stable. Energy dissipation efficiency refers to the percentage of energy consumed by the hydraulic jump in the total mechanical energy of the section before jump, as shown in Fig. 3.47b.

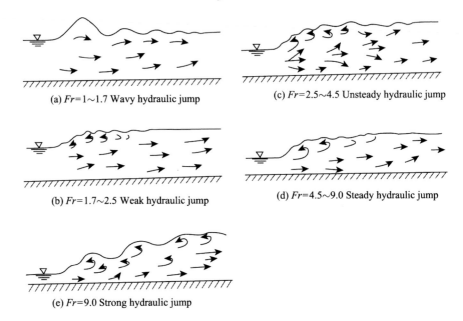

(a) *Fr*=1~1.7 Wavy hydraulic jump

(c) *Fr*=2.5~4.5 Unsteady hydraulic jump

(b) *Fr*=1.7~2.5 Weak hydraulic jump

(d) *Fr*=4.5~9.0 Steady hydraulic jump

(e) *Fr*=9.0 Strong hydraulic jump

Fig. 3.47 Type of hydraulic jump

When $2.5 \leq Fr_1 < 4.5$, it is called unstable hydraulic jump or swinging hydraulic jump. The bottom jet flows up intermittently, the rotation is unstable, the energy dissipation efficiency is 20–45%, and the water flow in the section after the jump fluctuates greatly, so auxiliary energy dissipators need to be set, as shown in Fig. 3.47c.

When $4.5 \leq Fr_1 \leq 9.0$, it is called a stable hydraulic jump. After the jump, the section water surface is stable, the energy dissipation effect is good, and the energy dissipation efficiency reaches 45–70%, as shown in Fig. 3.47d.

When $Fr_1 > 9.0$, it is called a strong water jump. The energy dissipation efficiency can reach 70%, but the intermittent water mass carried by the high-speed main stream rolls to the downstream continuously, resulting in large water surface fluctuation, so auxiliary energy dissipators need to be set, as shown in Fig. 3.7e.

In addition, it can be divided into free hydraulic jump and forced hydraulic jump according to whether there is a restriction to hydraulic jump or not.

(2) Conjugate depth equation of free hydraulic jump

The hydraulic jump occurs when the flat bottom rectangular channel passes through flow Q, the water depth of the section before the jump is H_1, the flow velocity is V_1, the water depth of the section after the jump is H_2,

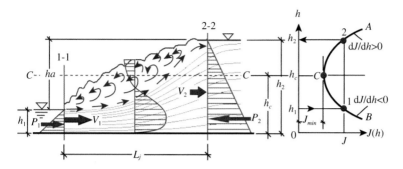

Fig. 3.48 Hydraulic jump in a rectangular channel with a flat bottom

the flow velocity is V_2, the hydrodynamic pressure acting on the two cross sections before and after the jump is P_1 and P_2, respectively, the distance between the two cross sections is the jump length L_j, and the hydraulic jump height is $ha = h_2-h_1$, as shown in Fig. 3.48.

If the friction resistance P_f on the contact surface between the water flow and the tank body is ignored, and the cross-section before and after the jump is assumed to be a gradual flow cross-section, and the hydrodynamic pressure obeys the law of hydrostatic pressure, the momentum equation along the water flow direction is

$$\frac{\gamma}{g} Q(\beta_2 V_2 - \beta_1 V_1) = P_1 - P_2$$

where γ is the unit weight of water, Q is the flow (= bh), and β is the momentum correction coefficient. For open channels with a rectangular section, the static pressure P on the section can be written as

$$P_1 = \frac{1}{2}\gamma h_1^2 b, \; P_2 = \frac{1}{2}\gamma h_2^2 b$$

Substituting the above formula gets

$$\frac{1}{2}\gamma h_1^2 b + \frac{\beta_1}{g}\gamma b h_1 V_1^2 = \frac{1}{2}\gamma h_2^2 b + \frac{\beta_2}{g}\gamma b h_2 V_2^2$$

The equation is the conjugate water depth equation in the famous rectangular open channel with a flat bottom. The formula shows that in the hydraulic jump zone, the sum of the momentum flowing into the pre jump section and the static pressure of the pre jump section in unit time is equal to the sum of the momentum flowing out of the post jump section and the static pressure of the post jump section in unit time. Because the momentum flowing into or out of a section in a unit time is equivalent to a force, in this case, it can be considered that the total thrust of the flow on the section before the jump is equal to the total resistance of the flow on the section after the jump. By substituting the continuous equation

$$q = Vh = V_1 h_1 = V_2 h_2, \text{we get}$$
$$\frac{h_1^2}{2} + \frac{\beta_1}{g} \frac{q^2}{h_1} = \frac{h_2^2}{2} + \frac{\beta_2}{g} \frac{q^2}{h_2}$$

where q is single width flow. The hydraulic jump function $J(h)$ can be expressed as

$$J(h) = \frac{h^2}{2} + \frac{\beta}{g} \frac{q^2}{h}$$

Its curve with water depth is shown in Fig. 3.48. In the rapid flow area, $dJ/dh < 0$; in the slow flow area, $dJ/dh > 0$; in the critical flow, $dJ/dh = 0$, where $h = h_c$, $J = J_{min}$.

If approximately $\beta_1 \approx \beta_2 \approx 1$ is taken, it is obtained from the above formula

$$\frac{h_2}{h_1} = \frac{1}{2}\left(\sqrt{1 + 8\frac{q^2}{gh_1^3}} - 1\right)$$

Because of formula $\frac{q^2}{gh_1^3} = \frac{V_1^2}{gh_1} = Fr_1^2$, Fr_1 is the Froude number of the flow in the section before the jump, which can be obtained by substituting the above formula

$$\frac{h_2}{h_1} = \frac{1}{2}\left(\sqrt{1 + 8Fr_1^2} - 1\right), \quad \eta = \frac{1}{2}\left(\sqrt{1 + 8Fr_1^2} - 1\right)$$

This is the conjugate depth equation of free hydraulic jump in horizontal rectangular channel derived by French hydraulician Blaise in 1860. Among them, $\eta = h_2/h_1$ is called the conjugate water depth ratio of hydraulic jump, and the verification of this formula and the experimental results are shown in Fig. 3.49. The formula shows that the conjugate depth ratio in the rectangular open channel is a function of Froude number (Fr_1) of the section before the jump. The water depth H_2 after the jump is an important basis for the design of hydraulic stilling basin depth.

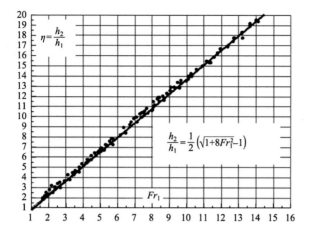

Fig. 3.49 Conjugate water depth ratio curve of horizontal rectangular channel

(3) Free hydraulic jump length in rectangular channel with flat bottom

Because of the complexity of hydraulic jump, the research of hydraulic jump mainly depends on experimental means. For example, on the basis of a large number of experiments, the momentum theorem is used to successfully solve the calculation of conjugate water depth of hydraulic jump, because in the hydraulic jump, most of the mechanical energy loss is caused by the rapid turbulent diffusion, strong vortex rupture and mixing process in the interface area between the main flow and the rolling of the hydraulic jump surface. In order to avoid the calculation of the high mechanical energy loss in the hydraulic jump, people have to give up the energy equation when analyzing the hydraulic jump, and the momentum theorem is successful. So far, there is no theoretical formula to follow the length of the hydraulic jump, and the empirical formula is still used in engineering. Through a large number of experiments, many empirical formulas have been put forward by scholars in various countries, and there are some differences between them in numerical value. The reason is that the definitions of hydraulic jump length are different from each other. Generally speaking, the horizontal distance between the start section and the end section of surface rolling is considered as the hydraulic jump length; the instability of the cross-section after the jump will also cause errors, especially in the unstable and strong water jump areas; the cross-section after the jump is accompanied by a large number of after waves, which are not stable enough to accurately measure the length of the hydraulic jump; the scale effect between the model and the prototype will also cause errors, because the model flow characteristics cannot completely reproduce the prototype flow field. The author (1993) first established a semi theoretical and semi empirical formula based on the principle of projectile. As shown in Fig. 3.50, for the highly diffused water flow in the hydraulic jump section, it can also be seen as a kind of diffused jet. During the flow process of deceleration along the path, the water particle quickly jumps up under the action of the pressure difference between the upstream and downstream. Analyze the movement process of any water particle on the dotted line as shown in Fig. 3.50 at the boundary between the main flow and the return area. Take the coordinate system shown in Fig. 3.50; take the surface particle of the section before the jump as the coordinate origin, vertically upward as the y-axis, horizontally along the path as the x-axis. In order to simplify the derivation, it is assumed that the lifting force caused by the pressure difference acting on the unit mass of the particle is f, and f changes along the path in the process of particle motion

$$\frac{d^2y}{dt^2} = f - g$$

By integrating the above formula twice for t and using the boundary conditions, we can get

$$y = \frac{1}{2}(f - g)t^2$$

At time t, the particle moves to y, and its horizontal velocity is assumed to be directly proportional to the average velocity V on the main flow section, namely

$$\frac{dx}{dt} = CV$$

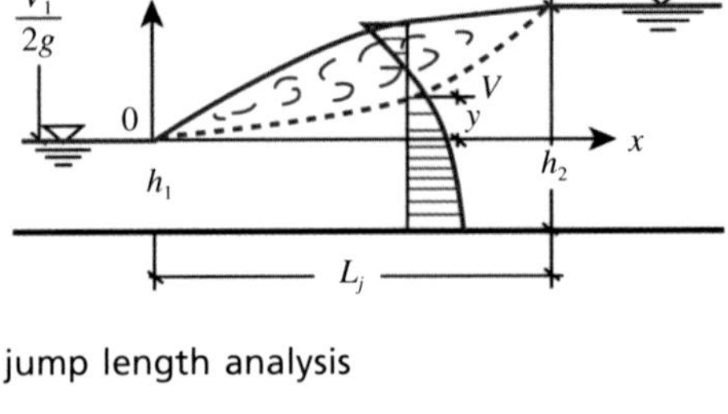

Fig. 3.50 Hydraulic jump length analysis

where C is the scale factor. For the average velocity V, it can be obtained from the total flow continuity equation

$$V = \frac{V_1 h_1}{h_1 + y}$$

The equation of motion of the particle in X-direction is

$$\frac{dx}{dt} = \frac{C V_1 h_1}{h_1 + \frac{1}{2}(f - g)t^2}$$

Integrate along the above formula along t, that is

$$L_j = C V_1 \frac{\sqrt{2h_1}}{f - g} tg^{-1}\left(\sqrt{\frac{f - g}{2h_1}} t_j\right)$$

As shown in Fig. 3.50, when $t = t_j$, $y = h_2\text{-}h_1$, we get

$$t_j = \sqrt{\frac{2(h_2 - h_1)}{f - g}}$$

Let $A = \frac{C\sqrt{2}}{\sqrt{f/g-1}}$, after finishing, get

$$\frac{L_j}{h_2} = A Fr_1 \frac{tg^{-1}\sqrt{\eta - 1}}{\eta}$$

where η is the conjugate depth ratio and the constant A is determined by experiments. Based on the analysis and comparison of the experimental curves given by USBR, it is suggested that a be 6.55, get

$$\frac{L_j}{h_2} = 6.55 Fr_1 \frac{tg^{-1}\sqrt{\eta - 1}}{\eta}$$

The comparison between the formula and the experimental results of the Bureau of reclamation of the United States Department of internal affairs is shown in Fig. 3.51. The results show that the relative error between the semi theoretical formula based on the principle of projectile and the experimental results is between 2% and 3% in the range of $Fr_1 = 2\text{--}13$.

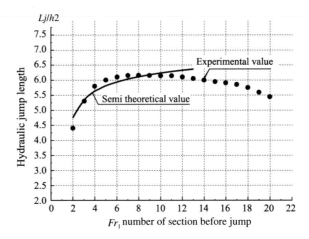

Fig. 3.51 Hydraulic jump length analysis

(4) Mechanical energy loss of hydraulic jump

It is found that the water flow in the hydraulic jump area can be divided into two parts: one is the surface water roll area of the hydraulic jump, the surface water body rolls violently and is full of air; the other is the mainstream under the surface water roll, the flow velocity changes from fast to slow, and the water depth changes from small to large. The shear effect between the mainstream and the surface water roll is great, and the turbulent mixing is intense. In the process of hydraulic jump development, there is strong turbulent shear in the flow, which consumes a lot of mechanical energy. From the particle of view of flow structure, in the process of the change in the hydraulic jump, the movement elements change dramatically. The velocity distribution of hydraulic jump section is S-shaped, and the velocity near the bottom is larger, but its value is smaller than that of the section before the jump; the velocity of the section after the jump will be further reduced, but the velocity near the bottom is still larger than that of the surface. In the post jump section, the velocity distribution will continue to adjust, the velocity near the bottom will gradually decrease, and the velocity at the upper part will gradually increase. Until the end of the post jump section, the velocity distribution of the cross section will show the velocity distribution of wall turbulence. The length of the post jump section is about 2–3 times of the length of the hydraulic jump, that is, $L_{JJ} = (2-3) \ L_J$. At the junction of the mainstream and the surface water roll in the hydraulic jump, the time average velocity gradient is very large, and the turbulent mixing is intense. This region is the main region for the generation of turbulent vortex. The greater the velocity gradient, the greater the turbulent shear stress, the greater the generation term of turbulent energy, and the more effective the turbulent mixing. In the process of diffusion, on the one hand, it makes its motion characteristics adjust along the depth of water and the direction of flow, which must be accompanied by the diffusion of momentum and energy along the transverse and longitudinal directions; On the other hand, the strong turbulent mixing produces a huge turbulent shear stress, which makes a part of the mechanical energy of the water flow quickly convert into turbulent pulsating energy and a part of the heat energy consumption. This part of mechanical energy loss is called the hydraulic jump energy loss. The interface area between the mainstream and the surface water roll is not only the generation area of turbulent vortex, but also the main area of mechanical energy dissipation. As shown in Fig. 3.52, the total head loss ΔE of the hydraulic jump should be the sum of the head loss E_J of the hydraulic jump section and the head loss E_{jj} of the post jump section.

$$\Delta E = E_j + E_{jj}$$

The total head (mechanical energy per unit weight of water body) of the section before and after the jump is

$$E_1 = h_1 + \frac{\alpha_1 V_1^2}{2g}, \quad E_2 = h_2 + \frac{\alpha_2 V_2^2}{2g}$$

The section before the jump is a gradually varied flow section and the kinetic energy correction coefficient $\alpha_1 \approx 1.0$. The section after the jump is not a gradually varied flow section and the kinetic energy correction

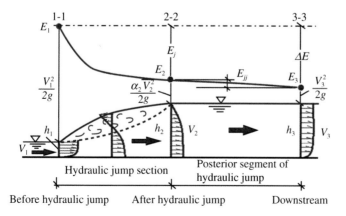

Fig. 3.52 Hydraulic jump mechanical energy loss

coefficient α_2 is larger than 1. The total head of the downstream section is

$$E_3 = h_3 + \frac{\alpha_3 V_3^2}{2g} \approx h_2 + \frac{\alpha_3 V_2^2}{2g}$$

where the section 3-3 is a gradually varied flow section and the velocity distribution tends to normal distribution, in which we can take $\alpha_3 \approx 1.0$. The head loss of the hydraulic jump section is

$$E_j = (h_1 + \frac{V_1^2}{2g}) - (h_2 + \frac{\alpha_2 V_2^2}{2g})$$

The head loss of the post jump section is defined as

$$E_{jj} = (h_2 + \frac{\alpha_2 V_2^2}{2g}) - (h_3 + \frac{V_3^2}{2g})$$

For the rectangular section, the kinetic energy correction coefficient of the section after the jump is calculated by empirical formula, namely

$$\alpha_2 = 0.85 Fr_1^{2/3} + 0.25$$

In the rectangular channel, the head loss of hydraulic jump section is

$$E_j = \frac{h_1}{4\eta}[(\eta - 1)^3 - (\alpha_2 - 1)(\eta + 1)]$$

The head loss of the post jump section is

$$E_{jj} = (\alpha_2 - 1)\frac{V_2^2}{2g} = \frac{h_1}{4\eta}(\alpha_2 - 1)(\eta + 1)$$

The total head loss is

$$\Delta E = E_j + E_{jj} = \frac{h_1}{4\eta}(\eta - 1)^3$$

The ratio of head loss of hydraulic jump section to total head loss is

$$\frac{E_j}{\Delta E} = 1 - (\alpha_2 - 1)\frac{\eta + 1}{(\eta - 1)^3}$$

The ratio of total head loss ΔE of hydraulic jump to total head E_1 of section before hydraulic jump is called energy dissipation efficiency of hydraulic jump. Namely

$$K = \frac{\Delta E}{E_1} = \frac{E_1 - E_3}{E_1} = \frac{E_j + E_{jj}}{E_1}$$

The energy dissipation rate of hydraulic jump section is defined as

$$K_j = \frac{E_1 - E_2}{E_1} = \frac{E_j}{E_1}$$

where the larger K_j is, the greater the energy dissipation efficiency of hydraulic jump is, and the better the energy dissipation effect is. For the hydraulic jump in the horizontal rectangular open channel, its energy dissipation efficiency is

$$K = \frac{\Delta E}{E_1} = \frac{\frac{h_1}{4\eta}(\eta - 1)^3}{h_1 + \frac{v_1^2}{2g}} = \frac{(\sqrt{1 + 8Fr_1^2} - 3)^3}{8(\sqrt{1 + 8Fr_1^2} - 1)(2 + Fr_1^2)}$$

The energy dissipation efficiency of hydraulic jump section is

$$K_j = \frac{E_j}{E_1} = \frac{(\sqrt{1 + 8Fr_1^2} - 3)^3 - 4(\alpha_2 - 1)(\sqrt{1 + 8Fr_1^2} + 1)}{8(\sqrt{1 + 8Fr_1^2} - 1)(2 + Fr_1^2)}$$

It can be seen that the energy dissipation efficiency K or K_j is a function of Froude number of the section before the jump, and the relation curve between them and Fr_1 is given in Fig. 3.53. It is obvious from the curve that in the wavy hydraulic jump zone, $1 < Fr_1 < 1.7$, the energy dissipation efficiency K is very low, $K < 5\%$; in the weak hydraulic jump zone, $1.7 < Fr_1 \leq 2.5$, the energy dissipation efficiency $K \approx 5-18\%$; in the unstable hydraulic jump zone (swing hydraulic jump), $2.5 < Fr_1 \leq 4.5$, the energy dissipation efficiency $K \approx 18-45\%$; in the stable hydraulic jump zone, $4.5 < Fr_1 \leq 9.0$, the energy dissipation efficiency $K \approx 45-70\%$; in the strong hydraulic jump zone, $Fr_1 > 9.0$, the energy dissipation efficiency $K > 70\%$.

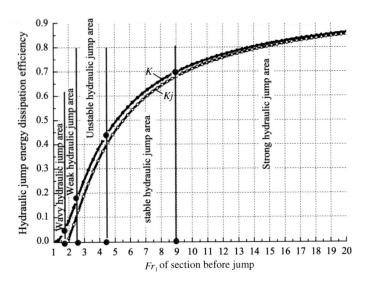

Fig. 3.53 Relationship between hydraulic jump energy dissipation efficiency and Froude number of cross section before jump

3.6 Unsteady Flow in a Pressure Pipeline

3.6.1 Overview

The unsteady flow of pressurized pipeline widely appears in water conservancy and hydropower, irrigation and water supply, energy and chemical engineering, hydraulic transmission, and other projects, such as the pressurized water delivery system of hydropower station and pump station, water supply pipe network, cooling water system of fire and nuclear power station, natural gas and oil transportation pipeline, hydraulic device pipeline system, etc. According to the time-varying speed of the hydraulic elements in the pipeline, the unsteady flow in the pressurized pipeline can be divided into water oscillation flow and water hammer phenomenon. If the water in the pipeline is incompressible and the hydraulic elements change slowly with time, it is called water oscillation flow, which is common in the water level of U-shaped pipeline and the water level fluctuation of the pressure regulating system of hydropower station, as shown in Figs. 3.54 and 3.55. Otherwise, the water is compressible, and the hydraulic elements change violently with time, resulting in the evolution of water flow in the pipeline in the form of pressure wave, which is called water hammer. This kind of situation often occurs in the pressure pipeline system. The working state of the flow passage element of the pipeline changes suddenly, which makes the

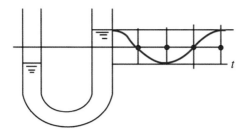

Fig. 3.54 Oscillatory flow of water body in U pipeline

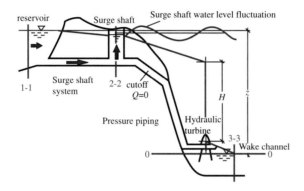

Fig. 3.55 Water level fluctuation of pressure regulating system of hydropower station

liquid flow rate in the system change rapidly, thus causing the liquid flow pressure to change rapidly and propagate in the pipeline in the form of pressure wave. For example, the sudden closing and opening of valves in the water diversion system, the sudden start-up and shutdown of the pump station, and the sudden change in generator load during the operation of the hydropower station will cause water hammer in the pipeline system, as shown in Figs. 3.56, 3.57, 3.58.

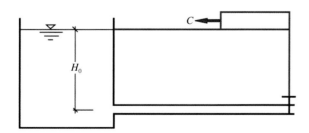

Fig. 3.56 Water hammer fluctuation process of sudden closing of water diversion pipeline valve

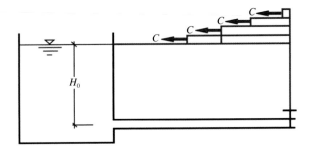

Fig. 3.57 Water hammer fluctuation process of gradual closing of water diversion pipeline valve

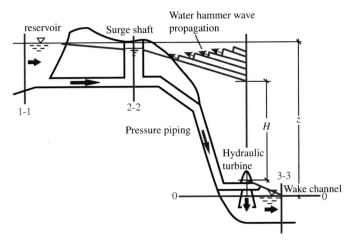

Fig. 3.58 Water hammer fluctuation process of gradual closing of turbine guide vane of hydropower station

From the physical essence, the unsteady flow in pressure pipeline is a kind of physical phenomenon of disturbance wave propagation in the pipeline. Wherever the wave goes, it will cause the hydraulic elements of the original steady flow state to change obviously with time. Because there is no free surface in the pressurized pipe, the water hammer phenomenon will cause the change in density and the elastic deformation of the pipe wall in addition to the change in velocity and pressure. From the mechanical particle of view, the oscillating flow of water in the pipeline mainly propagates in the form of gravity wave, and the inertial force and gravity play the main role in the flow process. In the water hammer phenomenon, the unsteady flow in the pipeline propagates in the form of pressure wave (elastic wave), and the inertial force and elastic force play the main role.

3.6.2 Basic Equation of One-Dimensional Unsteady Flow

The hydraulic elements of unsteady flow in pressure pipeline, such as the average velocity V and the average pressure p, can be expressed as the function of time t and flow path s, that is, $v = v\,(s,\,t)$, $p = p\,(s,\,t)$. The derivation of the basic equation can be obtained from the continuity equation, momentum equation, and energy equation. Considering the general requirements, it is assumed that both pipe section A and liquid density ρ are also functions of flow path s and time t. In any pipeline system, take the microelement control volume as shown in Fig. 3.59. According to the law of conservation of mass (continuity equation), in dt period, the mass difference of outflow and inflow microelement control volume should be equal to the reduction in the control volume mass in the same period, that is

$$\frac{\partial}{\partial s}(\rho A V dt)ds = -\frac{\partial}{\partial t}(\rho A ds)dt$$

We can obtain

$$\frac{\partial(\rho A)}{\partial t} + \frac{\partial(\rho A V)}{\partial s} = 0$$

This equation is the general form of one-dimensional unsteady flow continuous equation.

Take any microelement along the pipe, as shown in Fig. 3.60. Let the cross-sectional area of the microelement be a, perimeter P, length ds, D as the pipe diameter, the angle between the pipe axis and the horizontal direction be θ (the downward inclination of the pipe axis along the flow direction is positive, $\sin\theta = -\frac{\partial z}{\partial s}$), the shear stress acting on the pipe wall is τ_w,

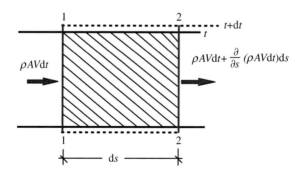

Fig. 3.59 Microelement control volume of unsteady flow in pressurized pipeline

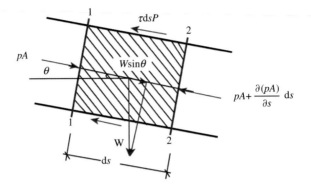

Fig. 3.60 Stress of micro segment liquid mass

according to Newton's second law, the motion equation established along the flow direction is

$$pA - \left(p + \frac{\partial p}{\partial s}ds\right)A + \gamma A ds \sin\theta - \tau_w ds\, P = \rho A ds\frac{dV}{dt}$$

After simplification, we get

$$\frac{1}{g}\frac{\partial V}{\partial t} + \frac{\partial z}{\partial s} + \frac{1}{\gamma}\frac{\partial p}{\partial s} + \frac{1}{g}V\frac{\partial V}{\partial s} + \frac{4\tau_w}{\gamma D} = 0$$

This equation is a differential equation of motion for one-dimensional unsteady gradually varied flow. Substituting the resistance equation $\frac{4\tau_w}{\gamma D} = \frac{dhw}{ds}$ of uniform flow into the above equation and integrating it from section 1-1 to section 2-2, the total flow energy equation is obtained as follows:

$$\int_1^2 \frac{\partial}{\partial s}\left(z + \frac{p}{\gamma} + \frac{V^2}{2g}\right)ds + \int_1^2 \frac{1}{g}\frac{\partial V}{\partial t}ds + \int_1^2 dhw = 0$$

$$z_1 + \frac{p_1}{\gamma} + \frac{V_1^2}{2g} = z_2 + \frac{p_2}{\gamma} + \frac{V_2^2}{2g} + \int_1^2 \frac{1}{g}\frac{\partial V}{\partial t}ds + hw_{1-2}$$

Compared with the energy equation of the constant total flow, there is one more term of water head due to the acceleration of the fluid in the right term of the equation, which is called inertia water head, expressed by hi. Namely

$$hi = \int_1^2 \frac{1}{g} \frac{\partial V}{\partial t} ds$$

In a word, the differential equations of unsteady flow in pressurized pipes can be obtained. Namely

$$\frac{\partial(\rho A)}{\partial t} + \frac{\partial(\rho A V)}{\partial s} = 0$$

$$\frac{1}{g} \frac{\partial V}{\partial t} + \frac{\partial z}{\partial s} + \frac{1}{\gamma} \frac{\partial p}{\partial s} + \frac{1}{g} V \frac{\partial V}{\partial s} + \frac{4\tau_w}{\gamma D} = 0$$

The equations are suitable for the unsteady flow in a gradual pipeline, including water hammer and oscillatory flow.

3.6.3   Water Hammer and Its Governing Equations

As early as 1898, the Russian scientist Jukowsky (as shown in Fig. 2.6) established the relationship between the pressure increment and the velocity change in elastic wave in the pipeline propagation based on the continuity equation and momentum theorem. For a simple pipe, the inlet end is connected with the water tank, and the outlet end is provided with a valve to regulate the flow, and the pipe length is L. In order to simplify the problem, it is assumed that the pipeline is horizontal and the influence of friction resistance is ignored. At this time, the piezometric headline under the condition of the constant flow of pipeline is a horizontal line. When the valve is fully opened, the flow in the pipe is constant, the flow rate is V_0, and the total head of the corresponding section is H_0. Now the valve is suddenly closed completely, resulting in water hammer wave (pressure wave) and wave velocity C from the valve to the upstream, where the wave will cause the flow rate to decrease and the pressure to increase. If the reference system is built on the wave crest, the relationship between the hydraulic elements before and after the liquid flow passing through the wave crest can be established according to the continuity equation and momentum theorem. The relation formula of hydraulic elements derived by Jukowsky is

$$\Delta p = -\rho C \Delta V, \ \Delta h = -\frac{C}{g} \Delta V$$

For example, the propagation velocity of water intake shock wave $C = 1435$ m/s, and the velocity $V_0 = 5$ m/s when the pipeline is in constant flow. Substituting the above formula, the water hammer pressure value generated by the sudden closing of the valve is $\Delta p = 7.175 \times 10^6$ pa.

The valve in the actual pipeline is not closed in a moment, but gradually closed in a certain period of time. The resulting water hammer pressure is much smaller than that generated by a sudden closing of the valve. The maximum water hammer pressure in front of the valve is the superposition of pressure waves generated by gradually closing the valve, which is called indirect water hammer. Relatively speaking, the water hammer caused by one-off shutdown is called direct water hammer. In addition to the water hammer generated by closing the valve, the sudden opening of the valve will also cause the water hammer wave in the pipeline, but the water hammer wave is negative at this time, causing the pressure to decrease.

If the elasticity of the pipe wall is not considered, the propagation velocity of the pressure wave caused by the compressibility of the water is the propagation velocity of the sound in the water $C = 1435$ m/s, but if the propagation velocity changes after considering the elasticity of the pipe, it can be derived using the continuity equation and momentum theorem

$$C = \frac{\sqrt{\frac{K_w}{\rho}}}{\sqrt{1 + \frac{DK_w}{\delta E}}} = \frac{1435}{\sqrt{1 + \frac{DK_w}{\delta E}}}$$

where, K_w is the bulk modulus of elasticity of water body $(= 2.1 \times 10^9$ Pa), D is the diameter of pipe, δ is the thickness of pipe wall, and E is the modulus of elasticity of pipe wall. Generally, the velocity of water hammer wave in Penstock of hydropower station is between 1000–1200 m/s.

Considering the compressibility of water and the elasticity of pipe wall, it can be assumed that the area of pipe is $A = A(s, t)$, $\rho = \rho (s, t)$, and the relationship of the derivative with the body can be used

$$\frac{dA}{dt} = \frac{\partial A}{\partial t} + V \frac{\partial A}{\partial s}, \frac{d\rho}{dt} = \frac{\partial \rho}{\partial t} + V \frac{\partial \rho}{\partial s}$$

Substituting into the continuous differential equation can be

$$\frac{1}{\rho}\frac{d\rho}{dt} + \frac{1}{A}\frac{dA}{dt} + \frac{\partial V}{\partial s} = 0$$

Bring relation

$$\frac{1}{\rho}\frac{d\rho}{dt} = \frac{1}{K_w}\frac{dp}{dt}, \frac{1}{A}\frac{dA}{dt} = \frac{d}{E\delta}\frac{dp}{dt}$$

Substituting into the above formula, we get

$$\frac{1}{\rho}\left(\frac{\partial p}{\partial t} + V\frac{\partial p}{\partial s}\right) + C^2\frac{\partial V}{\partial s} = 0$$

Combined with the equations of motion, the chain differential equations (wave equations of water hammer, first-order nonlinear partial differential equations) representing water hammer are formed, namely

$$\frac{1}{\rho}\left(\frac{\partial p}{\partial t} + V\frac{\partial p}{\partial s}\right) + C^2\frac{\partial V}{\partial s} = 0$$

$$\frac{1}{g}\frac{\partial V}{\partial t} + \frac{\partial z}{\partial s} + \frac{1}{\gamma}\frac{\partial p}{\partial s} + \frac{1}{g}V\frac{\partial V}{\partial s} + \frac{4\tau_w}{\gamma D} = 0$$

In the calculation of water hammer, the piezometric head $h(= z + p/\gamma)$ and pipeline velocity V are often used as independent variables, and the above equations are changed into

$$\frac{\partial h}{\partial t} + V\frac{\partial h}{\partial s} + V\sin\theta + \frac{C^2}{g}\frac{\partial V}{\partial s} = 0$$

$$\frac{\partial V}{\partial t} + V\frac{\partial V}{\partial s} + g\frac{\partial h}{\partial s} + \lambda\frac{V|V|}{2D} = 0$$

The wall shear stress of the pipeline is replaced by the following formula, where λ is the resistance coefficient along the pipeline.

$$\tau_w = \lambda\rho\frac{V^2}{8}$$

If the resistance is ignored and the partial derivative of (h, V) convection path s is far less than that of time, the following wave equation can be obtained after simplification.

$$\frac{\partial h}{\partial t} + \frac{C^2}{g}\frac{\partial V}{\partial s} = 0$$
$$\frac{\partial V}{\partial t} + g\frac{\partial h}{\partial s} = 0$$

The wave equations of H or P are obtained by solving the above equations, namely

$$\frac{\partial^2 h}{\partial t^2} = C^2\frac{\partial^2 h}{\partial s^2}, \frac{\partial^2 p}{\partial t^2} = C^2\frac{\partial^2 p}{\partial s^2}$$

The general solution of this wave equation is

$$h - h_0 = F(s - Ct) + f(s + Ct)$$
$$V - V_0 = \frac{g}{C}(F(s - Ct) - f(s + Ct))$$

$F(s\text{-}Ct)$ is the wave function of the forward water hammer wave and $f(s + Ct)$ is the wave function of the reverse water hammer wave. There is an analytic method, graphic method, and characteristic line method to solve the water hammer problem.

3.6.4 Water Oscillating Flow

For the fluctuation of U-tube water level and water level in surge shaft of hydropower station, the key point is to give the law of water level fluctuation, especially the surge height caused by turbine shutdown is an important parameter to determine the geometric size of surge shaft. Different from the water hammer problem, the compressibility of the water and the elasticity of the pipe wall are not considered here. The water is incompressible and the pipe is treated as rigid. The continuity obtained is $VA = f(t)$, and the differential equation of motion is

$$\frac{1}{g}\frac{\partial V}{\partial t} + \frac{V}{g}\frac{\partial V}{\partial s} + \frac{\partial h}{\partial s} + \lambda\frac{V|V|}{2gD} = 0$$
$$\frac{1}{g}\frac{dV}{dt} + \frac{\partial h}{\partial s} + \lambda\frac{V|V|}{2gD} = 0$$

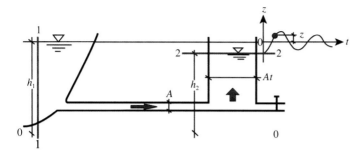

Fig. 3.61 Surge tank water level fluctuation process

As shown in Fig. 3.61, integration from 1-1 to 2-2 section can obtain

$$\int_1^2 \frac{1}{g}\frac{dV}{dt}ds + h_2 - h_1 + \int_1^2 \lambda\frac{V|V|}{2gD}ds = 0$$

If the length of the diversion pipeline L, the equal cross-sectional area A, and $z = h_2 - h_1$ (indicating the difference between the water level of the surge shaft and the water level of the reservoir), then

$$z + \frac{L}{g}\frac{dV}{dt} + \lambda\frac{V|V|}{2gD}L = 0$$

It is assumed that the sectional area of the surge shaft is A_t, and the continuous equation is as follows:

$$V = \frac{A_t}{A}\frac{dz}{dt}$$

Substituting the expression of V into the equation of motion, we get

$$\frac{LA_t}{gA}\frac{d^2z}{dt^2} + \frac{\lambda L}{2gD}\left(\frac{A_t}{A}\right)^2\frac{dz}{dt}\left|\frac{dz}{dt}\right| + z = 0$$

This is a damped wave equation, which belongs to the second-order nonlinear ordinary differential equation. Newton iterative method or fourth-order Runge–Kutta method is commonly used in the numerical solution.

3.7 Unsteady Gradually Varied Flow in Open Channel

3.7.1 Overview

Unsteady flow in open channel refers to the time-dependent flow of average velocity V and water depth h in open channel or natural channel. In water conservancy engineering, it is often seen in the evolution of flood in the natural river, the change in channel water depth caused by the opening and closing of the gate, and the tidal flow at the entrance to the sea. According to the theory of one-dimensional flow, the hydraulic elements of unsteady flow in open channel, such as the average velocity (or discharge) of cross section and water depth, can be expressed as the function of flow path s and time t, that is

$$V = V(s, t), h = h(s, t) \text{ or } Q = Q(s, t), h = h(s, t).$$

Different from the unsteady flow of pressure pipeline, there is a free water surface in the open channel flow, and the wave propagation is not a pressure wave but a gravity wave. In the flow process, the main forces of water are inertia force, gravity, and resistance. Although these waves belong to gravity waves, they are essentially different from the wind-driven waves on the surface of oceans and lakes. In the wind-driven wave motion, the water particles basically move in a circular motion along a certain track, almost no flow transfer, there is a certain phase difference between the particles, resulting in the propulsion of water surface wave, which is called propulsion wave. But for the unsteady flow in the open channel, not only the surface wave moves forward, but also the water particle moves forward. This wave is caused by the water particle moving, so it is called displacement wave or mass transfer wave. In the fluctuation area, the water depth and velocity of each section are not single value relations, but form the rope sleeve curve as shown in Fig. 3.62. For example, in the process of water rising, the upstream water level in the channel rises first, the water surface slope becomes steep, and the flow is larger than the constant flow; in the process of precipitation, the upstream water level falls first, the water surface slope becomes slower, and the flow is smaller than the constant flow. The wave of unsteady flow in open channel is generally a shallow water wave or long wave, and the ratio of water depth h to wavelength is less than 1/20. The flow resistance should be taken into account in the analysis. In addition, the unsteady flow in open channel can be divided into continuous wave and discontinuous wave according to the change in hydraulic elements with time. When the wave process evolves

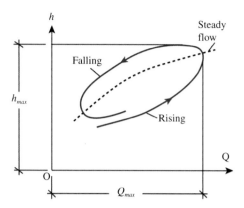

Fig. 3.62 Water depth relationship curve of non-constant flow in open channel (loop curve)

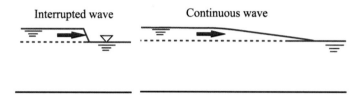

Fig. 3.63 Water wave motion in open channel

slowly and the instantaneous water surface gradient is not large (as shown in Fig. 3.63), the instantaneous streamline is close to a parallel straight line. This kind of unsteady flow has the characteristics of gradual flow. The pressure distribution along the vertical line is approximately in accordance with the hydrostatic pressure. The hydraulic element is the continuous differentiable function of location and time. This kind of wave is called a continuous wave, also known as open channel unsteady gradual flow, Such as the evolution of flood in open channels. On the contrary, if the hydraulic elements in the channel change rapidly with time, the instantaneous water surface slope is very steep, forming a discontinuous wave. For example, a dam break wave is a typical discontinuous wave, as shown in Fig. 3.63.

3.7.2 Differential Equation of Unsteady Gradually Varied Flow

In early 1848 and 1871, Saint Venant (as shown in Fig. 1.31), the French hydraulic scientist, established continuous equations and motion equations using mass conservation and momentum theorem and analyzed the solution methods of these equations.

(1) Continuous equation

As shown in Fig. 3.64, for prismatic channel without side flow, the differential equation of unsteady gradient flow in open channel can be obtained by taking the microelement control volume between section 1-2 and the conservation of mass theorem, that is

$$\left[\rho Q + \frac{\partial(\rho Q)}{\partial s}ds - \rho Q\right]dt = -\frac{\partial(\rho A)}{\partial t}dsdt$$

After simplification

$$\frac{\partial A}{\partial t} + \frac{\partial Q}{\partial s} = 0$$

For rectangular channels, substituting $A = bh$ and $q = bhV$ into the above equation, we can get

$$\frac{\partial h}{\partial t} + V\frac{\partial h}{\partial s} + h\frac{\partial V}{\partial s} = 0$$

This is the continuous equation of unsteady gradually varied flow in open channel. The equation shows that in DT micro segment, when the mass of the inflow micro control volume is greater than the outflow mass, the water

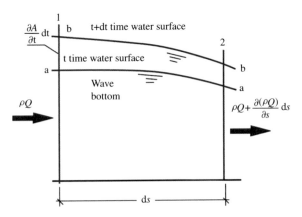

Fig. 3.64 Microelement control volume of unsteady gradient flow in open channel

level in the open channel will rise, that is, $\frac{\partial Q}{\partial s} < 0$, $\frac{\partial A}{\partial t} > 0$; otherwise, when the mass of the inflow micro control volume is less than the outflow mass, the water level in the open channel will fall, that is, $\frac{\partial Q}{\partial s} > 0$, $\frac{\partial A}{\partial t} < 0$.

(2) Equation of motion

In the gradually varied flow channel, any segment of the microelement with the length of ds is taken, as shown in Fig. 3.65. The cross-sectional area of the microelement is A, and the angle between the channel bottom wall and the horizontal direction is θ (the downward inclination along the flow direction is positive). According to Newton's second law, the motion equation established along the flow direction is

$$\frac{\partial V}{\partial t} + V \frac{\partial V}{\partial s} + g \frac{\partial (z_b + h)}{\partial s} + g \frac{\tau_w}{\gamma R} = 0$$

where z_b is the elevation of the canal bottom, $i = -\frac{\partial z_b}{\partial s}$; the shear stress of the canal wall is treated according to the uniform flow, that is $\tau_w = \gamma R J_f = \gamma \frac{V^2}{C^2}$, R is the hydraulic radius and C is the chezy coefficient. Put it into the above equation and get

$$\frac{\partial V}{\partial t} + V \frac{\partial V}{\partial s} + g \frac{\partial h}{\partial s} = g \left(i - \frac{V^2}{C^2 R} \right)$$

This equation is the energy equation of unsteady gradually varied flow in open channels. The Saint Venant equations are constructed together with the

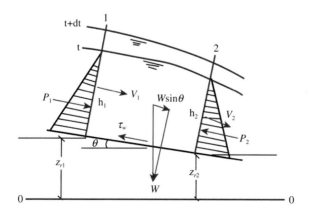

Fig. 3.65 Microelement of unsteady gradually varied flow in open channel

continuous equations to represent the unsteady gradually varied flows in open channels. For the rectangular open channel, the water depth and velocity can be used to represent the hydraulic elements. The corresponding Saint Venant equations are

$$
\begin{cases}
\dfrac{\partial h}{\partial t} + V\dfrac{\partial h}{\partial s} + h\dfrac{\partial V}{\partial s} = 0 \\[3mm]
\dfrac{\partial V}{\partial t} + V\dfrac{\partial V}{\partial s} + g\dfrac{\partial h}{\partial s} = g\left(i - \dfrac{V^2}{C^2 R}\right)
\end{cases}
$$

This is a set of first-order hyperbolic nonlinear partial differential equations. The unknown quantities V and h can be obtained by solving the equations with initial conditions and boundary conditions. At present, there is no general solution. In practice, the approximate calculation methods are the characteristic method, direct difference method, transient method, and finite element method.

3.8 Fundamentals of Water Wave Hydrodynamics

3.8.1 Overview

Wave phenomenon is a kind of water kinematics which is common in the ocean, lake, reservoir, and other broad water surfaces. The main characteristics of wave motion are regular undulation motion of liquid surface (as shown in Figs. 3.66 and 3.67), and periodic reciprocating oscillation of water particles. In the process of motion, water level and particle velocity are both functions of time, so water wave motion is a kind of unsteady motion.

It is found that any wave must meet the following three conditions:

(1) there must be an undisturbed equilibrium state (medium);
(2) there must be a disturbing force to break the balance;
(3) there must be a restoring force to reestablish equilibrium.

In wave motion, the disturbing forces include usually: wind force, tide force, ship force, earthquake force, etc.; the restoring forces are: gravity, surface tension, inertia force, etc.

As shown in Fig. 3.68, the main physical elements that characterize wave motion are: wave crest refers to the part above the static water surface; wave

Fig. 3.66 Wave motion (gravity wave)

Fig. 3.67 Wave motion (gravitationally progressive wave)

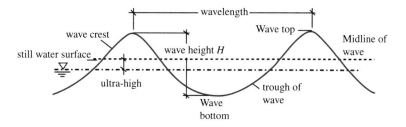

Fig. 3.68 Wave elements

crest refers to the highest particle of wave crest; wave trough refers to the part below the static water surface; wave bottom refers to the lowest particle of wave trough; wave height refers to the vertical distance h between the wave crest and wave bottom; wavelength refers to the horizontal distance λ between two adjacent wave crests; wave steepness refers to the ratio of wave height and wavelength.

The wave centerline is defined as the horizontal line of bisector wave height. In fact, due to the sharp crest and flat trough, the distance from the static water surface to the crest is greater than the distance from the static water surface to the trough, so the middle line of the wave is often located above the static water surface, and the height beyond it is called superelevation. The time that a complete wave process is repeated is called wave period T. In the process of wave motion, the wave whose waveform moves forward is called the progressive wave. In the progressive wave, the velocity of wave crest moving along the horizontal direction is called wave velocity, $a = \lambda / T$. Wave height, wavelength, wave steepness, wave velocity, and wave period are called wave elements.

If the water particle moves in a circle with the wave height h as the diameter, the average speed is $V_m = \pi h / T$, and the ratio between the water particle and the wave speed can be expressed as

$$\frac{V_m}{a} = \frac{\pi h / T}{\lambda / T} = \pi \frac{h}{\lambda}$$

Generally, the wave height is much smaller than the wavelength, usually between 0.15 and 0.02. According to statistical records, the relationship between wave height and frequency of occurrence of ocean waves is shown in Table 3.1.

45% of the wave height is below 1.3 m, 80% of the wave height is below 4 m, and 10% of the wave height is above 6.7 m. According to the nautical records, from 1836 to 1839, the wave height of 7.6 m was measured in South America; in 1894, the wave height of at least 12 m was encountered in the Atlantic Ocean, and 20 m was encountered in the Pacific Ocean; in 1933, the wave height of 35 m was encountered in the Pacific Ocean. The periodic range of ocean waves is given in Table 3.2. Generally speaking, viscosity has a greater attenuation effect on the surface tension wave, but a smaller attenuation effect on the gravity wave, so the gravity wave can be maintained for a long time. In the process of wave propagation, the direction is constant, but the situation of encountering obstacles and entering shallow water is different. If we encounter physical waves, we will reflect them. If we encounter shallow

Table 3.1 Frequency of wave height

Wave height (h(m))	0–1	1–1.3	1.3–2.3	2.3–4.0	4–6.7	>6.7
Frequency (Hz)	20	25	20	15	10	10

Table 3.2 Ocean wave period

Wind velocity (m/s)	5	10	20
Average period (s)	2.86	5.0	11.4
Cycle upper limit (s)	1.0	3.0	6.5
Lower period limit (s)	6.0	11.1	21.7

water waves, we will refract them. The influence of wave on depth is attenuated exponentially. In deepwater, a wave with a height of 5 m and a length of 150 m can only cause a movement with a diameter of 60 cm at 50 m below the water surface, and the water particle velocity is reduced from 1.6 m/s to 0.2 m/s. Under 15 m, 2–3 m waves have little effect.

3.8.2 Basic Characteristics of Wave Motion

There is two-dimensional wave motion, which propagates on the water surface with constant water depth. Take the x-axis as the direction of wave advance and the z-axis as vertical upward. The height of the equilibrium water surface is z = 0, and the height of the water surface where the wave travels is, as shown in Fig. 3.69.

At the same time, the acceleration in the Z-direction is ignored, and the pressure obeys the distribution of hydrostatic pressure, that is

$$p = p_a + \gamma(\eta - z)$$

Among them, p_a is the atmospheric pressure and γ is the bulk density of the water body. The following can be obtained:

$$\frac{\partial p}{\partial x} = \rho g \frac{\partial \eta}{\partial x}$$

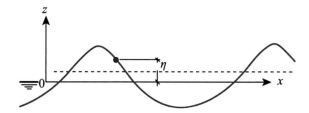

Fig. 3.69 Two-dimensional wave

From the horizontal equation of motion (Euler equation), we get

$$\frac{\partial u}{\partial t} + u\frac{\partial u}{\partial x} = -\frac{1}{\rho}\frac{\partial p}{\partial x}$$

The z-direction velocity and acceleration are ignored. If the nonlinear term is not included, there are

$$\frac{\partial u}{\partial t} = -g\frac{\partial \eta}{\partial x}$$

In a certain period of time, the horizontal displacement of water particle movement is

$$\xi = \int u dt$$

From the principle of conservation of mass, we can get

$$\frac{\partial}{\partial x}\left[\xi H + \int \eta dx\right] = 0, \ \eta = -H\frac{\partial \xi}{\partial x}$$
$$\frac{\partial^2 \xi}{\partial t^2} = gH\frac{\partial^2 \xi}{\partial x^2},$$

Eliminate ξ and get

$$\frac{\partial^2 \eta}{\partial t^2} = gH\frac{\partial^2 \eta}{\partial x^2}, \ \frac{\partial^2 \eta}{\partial t^2} = a^2\frac{\partial^2 \eta}{\partial x^2}$$

This is the famous wave equation, which was deduced by the French scientist D'Alembert (as shown in Fig. 1.14) in 1747. Among them, water surface wave velocity $a = \sqrt{gH}$, H is water depth.

If we set $x_1 = x - at$, $x_2 = x + at$ and substitute it into the wave equation about wave height, we can get

$$\frac{\partial^2 \eta}{\partial x_1 \partial x_2} = 0$$

Its solution is

$$\eta = F(x - at) + G(x + at)$$

where F and G are waveform functions. On the surface of the water

$$u = \frac{\partial \xi}{\partial t} = \frac{\partial \xi}{\partial x} \frac{\partial x}{\partial t} = a \frac{\partial \xi}{\partial x}, \frac{u}{a} = \frac{\partial \xi}{\partial x}$$

Substituting

$$\eta = -H \frac{\partial \xi}{\partial x}$$

can get

$$\frac{u}{a} = -\frac{\eta}{H}$$

The above formula shows that the waveform remains unchanged in the process of wave propagation.

Wave energy includes wave potential energy and wave kinetic energy. The potential energy can be expressed as

$$E_p = \rho g \int \left(\int_0^\eta z \, dz \right) dx = \frac{1}{2} \rho g \int \eta^2 dx$$

Wave kinetic energy is

$$E_k = \frac{1}{2} \rho H \int u^2 dx = \frac{1}{2} \rho H \int a^2 \frac{\eta^2}{H^2} dx = \frac{1}{2} \rho g \int \eta^2 dx$$

It can be seen that $E_T = E_p + E_k = 2E_k = 2E_p$. This conclusion was first proposed by Rayleigh (as shown in Fig. 1.47) in 1876. If sine wave is assumed, the wave function is

$$\eta = A \cos(x - at)$$

where A is the amplitude. Substituting E_P and E_K, we get

$$E_p = E_k = \frac{1}{2} \rho g \int_0^\lambda \eta^2 dx = \frac{1}{2} \rho g \int_0^\lambda A^2 \cos^2(x - at) dx = \frac{1}{4} \rho g A^2 \lambda$$

The total energy per wavelength is

$$e_T = \frac{E_P + E_K}{\lambda} = \frac{1}{2}\rho g A^2$$

If the potential flow of an ideal fluid is considered, then

$$u = \frac{\partial \phi}{\partial x}, v = \frac{\partial \phi}{\partial y}$$

Kinetic energy is

$$E_k = \frac{1}{2}\rho \int\limits_0^\lambda \int\limits_{-H}^\eta \left[\left(\frac{\partial \phi}{\partial x}\right)^2 + \left(\frac{\partial \phi}{\partial y}\right)^2 \right] dxdz$$

3.8.3   Types of Waves

Waves in the ocean are of various types, shapes, and names. The disturbing forces of waves are: sun, moon, storm, earthquake, wind, etc.; the restoring forces are: Coriolis force, gravity, surface tension, etc. Among the ocean waves, the wind-induced gravity wave is the main one, also known as wind wave.

(1) according to the cause of formation and frequency classification, there are: surface tension wave with frequency less than 10 Hz; short period gravity wave with frequency 1–10 Hz; gravity wave with frequency 1/30–1 Hz; long period gravity wave with frequency 1/300–1/30 Hz; long wave with frequency 1/8.64 × 10-4-1/300 Hz; inertial wave and Star wave with frequency greater than 1/8.64 × 10-4 Hz.
(2) according to the interference force, there are wind waves, tide waves, ship waves, etc.
(3) according to the excitation force, there are free wave and forced wave. For example, the waves in still water caused by rocks are free waves. It refers to the wave after the cancelation of the interference force. The propagation and evolution of the wave are only restricted by the nature of water. The forced wave refers to the wave which is continuously acted by the interference force. The operation of the forced wave is restricted by the interference force and the nature of water. Tidal wave is a kind of forced wave.

(4) according to the classification of mass transport, there is transport wave (tide wave, flood wave, etc.) and vibration wave. If there is a mass transport (generating flow) in the average sense when the wave moves forward, it is called a transport wave (as shown in Figs. 3.70, 3.71, 3.72). Otherwise, if there is no flow when the wave moves forward (the average flow is zero), it is called vibration wave (as shown in Fig. 3.73). Wind wave is a kind of vibration wave. Particles move in a circular way without producing flow. In the vibration wave, according to whether the wave shape has horizontal movement relative to the medium, it can be divided into progressive wave and standing wave. Obviously, the horizontal motion of the wave is propulsion wave; the horizontal motion is

Fig. 3.70 Transport wave (tide wave)

Fig. 3.71 Ebb wave (different wave speed)

Fig. 3.72 High tide waves

Fig. 3.73 Vibration wave

standing wave (the water particle vibrates in the vertical direction). The movement of wave shape is different from that of water particle. If the movement direction of wave shape is the same as that of water particle, it is called longitudinal wave. If the direction of motion is vertical, it is called transverse wave. Generally speaking, sea surface waves are neither longitudinal waves nor transverse waves. The water particle moves in a circle or ellipse on a vertical plane.

(5) According to water depth, there are deepwater waves, shallow water waves, and intermediate water waves (as shown in Figs. 3.74 and 3.75). Let the water depth be H and the wavelength be λ, then H/λ is the ratio

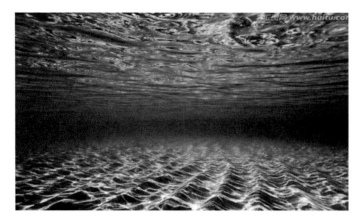

Fig. 3.74 Deepwater waves

Fig. 3.75 Shallow water waves

of water depth to wavelength. Whether the bottom of the riverbed affects the wave motion depends on the relative water depth ratio. It is shown that when $H/\lambda > 1/2$, the effect of the bottom on the waves is negligible, which is called deepwater. For example, using a more complete wave speed a formula

$$a^2 = \frac{g}{k}\tanh(kH)$$

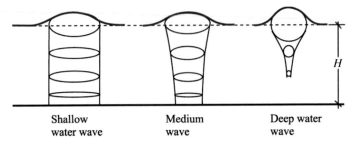

Fig. 3.76 Particle trajectory shape under different water depth ratio

When $H/\lambda > 1/2$, $kH = \frac{2\pi H}{\lambda} > \pi$, from which, $\tanh(kH) = 0.9963$ can be obtained, and the wave velocity is

$$a^2 = \frac{g}{k} = \frac{2\pi g}{\lambda}$$

It has nothing to do with the depth of water. If the limit of shallow water is determined in the same way, it can be reached only when $kH = 0.05$

$$\tanh(kH) = kH$$

It can be obtained.

$$\frac{H}{\lambda} = \frac{0.05}{2\pi} = 0.008 = \frac{1}{125}$$

Then, at $H/\lambda < 1/125$, the wave velocity is

$$a^2 = \frac{g}{k}\tanh(kH) = gH$$

Generally (as shown in Fig. 3.76):
When $H/\lambda > 1/2$ (deepwater wave); $\frac{H}{\lambda} \le \frac{1}{200}$ (shallow water wave); $\frac{1}{200} < \frac{H}{\lambda} \le \frac{1}{2}$ (medium water wave)

3.8.4 Linear Wave Theory (Micro Amplitude Wave Theory)

The research of wave theory has gone through the process from regular wave to random wave. The characteristic of regular wave theory is to take wave motion as a definite function form and to study the dynamic properties and

Fig. 3.77 Micro amplitude wave

motion rules of wave in various cases through hydrodynamics analysis. Since the nineteenth century, the research of regular wave theory has gone through a process from linear theory to nonlinear theory and turbulence theory. It mainly includes: micro amplitude wave theory (as shown in Fig. 3.77), Stokes (as shown in Fig. 1.32) high-order wave theory, elliptic cosine wave theory, solitary wave theory, etc., in which the micro amplitude wave theory was put forward by the British mathematician and astronomer G.B.Airy (1801–1892, as shown in Fig. 3.78) in 1845. This theory is a linear wave theory that applies the velocity potential function to study wave motion. As the most basic and important content of wave theory, it is widely used in offshore engineering. In 1887, Stokes, the British mathematician, put forward the high-order wave theory, which is often used to calculate the maximum wave height in offshore engineering calculation. Because Stokes' high-order wave theory does not consider the influence of the change in water depth, it is only suitable for the case of deepwater. In the case of shallow water, the theoretical error of Stokes wave is large, but if we adopt the elliptic cosine wave theory which can reflect the main law of wave motion, we can get high precision. The theory of elliptic cosine wave was first proposed by D.J. Korteweg (1848–1941, as shown in Fig. 3.79) in 1895, Another famous achievement of Koteweg's wave theory is that he and his student, G.de Vries (1866–1934, as shown in Fig. 3.80), worked together in 1895 to study the small amplitude long wave motion in shallow water, This paper proposes a partial differential equation (i.e., the famous KdV equation) for shallow water waves in one direction. The solution is a cluster of solitons (solitary waves). Results of the comparison of various wave theories, due to different criteria used, are quite different. In terms of qualitative analysis, at present, it can only be determined that elliptic

Fig. 3.78 George Biddell Airy (1801–1892, British mathematician and astronomer)

Fig. 3.79 D.J. Korteweg (1848–1941, Dutch mathematician)

Fig. 3.80 G. de Vries, (1866–1934, Dutch mathematician)

cosine wave is generally used in shallow water, solitary wave is generally used in nearshore shallow water and the peak energy of periodic wave accounts for more than 90% of the total wave energy, and micro amplitude wave is generally used in deepwater, However, for the limited water depth area, the situation is more complex, and the application range of various wave theories intersects here, so it is necessary to analyze according to the actual conditions to select the appropriate wave theory.

(1) Potential wave equation

Micro amplitude wave theory is a linear wave theory which studies wave motion from potential function. The basic assumption is that the water is ideal and incompressible (most wave motion, this assumption is reasonable); the mass force acting on the particle is only gravity (only to discuss the gravity wave). When the free surface is in the equilibrium position, set it as xoy plane, and Z is vertical upward, then the Euler equations representing the movement of water particle are

$$\frac{\partial u}{\partial t} + u\frac{\partial u}{\partial x} + v\frac{\partial u}{\partial y} + w\frac{\partial u}{\partial z} = -\frac{1}{\rho}\frac{\partial p}{\partial x}$$

$$\frac{\partial v}{\partial t} + u\frac{\partial v}{\partial x} + v\frac{\partial v}{\partial y} + w\frac{\partial v}{\partial z} = -\frac{1}{\rho}\frac{\partial p}{\partial y}$$

$$\frac{\partial w}{\partial t} + u\frac{\partial w}{\partial x} + v\frac{\partial w}{\partial y} + w\frac{\partial w}{\partial z} = -\frac{1}{\rho}\frac{\partial p}{\partial z} - g$$

$$\frac{\partial u}{\partial x} + \frac{\partial v}{\partial y} + \frac{\partial w}{\partial z} = 0$$

where u, v, and w represent the velocity components in three coordinate directions, respectively, and g is the acceleration of gravity. Only gravity wave is considered here, and the recovery force of wave is gravity. That is to say, the original water body in equilibrium deviates from the equilibrium state after being disturbed, and returns to the equilibrium position under the action of gravity, which forces the water to oscillate. There are two kinds of water body disturbed: one is the free surface disturbed (wind wave); the other is the particle velocity disturbed (earthquake wave). Since the water is inviscid, it is in a static state before being disturbed, and the motion is still irrotational after being disturbed. Therefore, it can be assumed that the wave is irrotational, so the velocity of water particle has velocity potential function ϕ (x, y, z, t), which satisfies

$$u = \frac{\partial \varphi}{\partial x}, v = \frac{\partial \varphi}{\partial y}, w = \frac{\partial \varphi}{\partial z}$$

So we can get it from the continuous equation

$$\frac{\partial^2 \varphi}{\partial x^2} + \frac{\partial^2 \varphi}{\partial y^2} + \frac{\partial^2 \varphi}{\partial z^2} = 0$$

Therefore, to solve the wave equation problem is essentially to solve the potential flow problem (also known as potential wave). Then, the energy equation is used to calculate the pressure field, namely

$$\frac{\partial \phi}{\partial t} + \frac{p}{\rho} + \frac{1}{2}V^2 + gz = f(t)$$

If the solid wall equation is $z = -H(x, y)$, the solid wall boundary condition of an ideal fluid is

$$\frac{\partial \varphi}{\partial n} = \vec{n} \bullet \nabla \varphi = 0, u\frac{\partial H}{\partial x} + v\frac{\partial H}{\partial y} + w = 0$$

At the free surface, the dynamic and kinematic conditions need to be satisfied. Let the free surface equation be

$$z = \eta(x, y, t)$$

On the free surface, $p = p_a$, the kinetic condition of the free surface is obtained using the energy equation

$$\frac{\partial \varphi(x, y, \eta, t)}{\partial t} + \frac{1}{2}V^2 + g\eta = 0$$

Since the free surface is a flow surface, its kinematic condition is

$$w - \frac{\partial \eta}{\partial t} - u\frac{\partial \eta}{\partial x} - v\frac{\partial \mu}{\partial y} = 0$$

The boundary condition of free surface includes nonlinear term. In the theory of micro amplitude wave, it is assumed that the velocity of water particle is very small and $\frac{1}{2}V^2$ terms are ignored; the deviation of free surface from the horizontal plane is very small, so it can be replaced by the physical quantity $z = 0$ on the horizontal plane; the tangent plane on the free surface is almost the same as the horizontal plane, that is, it is assumed that $\frac{\partial \eta}{\partial x}, \frac{\partial \eta}{\partial y}$ is also a small quantity. It can be concluded that the dynamic condition of the free surface is

$$\eta = -\frac{1}{g}\frac{\partial \varphi(x, y, 0, t)}{\partial t}$$

Kinematic conditions become

$$\frac{\partial \eta}{\partial t} = \frac{\partial \varphi(x, y, 0, t)}{\partial z}$$

The above two equations can combine dynamics and kinematics, that is

$$\frac{\partial^2 \phi}{\partial t^2} + g\frac{\partial \phi}{\partial z} = 0$$
$$z = 0$$

The dynamic calculation is

$$\frac{p - p_a}{\rho} = -\frac{\partial \phi}{\partial t} - gz$$

(2) Basic solution

In the wave problem, we mainly study the shape of free surface, the propagation speed of wave, the velocity and trajectory of water particle, wave energy, and so on. For two-dimensional problems, the Laplace equation is solved by the method of separating variables. The general solution of its velocity potential function is

$$\varphi(x, z) = \cosh k(z + H)(A^* \sin kx + B^* \cos kx)$$

where k is the wave number and H is the water depth. The constants A^* and B^* are determined by free surface conditions, which are functions of time in wave problems. If the problem of surface tension is not considered, the free surface condition is substituted and a basic solution is obtained

$$\varphi(x, z) = A \cosh k(z + H) \sin(kx - \omega t)$$

Among them, $\omega^2 = gk \tanh kH$. This is a typical velocity potential function of progressive wave. If the wave velocity is $a = \omega/k = \lambda f = \sqrt{\frac{g\lambda}{2\pi}}$, the wave form remains unchanged at the constant phase angle ($= kx - \omega t$), as shown in Fig. 3.81.

In wave dynamics, the wave number k is the number of internal waves in 2π, and the circular frequency ω is the number of waves in 2π. The function relationship between ω and k is called the dispersion relationship during wave group propagation. If the wave group meets the isophase condition during the propagation process, that is, $kx - \omega t = $ constant, the propagation speed of each wave $a = \frac{dx}{dt} = \frac{\omega}{k} = $ constant, which is called a monochromatic wave; otherwise, if the ω/k of each wave is not constant, it is called a dispersion wave. Intuitively, this wave group has different frequencies and propagation speeds

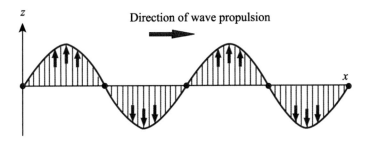

Fig. 3.81 Progressive wave

during the propagation process, which will cause dispersion. For gravity waves, the dispersion relationship is $\omega^2 = gk \tanh kH$ and the wave speed varies with wavelength. In nonuniform media, wave velocity is also related to the direction of wave propagation. This formula can further show that for deepwater waves, the dispersion relationship is $\omega = \sqrt{gk} = \sqrt{g\frac{2\pi}{\lambda}}$, and the wave frequency is inversely proportional to the square root of the wavelength; for shallow water waves, the dispersion relationship is $\omega = \sqrt{gh}k = \sqrt{gh\frac{2\pi}{\lambda}}$, where the wave frequency is inversely proportional to the wavelength and proportional to the water depth. As shown in Fig. 3.82.

Another basic solution is

$$\varphi = B \cosh k(z + H) \cos kx \cos \omega t$$

This is a typical standing wave velocity potential function. In this kind of wave, the wave function does not propagate along the x-axis, but makes periodic movement in place. As shown in Fig. 3.83, the shape of the standing wave is fixed, and there are antinodes and nodes, and the distance between adjacent belly (node) particles is $\lambda/2$, and the distance between adjacent belly and nodes is $\lambda/4$.

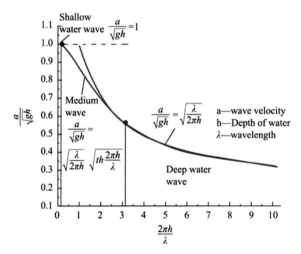

Fig. 3.82 Gravity wave dispersion relationship

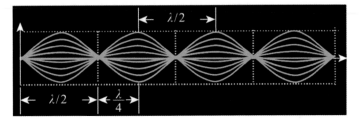

Fig. 3.83 Standing wave

(3) Plane progressive wave

Plane progressive wave is a simple wave phenomenon. Its waveform profile is a sine curve, and the waveform advances at a certain speed on the water surface, and the waveform remains unchanged during the advance, but no quality migration occurs.

For deepwater wave ($\omega^2 = gk$), in order to ensure that $z \to -\infty$ has a limit at $u = \frac{\partial \phi}{\partial x}$, we get

$$\varphi(x, z, t) = Ae^{kz} \sin(kx - \omega t)$$

The free surface equation is

$$\eta = \frac{h}{2} \cos(kx - \omega t)$$

where h is the free surface wave height, $H = 2A\omega/g$. At a certain time, the contour of the free surface is a cosine wave curve. The velocity of water particle is very small, which is divided into

$$\frac{dx}{dt} = u = \frac{\partial \varphi}{\partial x} = \frac{h}{2}\omega e^{kz} \cos(kx - \omega t)$$
$$\frac{dz}{dt} = w = \frac{\partial \varphi}{\partial z} = \frac{h}{2}\omega e^{kz} \sin(kx - \omega t)$$

The x and z in the right term of the above formula are replaced by the x_0 and z_0 of the initial position of the particle.

$$(x - x_0)^2 + (z - z_0)^2 = \left(\frac{h}{2}\right)^2 e^{2kz_0}$$

This formula indicates that the trajectory of the water particle in the deep-water area is a circle with the equilibrium position as the center and the radius. The radius on the free surface is h/2. When the progressive wave propagates in a forward direction, each particle moves in a clockwise direction in a circular motion. Because the radius is the amplitude and velocity of particle fluctuations decay in the water depth direction by exponent e. The deeper the free surface, the smaller the particle velocity and amplitude.

For the case of shallow water waves, it is assumed that the water depth in the water area is $z = H = $ constant, and the velocity potential function of the water particle is

$$\varphi(x, z, t) = A \cosh k(z + H) \sin(kx - \omega t)$$

The relationship between wave speed and wavelength is

$$\omega^2 = gk \tanh kH$$

Compared to the infinite depth case ($\omega^2 = gk$), the depth of the water affects the frequency of the wave. Free surface shape is

$$\eta = \frac{A\omega}{g} \cosh kH \cos(kx - \omega t) = \frac{h}{2} \cos(kx - \omega t)$$

Let $\frac{h}{2} = \frac{A\omega}{g} \cosh kH$ have a velocity potential of

$$\varphi = \frac{h}{2} \frac{g}{\omega} \frac{\cosh k(z + H)}{\cosh kH} \sin(kx - \omega t)$$

Among them, the propulsion wave speed is

$$a = \frac{\omega}{k} = \sqrt{\frac{g \tanh kH}{k}} = \sqrt{\frac{g\lambda}{2\pi} \tanh \frac{2\pi H}{\lambda}}$$

When $kH >> 1$, $\tanh kH \approx 1$ The above formula becomes deepwater wave condition. When the water depth is small, $\tanh kH \approx kH$, then

$$a = \sqrt{gH}$$

The speed of the particle motion is

$$\frac{dx}{dt} = u = \frac{\partial \varphi}{\partial x} = \frac{h}{2} \frac{gk}{\omega} \frac{\cosh k(z + H)}{\cosh kH} \cos(kx - \omega t)$$

$$\frac{dz}{dt} = w = \frac{\partial \varphi}{\partial z} = \frac{h}{2} \frac{gk}{\omega} \frac{\sinh k(z+H)}{\cosh kH} \sin(kx - \omega t)$$

The approximate trajectory equation of particle is

$$\frac{(x - x_0)^2}{\left[\frac{h}{2} \frac{\cosh k(z_0+H)}{\sinh kH}\right]^2} + \frac{(z - z_0)^2}{\left[\frac{h}{2} \frac{\sinh k(z_0+H)}{\sinh kH}\right]^2} = 1$$

This is an elliptic equation. The above formula shows that the trajectory of water particles in shallow water area is an ellipse, and the horizontal and vertical axes of the ellipse decrease with the increase in water depth. At the bottom of the water, the vertical axis of the ellipse is zero, and the particles move back and forth horizontally.

(4) Plane standing wave

The plane standing wave is also a simple waveform. At the same time, the distance between each particle on the wave surface and the equilibrium position is the harmonic function of x. At the same x, the displacement of the particle is a periodic function of time. There is a particle in the wave where the height of the free surface is always zero, which is a node. In a standing wave, the wave surface does not move forward, but moves up and down periodically.

For the plane standing wave of the sine curve of the deepwater wave, in order to ensure the $u = \frac{\partial \phi}{\partial x}$ finite value at $z \to -\infty$, the obtained particle velocity potential function is

$$\varphi(x, z, t) = \frac{gh}{2\omega} e^{kz} \sin kx \cos \omega t$$

According to the free surface boundary conditions, $\omega^2 = gk$ is known, which means that the frequency of the wave is determined after the constant k is given.

The shape function of the free surface is

$$\eta = \frac{h}{2} \sin(\omega t) \sin(kx)$$

At a certain time, the free surface is a sine curve on the *xoz* plane, and the height of the free surface at each particle is determined by the coordinate of particle X. The intersection of free surface and *ox* is

$$x_0 = \frac{n\pi}{k}, k = 0, \pm 1, ...,$$

The positions of these intersections do not change with time, so they are called nodes. The horizontal distance between two adjacent peaks is called the wavelength. Wave period is

$$T = \frac{2\pi}{\omega} = \frac{\kappa\lambda}{\omega} = \sqrt{\frac{2\pi\lambda}{g}} \left(\omega^2 = kg\right)$$

Vibration speed of water particle

$$u = \frac{\partial\varphi}{\partial x} = \frac{h}{2}\frac{gk}{\omega}e^{kz}\cos kx \cos \omega t$$

$$w = \frac{\partial\varphi}{\partial z} = \frac{h}{2}\frac{gk}{\omega}e^{kz}\sin kx \cos \omega t$$

The approximate trajectory of a water particle is

$$(z - z_0) = (x - x_0)tgkx_0$$

For shallow standing wave standing waves, a constant water depth is H. Let $z = -H ==$ constant, then

$$\omega^2 = kg \tanh kH$$

The velocity potential function is

$$\varphi = \frac{h}{2}\frac{g}{\omega}\frac{\cosh k(z + H)}{\cosh kH}\sin kx \cos \omega t$$

The shape of the free surface is

$$\eta = \frac{h}{2}\sin kx \sin \omega t$$

The water particle velocity is

$$u = \frac{\partial \varphi}{\partial x} = \frac{h}{2} \frac{gk}{\omega} \frac{\cosh k(z+H)}{\cosh kH} \cos kx \cos \omega t$$

$$w = \frac{\partial \varphi}{\partial z} = \frac{h}{2} \frac{gk}{\omega} \frac{\sinh k(z+H)}{\cosh kH} \sin kx \cos \omega t$$

The trajectory of water particle is

$$x = x_0 + \frac{h}{2} \frac{\cosh k(z_0 + H)}{\sinh kH} \cos kx_0 \sin \omega t$$

$$z = z_0 + \frac{h}{2} \frac{\sinh k(z_0 + H)}{\sinh kH} \sin kx_0 \sin \omega t$$

3.8.5   Wave with Finite Amplitude

For waves with limited amplitude, because the nonlinear convection term in the equation of motion cannot be ignored, it is more difficult to directly obtain the solution of the governing equation, and some theoretical results have certain limitations. Experiments have found that the shape of this type of wave is no longer a sine (or cosine) curve, but a shape with a steeper wave peak and a more flat wave surface. The shape of this wave is similar to the shape of a trochoid curve. Therefore, in engineering practice, for the sake of simplicity, we use the trochoid theory as an approximation. Compared with the potential flow theory, this theory is different in that the motion characteristics of the water mass particle are given first, and then the correctness of the hypothesis is verified according to the basic equations of water flow, so as to establish the fluctuation law.

(1) Two-dimensional Lagrangian continuous equation and equation of motion

When solving with the cycloid theory, it is required to track individual water particles to study its motion law, so the continuous and motion equations expressed by the Lagrangian method are needed. In a two-dimensional flow space of incompressible fluid, we take a fixed rectangular coordinate system. Then, the position of any water particle expressed in Lagrangian at time t is

$$x = x(a, b, t), z = z(a, b, t)$$

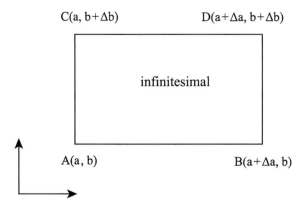

Fig. 3.84 Microelement

where, a and b are the position of the water particle at the initial moment, and its value does not change with time. It is used to distinguish different water particles. Suppose that at time t, any micro-facet (such as the rectangle shown in Fig. 3.84) is taken from the two-dimensional moving liquid.

$$\Delta A = \Delta a \Delta b$$

Tracking this microelement surface, after a certain period of time, it moves to a new position. Then the coordinates of each particle can be obtained. The position of particle A is

$$x = x(a, b, t), z = z(a, b, t)$$

The position of particle B is

$$x_B = x(a, b, t) + \frac{\partial x}{\partial a} \Delta a, z_B = z(a, b, t) + \frac{\partial z}{\partial a} \Delta a$$

The position of particle C is

$$x_C = x(a, b, t) + \frac{\partial x}{\partial b} \Delta b, z_C = z(a, b, t) + \frac{\partial z}{\partial b} \Delta b$$

The microelement area of the new location is

$$\Delta A^* = \begin{vmatrix} \dfrac{\partial x}{\partial a} & \dfrac{\partial z}{\partial a} \\ \dfrac{\partial x}{\partial b} & \dfrac{\partial z}{\partial b} \end{vmatrix} \Delta a \Delta b$$

Because the liquid is incompressible, the volume of the microelements does not change at any time. Based on that, we can obtain

$$
\begin{vmatrix} \dfrac{\partial x}{\partial a} & \dfrac{\partial z}{\partial a} \\[2mm] \dfrac{\partial x}{\partial b} & \dfrac{\partial z}{\partial b} \end{vmatrix} = 1
$$

This is the uncompressed Lagrange continuous equation. Usually, we use

$$
\frac{\partial}{\partial t} \begin{vmatrix} \dfrac{\partial x}{\partial a} & \dfrac{\partial z}{\partial a} \\[2mm] \dfrac{\partial x}{\partial b} & \dfrac{\partial z}{\partial b} \end{vmatrix} = 0
$$

Using Newton's second law, the equation of motion established under gravity is

$$
\frac{\partial^2 x}{\partial t^2} = -\frac{1}{\rho}\frac{\partial p}{\partial x}
$$

$$
\frac{\partial^2 z}{\partial t^2} = g - \frac{1}{\rho}\frac{\partial p}{\partial y}
$$

Based on the derivation of compound functions

$$
\frac{\partial p}{\partial a} = \frac{\partial p}{\partial x}\frac{\partial x}{\partial a} + \frac{\partial p}{\partial z}\frac{\partial z}{\partial a}
$$

The equation of motion can be expressed as

$$
\left(-\frac{\partial^2 x}{\partial t^2}\right)\frac{\partial x}{\partial a} + \left(g - \frac{\partial^2 z}{\partial t^2}\right)\frac{\partial z}{\partial a} - \frac{1}{\rho}\frac{\partial p}{\partial a} = 0
$$

$$
\left(-\frac{\partial^2 x}{\partial t^2}\right)\frac{\partial x}{\partial b} + \left(g - \frac{\partial^2 z}{\partial t^2}\right)\frac{\partial z}{\partial b} - \frac{1}{\rho}\frac{\partial p}{\partial b} = 0
$$

Substituting a and b with x_0 and z_0, respectively, we can obtain the Lagrangian-type continuous equation and motion equation that characterizes the motion of the particle as follows:

$$\frac{\partial}{\partial t}\begin{vmatrix} \dfrac{\partial x}{\partial x_0} & \dfrac{\partial z}{\partial x_0} \\ \dfrac{\partial x}{\partial z_0} & \dfrac{\partial z}{\partial z_0} \end{vmatrix} = 0$$

$$\left(-\frac{\partial^2 x}{\partial t^2}\right)\frac{\partial x}{\partial x_0} + \left(g - \frac{\partial^2 z}{\partial t^2}\right)\frac{\partial z}{\partial x_0} - \frac{1}{\rho}\frac{\partial p}{\partial x_0} = 0$$

$$\left(-\frac{\partial^2 x}{\partial t^2}\right)\frac{\partial x}{\partial z_0} + \left(g - \frac{\partial^2 z}{\partial t^2}\right)\frac{\partial z}{\partial z_0} - \frac{1}{\rho}\frac{\partial p}{\partial z_0} = 0$$

(1) Deepwater propulsion wave (circle trochoid theory)

For finite-amplitude deepwater propulsion waves, a commonly used approximation theory is the circle trochoid theory proposed by German physicist F. Gerstner (1756–1832, as shown in Fig. 3.85) in 1802. Let us take a two-dimensional deepwater wave as an example. It is assumed in the analysis that the water is an ideal incompressible liquid without considering the influence of viscosity. The water depth is infinite and the wave motion is not affected by

Fig. 3.85 F. Gerstner (1756–1832, German physicist)

the seafloor. The water particle is on a vertical plane, with a uniform circular motion. The center of the circle is a certain distance above the position of the particle at rest. Water particles located on the same horizontal plane at rest forms the wave surface during waving. Water particles on the same wave surface have equal radius r of circular motion and at the water surface $r = h/2$. But in the vertical direction, the r decreases sharply from the water surface. When the water particle moves in a circle, the angle between the radial line and the upward vertical line is the phase angle θ. At the same time particle, on any wave surface, the phase angle decreases along with the direction of the wave as the distance increases. At the same time particle, the phase angles of the water particles with the center of the circle on the same vertical line are equal. This is shown in Fig. 3.86.

As shown in Fig. 3.87, we take the midline of the wave as x-axis, the positive direction is the wave propulsion direction. The positive direction of the vertical axis z is along the downside. The center of the water mass particle A is (x_0, z_0). A moves at a constant speed around the center particle with radius r, and its trajectory is

$$x = x_0 + r \sin \theta$$
$$z = z_0 - r \cos \theta$$

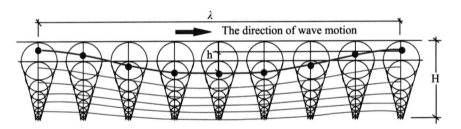

Fig. 3.86 The wave shape of circle trochoid curve with different water depth

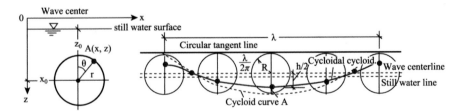

Fig. 3.87 The establishment of trochoid curve

In above equation, we take $r = f(z_0)$. Then the phase angle θ can be rewritten

$$\theta = \sigma t - kx_0$$

where σ is angular velocity of particle; k is the curvature. According to the periodicity of the waves, for wave period T, the phase angle increased by 2π when the particle rotates once. Then we can obtain

$$\theta = \sigma(t + T) - kx_0 = \sigma t + 2\pi - kx_0$$

Therefore, we take $\sigma = 2\pi/T$. Similarly in x_0 direction, for each increase in wavelength λ, the phase angle decreases 2π correspondingly. We have

$$\theta = \sigma t - k(x_0 + \lambda) = \sigma t - kx_0 - 2\pi$$

Therefore, we $k = 2\pi/\lambda$. Then the water particle motion equation of deepwater propulsion wave can be obtained. If we take $z_0 = 0$, then the water surface curve can be derived easily.

$$x = x_0 + r\sin\left(\frac{2\pi}{T}t - \frac{2\pi}{\lambda}x_0\right)$$
$$z = z_0 - r\cos\left(\frac{2\pi}{T}t - \frac{2\pi}{\lambda}x_0\right)$$

Substituting the above-mentioned motion trajectory equation of the water mass particle into a Lagrange-type continuous equation, we can obtain

$$\frac{\partial}{\partial t}\left[1 + kr\frac{\partial r}{\partial z_0} - \left(kr + \frac{\partial r}{\partial z_0}\right)\cos(\sigma t - kx_0)\right] = 0$$

The coefficient before the function $\cos(\sigma t - kx_0)$ must be zero before the above equation exists. Along with the water surface condition, we can obtain

$$r = \frac{h}{2}e^{-kz_0} = \frac{h}{2}e^{-\frac{2\pi}{\lambda}z_0}$$

This equation shows that for deepwater propulsion waves, the water particle trajectory circle radius r decreases in the vertical direction according to the e-exponential law, and the particle with smaller wavelength attenuates quicker. This is consistent with physical phenomena. Similarly, substituting the particle trajectory equation into the Lagrange-type equation of motion, and the pressure at the free surface after integration is

$$\frac{p_a}{\rho} = \left(\frac{\sigma^2}{k} - g\right) r_0 \cos(\sigma t - kx_0) + \frac{1}{2}\sigma^2 r_0^2 + C$$

where p_a is atmospheric pressure; C is the free constant. Obviously, for the existence of the above formula, it must be satisfied

$$\frac{\sigma^2}{k} - g = 0$$

This shows that in the case of deepwater propulsion waves, the square value of the angular velocity of the particle motion is equal to the product of the curvature of the circle and the acceleration of gravity (Dispersion relation of deepwater wave). Substituting $\sigma = 2\pi/T$ and $k = 2\pi/\lambda$ into the dispersion equation, and we can obtain the wave velocity and the period

$$a = \frac{\lambda}{T} = \sqrt{\frac{g\lambda}{2\pi}}, \ T = \sqrt{\frac{2\pi\lambda}{g}}$$

The wave velocity and wave period of a deepwater propulsion wave are directly proportional to the wavelength. That is, the longer the wavelength, the greater the wave speed and wave period.

For water surface wave, we have $t = 0$, $z_0 = 0$, $r = h/2$, as well as $\theta = \sigma t - kx_0 = -kx_0$. The water surface wave equation can be obtained by substituting the front condition into the motion equation of the particle. θ is a changeable parameter.

$$x = -\frac{\lambda}{2\pi}\theta + \frac{h}{2}\sin\theta$$
$$z = -\frac{h}{2}\cos\theta$$

The curve obtained from this equation is the circle trochoid curve. The circle trochoid curve is a curve drawn by a particle inside a circle as each circle rolls along its tangent. If it is a curve drawn on a circle, it is called a cycloid, which is plotted in Fig. 3.88. Similarly, we can obtain a curve equation of wave surface at an arbitrary water depth

$$x = -\frac{\lambda}{2\pi}\theta + \frac{h}{2}e^{-kz_0}\sin\theta$$

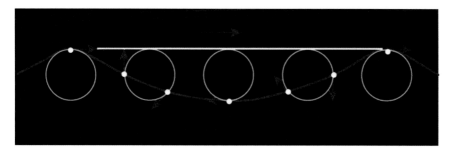

Fig. 3.88 The cycloid curve

$$z = z_0 - \frac{h}{2}e^{-kz_0}\cos\theta$$

Based on the above equation, the wave shape at different water depths can be obtained, which is plotted in Fig. 3.89. The particle circular motion speed at a different layer is

$$u = \frac{2\pi r}{T} = r\sqrt{\frac{2\pi g}{\lambda}} = \frac{h}{2}e^{-kz_0}\sqrt{\frac{2\pi g}{\lambda}}$$

(2) Shallow water propulsion wave (ellipse trochoid theory)

Considering that the characteristics of deepwater propulsion waves are greatly affected by water depth, the French scientist J.V. Boussinesq (shown as

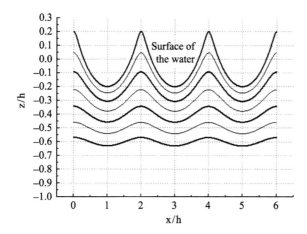

Fig. 3.89 The circle trochoid wave shape at different water depth

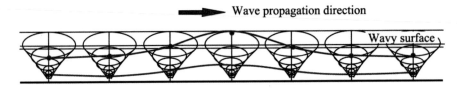

Fig. 3.90 The water surface curve of ellipse trochoid curve

Fig. 1.71) proposed the ellipse trochoid theory. The assumption and derivation of the ellipse trochoid theory is similar to the circle trochoid theory. The main difference is that the water particle moves elliptically. The long axis of ellipse is the horizontal axis (the direction of wave propagation). The short axis of ellipse is the vertical axis. The long and short axes of ellipse decrease along the vertical direction. Water particles no longer move at a constant speed on elliptical trajectories. But their phase velocity is still constant. This is plotted in Fig. 3.90

As shown in Fig. 3.91, suppose the wave centerline as x-axis; the positive direction of z-axis is the downside of the vertical line; the water depth is H. Take a particle (x_0, z_0) arbitrarily in the waters as the center of an ellipse whose long axis is a, short axis is b, and phase angle is θ. Then the trajectory of the water mass particle M on the ellipse is

$$x = x_0 + a\sin\theta = x_0 + a\sin(\sigma t - kx_0)$$
$$z = z_0 - b\cos\theta = z_0 - b\cos(\sigma t - kx_0)$$

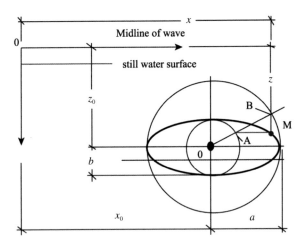

Fig. 3.91 The definition of ellipse trochoid curve

Correspondingly, the velocity of the mass particle is

$$u = \frac{\partial x}{\partial t} = \sigma a \cos \theta, w = \frac{\partial z}{\partial t} = \sigma b \sin \theta$$

But according to the definition, the phase angle θ is not the angle of the line connected by particle M and center of trajectory and the vertical line, but the angle of the angle between the radial and vertical lines of the inner and outer auxiliary circle. Drawing horizontal and vertical lines from the intersection particles A and B of the radial line and the inner and outer auxiliary circles, respectively, the position of water mass particle M is their intersection. $\theta = \sigma t\text{-}kx_0$, where σ and k represent the phase angular velocity and round curvature, respectively. The derivation process is exactly the same. Using the Lagrangian continuous equation, we can obtain that the long and short semi-axes of the particle locus circle are

$$a = \frac{h}{2}\frac{\cosh k(H - z_0)}{\sinh kH}, b = \frac{h}{2}\frac{\sinh k(H - z_0)}{\sinh kH}$$

And the following condition should be satisfied:

$$k^2(a^2 - b^2) = \left(\frac{h}{\lambda}\right)^2 \frac{\pi^2}{\sinh^2 kH} = 0$$

For the ratio of wave height h to wavelength λ is very small, the above formula can be established. When shallow water waves generally occur, the water depth is small. Under the circumstance, when the kH is very small, we have $\sinh kH \approx kH$. Therefore, we can obtain

$$k^2(a^2 - b^2) = \frac{1}{4}\left(\frac{h}{H}\right)^2$$

It shows that the larger the ratio of wave height to water depth, the less satisfying the continuity equation. The magnitude of the error is proportional to the square of the ratio of wave height to water depth. When h/H is very small, $a \approx b$. In this case, water particles will make circular motions, called deepwater propulsion waves. a and b are the function of z_0. When on the water, $z_0 = 0$, and we can obtain

$$a_0 = \frac{h}{2}\coth kH, b_0 = \frac{h}{2}$$

When under the water, $z_0 = H$, and we can obtain

$$a_H = \frac{h}{2}\frac{1}{\sinh kH}, b_H = 0$$

The elliptical focal length is

$$a_H = \frac{h}{2}\frac{1}{\sinh kH}, b_H = 0$$

Obviously, the ellipse of the trajectory of the water particle is gradually flattened down from the water surface, but the focal length remains unchanged, and the water mass particle at the bottom of the water vibrations horizontally between the two focal particles.

Substituting the motion trajectory equation of the water mass particle into the Lagrange equation of motion, and then simplifying it to obtain the condition that satisfies the motion is

$$gb_0 - \frac{\sigma^2 a_0}{k} = 0$$

Substituting $\sigma = 2\pi/T$ and $k = 2\pi/\lambda$ into the above equation, the wave speed and wave period can be obtained

$$a = \frac{\lambda}{T} = \sqrt{\frac{g\lambda}{2\pi}}\sqrt{thkH}, \quad T = \sqrt{\frac{2\pi\lambda}{g}\frac{a_0}{b_0}} = \sqrt{\frac{2\pi\lambda}{g}}\sqrt{cthkH}$$

It can be seen that compared with deepwater propulsion waves, the speed and period of shallow-water propulsion waves are not only related to the wavelength, but also change with the depth of water. At the same wavelength, the period of shallow-water wave is larger than that of deepwater wave, and the speed of shallow-water wave is smaller than that of deepwater wave. Because when $H/\lambda > 1/2$, $thkH \approx 1$, so substitute it into the above formula and get the same result as the wave velocity and period of the deepwater wave. Therefore, in actual calculations, $H/\lambda = 1/2$ is often used as the dividing line for deep and shallow water waves.

For waveform of shallow water waves, because $t = 0$, $z_0 = 0$, $a = a_0$, $b = b_0 = h/2$, and $\theta = \sigma t - kx_0 = -kx_0$, the water surface wave equation can be obtained by substituting front conditions into particle motion equation. θ

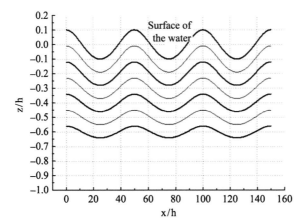

Fig. 3.92 The shape of ellipse trochoid curve under different water depth

is a changeable parameter in the equation.

$$x = -\frac{\lambda}{2\pi}\theta + a_0 \sin\theta = -\frac{\lambda}{2\pi}\theta + \frac{h}{2} cthkH \sin\theta$$

$$z = -b_0 \cos\theta = -\frac{h}{2}\cos\theta$$

It can also be found that the wave equation of the underwater arbitrary wave surface is

$$x = -\frac{\lambda}{2\pi}\theta + a\sin\theta = -\frac{\lambda}{2\pi}\theta + \frac{h}{2}\frac{\cosh k(H-z_0)}{\sinh kH}\sin\theta$$

$$z = z_0 - b\cos\theta = z_0 - \frac{h}{2}\frac{\sinh k(H-z_0)}{\sinh kH}\cos\theta$$

Compared with the circle trochoid curve, the water peaks of the ellipse trochoid curve are sharper and troughs are calmer. The shape curve under different water depth is plotted at Fig. 3.92.

3.8.6 Solitary Wave

A solitary wave is a wave of limited amplitude that propagates as a single crest or trough and can appear in shallow waters, as shown in Fig. 3.93. This kind of fluctuation was first discovered in the laboratory by the British scientist and shipbuilding engineer John Scott Russell (1808–1882, as shown in Fig. 3.94) in 1844, and theoretically carried out research to propose the solitary wave

Fig. 3.93 Solitary wave

Fig. 3.94 John Scott Russell (1808–1882, British scientist and shipbuilding engineer)

theory (As shown in Fig. 3.95). Since the waves are similar to solitary waves after they are introduced into shallow flat areas with flat bottom slopes, the results of solitary wave studies are often used to analyze nearshore waves. According to records, in the fall of 1834, Russell noticed on a canal that a fast-moving ship with two horses and horses suddenly stopped, and a large

Fig. 3.95 Solitary wave in the river

body of water pushed by the ship did not stop, but formed a smooth and a well-defined large water bag with a height of about 0.3 to 0.5 m, a length of about 1 m, and spread forward along the river at a speed of about 13 km per hour. When Russell tracked the water bag along the canal, he found that its size, shape, and speed changed slowly. It did not disappear on the river until 3 to 4 km. Later, Russell conducted a large number of tank tests and called this strange wave packet a solitary wave (as shown in Figs. 3.96 and 3.97). Russell reproduced this solitary wave in the sink and tried to find a solution, but failed.

Later in 1895, the Dutch mathematicians Kottweig and Devries proposed a partial differential equation to characterize single wave propagation, known as the KdV equation. Their research showed that the solitary wave observed by Russell is the result of the balance between nonlinear effects and dispersion in the wave process, and the KdV equation characterizes the phenomenon of weak nonlinearity and weak dispersion. The basic characteristic is that a solitary wave is a traveling wave that propagates in one direction in a small area, and its waveform does not change with time. When two solitary waves collide, they penetrate each other and maintain the original waveform and speed.

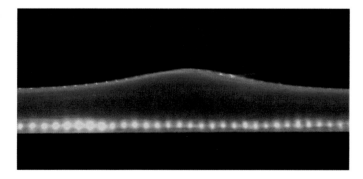

Fig. 3.96 Solitary wave packet

Fig. 3.97 Solitary wave on the coast

For the sake of solution aspect, in the classical micro-amplitude linear theory, the vertical acceleration of water particles, the nonlinear convection term in the equation of motion, and the nonlinear kinetic energy (free surface isobaric condition) term in the energy equation are all ignored. The sine or cosine wave theory is derived from this. The theory ignores the vertical velocity and acceleration of the water mass particle to obtain the static pressure distribution of the water mass particle that satisfy the hydrostatic pressure distribution rule; ignores the nonlinear convection term in the motion equation to obtain the linear wave equation; ignores the kinetic energy term in the free surface isobaric conditional energy equation to obtain the linearized boundary condition. The theoretical solutions of micro-amplitude waves established using these assumptions are suitable for deepwater waves and only obtain small errors. But for long wave motion problems in shallow water, these assumptions are not suitable. In order to improve the accuracy of theoretical predictions, corrections are needed. In 1871, the French scientist Boussinesq studied the motion of shallow water waves. First, he considered the influence of convection nonlinear terms and vertical acceleration on the wave and proposed a correction relationship for the distribution of hydrostatic pressure. The vertical line is set as z-axis. The x-axis is set in the balanced water surface. Based on the vertical motion equation (the vertical velocity is w), we can obtain the following equation by ignoring the nonlinear convection term in the vertical equation:

$$\frac{\partial w}{\partial t} = -\frac{1}{\rho}\frac{\partial p}{\partial z} - g$$

In case with horizontal bottom, the water depth H is constant and the normal velocity at the bottom is zero. That is $w_b = 0$. However, on the water surface, the vertical velocity of the water mass particle is $w_s \approx \partial\eta/\partial t$. Assume that the vertical velocity varies linearly with water depth, then

$$w(x, z, t) = \frac{z}{H + \eta}\frac{\partial\eta}{\partial t}$$

where η is the height of the wave. Differentiate the above equation with time t, then ignore the high-order terms and substitute them into the previous formula to obtain

$$\frac{z}{H + \eta}\frac{\partial^2\eta}{\partial t^2} = -\frac{\partial}{\partial z}\left(\frac{p}{\rho} + gz\right)$$

Integral above equation from z to $(H + \eta)$ is obtained

$$\frac{p}{\rho} = g(H + \eta - z) + \frac{(H + \eta)^2 - z^2}{2(H + \eta)}\frac{\partial^2\eta}{\partial t^2}$$

This formula is the correction of the distribution law of hydrostatic pressure at water particles by considering vertical acceleration derived by Businniques. The equation is substituted into the equation of motion in the horizontal direction and the small terms are omitted to obtain

$$\frac{du}{dt} = -g\frac{\partial\eta}{\partial x} - \frac{(H + \eta)^2 - z^2}{2(H + \eta)}\frac{\partial^3\eta}{\partial t^2\partial x}$$

Considering the characteristics of shallow water waves, we Integrate the above equation along the water depth and change $u(x, z, t)$ into the average velocity of water depth $v(x, t)$, that is

$$v(x,t) = \frac{1}{H + \eta}\int_0^{H+\eta} u(x, z, t)dz$$

The famous Boussinesq equation about nonlinear water wave motion is obtained as

$$\frac{\partial v}{\partial x} + v\frac{\partial v}{\partial x} + g\frac{\partial \eta}{\partial x} + \frac{H + \eta}{3}\frac{\partial^3 \eta}{\partial t^2 \partial x} = 0$$

The last term in this equation is the additional term considering the effect of vertical acceleration.

According to the idea of asymptotic approximation, the famous KdV equation was proposed by the Dutch mathematician Kottweig and his student Devries in 1895 when they were studying the motion of shallow water waves. For the two-dimensional wave problem, taking the origin of the vertical coordinate z at the bottom wall, then the definite solution problem of the velocity potential function is

$$\frac{\partial^2 \varphi}{\partial x^2} + \frac{\partial^2 \varphi}{\partial z^2} = 0$$

$$\frac{\partial \varphi}{\partial z}\bigg|_{z=0} = 0$$

$$\frac{\partial \eta}{\partial t} + \frac{\partial \varphi}{\partial x}\frac{\partial \eta}{\partial x} - \frac{\partial \varphi}{\partial z} = 0, z = H + \eta$$

$$g\eta + \frac{\partial \varphi}{\partial t} + \frac{1}{2}\left[\left(\frac{\partial \varphi}{\partial x}\right)^2 + \left(\frac{\partial \varphi}{\partial z}\right)^2\right] = 0, z = H + \eta$$

In this set of equations, the governing equations are linear and the free surface conditions (flow surface and isobaric conditions) are nonlinear, so it is difficult to solve directly. However, the long wave problem in shallow water can be solved approximately. Let the wavelength be λ, the amplitude of the surface wave be A. The parameters defined by the scale ratio in the x and z-direction is

$$\alpha = \frac{A}{h}, \beta = \frac{h^2}{\lambda^2}$$

where α and β are small terms and the approximate solution parameters. In order to nondimensionalize above definite problem, we define

$$x = \lambda x', z = Hz', t = t'\frac{\lambda}{\sqrt{gH}}, \eta = A\eta', \varphi = \frac{g\lambda A}{\sqrt{gH}}\varphi'$$

Then definite solution problem for the dimensionless velocity potential function is obtained.

$$\beta^2 \frac{\partial^2 \varphi'}{\partial x'^2} + \frac{\partial^2 \varphi'}{\partial z'^2} = 0$$

$$\left.\frac{\partial \varphi'}{\partial z'}\right|_{z=0} = 0$$

$$\frac{\partial \eta'}{\partial t'} + \alpha \frac{\partial \varphi'}{\partial x'} \frac{\partial \eta'}{\partial x'} - \frac{1}{\beta^2} \frac{\partial \varphi'}{\partial z'} = 0, z' = 1 + \alpha \eta$$

$$\eta' + \frac{\partial \varphi'}{\partial t'} + \frac{1}{2}\alpha \left[\left(\frac{\partial \varphi'}{\partial x'}\right)^2 + \frac{1}{\beta}\left(\frac{\partial \varphi'}{\partial z'}\right)^2 \right] = 0, z' = 1 + \alpha \eta$$

Approximating the above equation set, keeping the first-order small terms of α and β, then the approximate solution is

$$\varphi' \approx -\frac{z'^2}{2}\frac{\partial V'}{\partial x'}\beta, V' = \frac{\partial \varphi'}{\partial x'} = \eta' - \frac{1}{4}\alpha\eta'^2 + \frac{1}{3}\beta\frac{\partial^2 \eta'}{\partial x'^2}$$

The nonlinear differential equation obtained from the free surface condition is

$$\frac{\partial \eta'}{\partial t'} + \frac{\partial \eta'}{\partial x'} + \frac{3}{2}\alpha\eta'\frac{\partial \eta'}{\partial x'} + \frac{1}{6}\beta\frac{\partial^3 \eta'}{\partial x'^3} = 0$$

This equation is the dimensionless KdV equation. The third term in the equation is a nonlinear effect and the fourth term is a dispersion effect. Transform it into a dimensional form as

$$\frac{\partial \eta}{\partial t} + \sqrt{gH}\left(1 + \frac{3}{2}\frac{\eta}{H}\right)\frac{\partial \eta}{\partial x} + \sqrt{gH}\frac{H^2}{6}\frac{\partial^3 \eta}{\partial x^3} = 0$$

The solution of this equation has typical solitary wave properties, the propagation direction is single, and the propagation speed is

$$a = \sqrt{gH}\left(1 + \frac{1}{2}\frac{A}{H}\right)$$

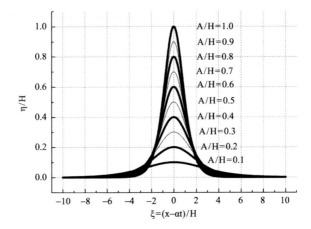

Fig. 3.98 Solitary wave solution of KdV equation

where A is the height of the isolated wave. Its solution is

$$\frac{\eta}{H} = \frac{A}{H} \sec h^2 \left(\sqrt{\frac{3A}{4H}} \frac{x - at}{H} \right)$$

The solitary wave surface distribution curve calculated by this formula is plotted in Fig. 3.98.

3.9 Applications in Hydraulics

Hydraulics is an applied science. The topics studied are from production practices and are closely related to engineering technology. The main application areas are as follows:

3.9.1 Water Resources and Hydropower Engineering

Water conservancy and hydropower engineering is one of the oldest engineering sciences. In flood control projects, data such as flood storage capacity, flood discharge capacity, and levee height need to be determined; flood forecasting needs to know the flood operation law; industrial development must prevent pollution of rivers, and these problems can be solved by studying open channel flow (as shown in Figs. 3.99 and 3.100). The aerated water flowing down through the high dam has great kinetic energy and will cause erosion. Various different forms of energy dissipaters must be established (as

Fig. 3.99 South-to-North Water Transfer Project (open channel flow)

Fig. 3.100 Yangtze River (Clear Water) and Yellow River (Sand Flow, River Flow)

Fig. 3.101 Free water jump and typical underflow energy dissipation

shown in Figs. 3.101, 3.102, 3.103, 3.104, 3.105). Sedimentation of rivers, estuaries, and reservoirs may affect waterways or render existing works ineffective. These problems can be solved by studying hydraulics and sediment movement. When constructing hydroelectric power stations and pumping projects (as shown in Fig. 3.106), it is necessary to study the output of hydraulic machinery, the conditions under which vibration occurs, and the changes in characteristics during the opening and closing process, mainly to prevent or reduce cavitation damage. These aspects are all applications of hydraulics.

3.9.2 Ship Engineering

Due to the needs of shipbuilding technology, people already had a certain understanding of ship mechanics in ancient times. The resistance encountered by a ship moving at a constant speed and acceleration and the safety during navigation is always the most important issues in shipbuilding engineering. The propeller output (as shown in Fig. 3.107), wave resistance, attachment quality, and airworthiness that have been studied for a long time are all aimed at solving these problems. The innovation of shipbuilding technology, the emergence of hydrofoil and hovercraft (see hydrofoil, air cushion),

Fig. 3.102 Wind-tank dissipative energy of Fengtan Hydropower Station and over-flow dam of Baishan Hydropower Station

put forward higher requirements on hydrodynamic. Cavitation flow from hydrofoil, torpedo, etc. running at high-speed in water; speedboat, rowing boat, seaplane floating boat gliding on the water (as shown in Figs. 3.108 and 3.109); The vibration of elastomers such as ships, gates, pipelines, and hydrodynamic noise generated by surface ships, submarines, torpedoes, etc., are all important research topics in hydrodynamics. Modern weapons (such as submarines, torpedoes, and anti-submarine missiles, as shown in Figs. 3.110 and 3.111) are weapons that are closely related to hydrodynamic research. The need for underwater launches results in research on water outflow; water ingress of torpedoes, anti-submarine missiles, instrument cabins, and cockpits of spacecraft result in research on water collision and ingression.

Fig. 3.103 Geheyan Hydropower Station's wide tail pier flow energy dissipation and Dachaoshan wide tail pier dam face step flow energy dissipation

3.9.3 Lubrication and Hydraulic Transmission

Lubrication and hydraulic transmission in mechanical engineering is one of the research topics of fluid dynamics (as shown in Fig. 3.112). Lubrication is an important part of tribology research. Improving the friction state is a technical measure to reduce friction resistance and slow down wear. Lubrication is generally achieved through lubricants. In addition, the lubricant also has the functions of rust prevention, vibration reduction, sealing, and power transmission. Making full use of modern lubrication technology can significantly improve the performance and life of the machine and reduce energy consumption.

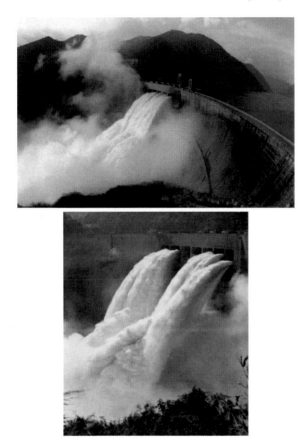

Fig. 3.104 Energy Dissipation in Ertan Hydropower Station

Hydraulic transmission is a transmission method that uses liquid as the working medium to transfer energy and control. Hydraulic transmission and pneumatic transmission are called fluid transmission, which is an emerging technology developed based on the principle of hydrostatic transmission proposed by Pascal in the seventeenth century. It is a widely used technology in industrial and agricultural production. Today, the level of fluid transmission technology has become an important symbol of a country's industrial development level.

Fig. 3.105 Flood relief at the orifice of the dam of the Three Gorges hydropower station (prototype)

3.9.4   Marine and Coastal Engineering

Due to the needs of the shipbuilding industry and hydraulic engineering, research on free surface flow has already begun. The development of ocean engineering has put forward new requirements for this aspect of research. Therefore, the study of wave motion has been the focus of this field. The wave motion is complex, especially for the movement of ocean waves (wind-generated waves), there are fewer regular waves, which usually appear as irregular random waves, especially when encountering strong winds, waves being impacted by the shore or ships, will cause complex Breaking wave motion. Formally a "white hat" at the tip of the wave will be generated. These waves will have an important impact on shipping, port, and ocean engineering, as shown in Figs. 3.113, 3.114, 3.115, 3.116, 3.117.

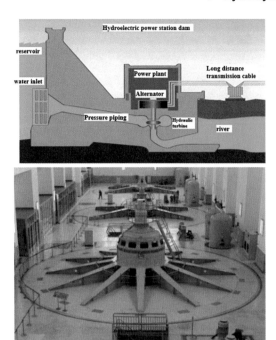

Fig. 3.106 Hydroelectric power station (turbine)

Fig. 3.107 Cavitation stream (baike.baidu.com)

Fig. 3.108   Ship traveling wave resistance

Fig. 3.109   Seaplane and ground effect aircraft

Fig. 3.110 Speed boat and torpedo

Fig. 3.111 Julang-2 Intercontinental Missile

Fig. 3.112 Lubrication and hydraulic transmission

Fig. 3.113 Coastal wave shock

Fig. 3.114 Wave breaking

Fig. 3.115 Wave breaking (Rapidly changing water flow, dynamic viscosity coefficient $\mu = 1.005 \times 10^{-3}$ Pa · s)

Fig. 3.116 Wave at minus 7° Celsius (dynamic viscosity coefficient $\mu = 3.5 \times 10^{-3}$ Pa · s)

Fig. 3.117 Wave at minus 7° Celsius (dynamic viscosity coefficient $\mu = 3.5 \times 10^{-3}$ Pa · s)

4

Computational Fluid Dynamics

4.1 Derivation of Computational Fluid Dynamics

Computational fluid dynamics is a science rapidly evolving since 1960s, which constitutes three branches of modern fluid dynamics with experimental fluid dynamics as well as theoretical fluid dynamics. Moreover, computational fluid dynamics play a more prominent role in industries, especially for aircraft design, computational fluid dynamics is called an important flow simulation and analysis tool.

The combination of computer technology and discrete numerical method constitutes the basis of computational fluid dynamics. At the beginning of the twentieth century, L.f. Richardson (1881–1953), the British meteorologist, first tried to make numerical weather forecast. In 1922, in his book *numerical method for weather forecast*, he discussed the principle and possibility of numerical forecast, and applied complete primitive equations to forecast the surface pressure field in Europe for six hours. But the result is not ideal. He predicted that the pressure in the area would change by 154 mba (mill bar, or 100 Pa) in six hours, while the actual pressure would hardly change. At that time, Richardson attributed the failure to the inaccuracy of the initial value. His failure once made people doubt the possibility of numerical weather forecast. Until the end of World War II, due to the emergence of computers and the development of meteorological observation network, especially the high-altitude observation, the meteorological data have been greatly improved, and the numerical weather forecast has attracted people's attention again. In particular, people realize that Richardson's failure is mainly

© Science Press 2021
P. Liu, *A General Theory of Fluid Mechanics*,
https://doi.org/10.1007/978-981-33-6660-2_4

due to the solution of the equations used by him, which not only includes the slow process such as long wave, but also includes the high-speed sound wave and gravity wave. The actual amplitudes of these high-speed waves are very small, but they are often enlarged in the calculation process, so as to cover up the meteorological meaningful disturbance. In 1948, the American meteorologist Jule Gregory Charney (1917–1981, as shown in Fig. 4.1, one of the greatest meteorologists in the twentieth century) put forward the filtering theory based on the work of the American meteorologist Rossby (Carl Gustaf Arvid, 1898–1957, as shown in Fig. 4.2), and proved that static balance and geostrophic balance approximation can eliminate gravity wave and sound wave. The simplified equations could avoid the influence of sound wave and gravity wave. In 1950, the American mathematician J. von Neumann (1903–1957, as shown in Fig. 4.3) and other mathematicians first succeeded in using quasi-geostrophic barotropic model to predict the pressure field of 500 hPa in North America, for the next 24 h with the computer.

Since the computer came out in 1946, numerical simulation began to develop rapidly. In 1963, the American scientists F. H. Harlow and J. E. Fromm successfully solved the flow around the two-dimensional rectangular cylinder with the IBM 7090 computer at that time, and gave the formation and evolution process of wake vortex-street, which has attracted widespread attention. In 1965, Harlow and Fromm published the paper *Computer experiment of hydrodynamics*, which introduced ceremoniously the great role of computers in hydrodynamics. Since then, the mid-1960s has been regarded as a sign of the rise of computational fluid dynamics.

Fig. 4.1 Jule Gregory Charney (1917–1981, American meteorologist)

Fig. 4.2   Rossby (1898–1957, American meteorologist)

Fig. 4.3   J. Von Neumann (1903–1957, American mathematician)

Although the history of computational fluid dynamics is not long, it has been widely used in various fields of fluid mechanics, and various numerical solutions have been formed accordingly. At present, the main methods are finite difference method and finite element method. The finite difference method has been widely used in fluid mechanics, while the finite element method is developed from solving the problems of solid mechanics. In recent years, there have been many applications in dealing with the problem of low-speed fluid, and they are developing rapidly (a large number of weather forecasts are included, as shown in Fig. 4.4).

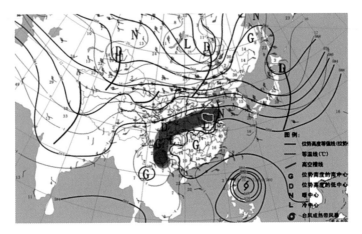

Fig. 4.4 Meteorological forecast (isobaric distribution) (www.chinabaike.com)

4.2   Discrete Techniques and Iterative Methods

Computational fluid dynamics is a combination of modern fluid mechanics, numerical mathematics, and computer science. It uses computer and various discrete numerical methods to carry out numerical experiments, numerical simulation, and analysis for various problems in modern fluid mechanics. Although it is only more than 50 years old, it has developed rapidly and is active in many fields.

As early as the beginning of the twentieth century, Runge (1909), Richardson (1910), and Liebmann proposed a five point difference discrete scheme and an iterative method for solving harmonic equations. In 1928, courant, Friedrichs, and Lewy first proposed the convergence of the difference method in their famous paper *On the Partial Differential Equation of the Mathematical Physical Equation*, and proved the CFL condition for the convergence of the hyperbolic equation, which raised the understanding of the difference method to a new height. Although Poincare had pointed out for a long time that the approximation of the discrete simple algorithm could achieve arbitrary accuracy, because of the backward calculation means and the extremely complex problems for fluid mechanics, it was difficult to calculate the satisfactory approximate solution even for the extremely simple hydro-dynamic model at that time. In 1946, the first electronic computer ENIAC came out. At the same time, von Neumann predicted in a report that the numerical method could replace the analytical method to solve the nonlinear problem of fluid. In the following ten years, a large number of numerical examples appeared rapidly, and related numerical methods and theoretical research began to expand in many aspects, such as the arithmetic mean implicit scheme proposed by Crank–Nicolson (1947), the artificial viscosity method for calculating shock tube problems proposed by von Neumann and Richtmyer (1950), conservative scheme of lax (1954), the alternating direction method and characteristic line assembly method of Peaceman, Rachford (1955), and Douglas (1956), particle method in the lattice of Harlow et al. (1957), etc. Theoretical research had made fruitful progress in the compatibility, convergence, and stability of difference schemes. Lax equivalence theorem, von Neumann stability analysis method, and the analysis and formality of artificial viscosity term had acquired quite perfect results. A variety of problems such as the problems of incompressible viscous flow and shock tube were calculated. However, due to the limitation of computer function and the difficulty of man–machine conversation, most of the fluid dynamic models that could be realized were ideal fluid models. During this period, the main purpose of the work was to lay the foundation of computational fluid dynamics.

In the 1960s, the wide application of high-speed, large capacity, and multifunctional computers promoted the rapid development of various numerical methods of fluid dynamics, such as lattice method (MAC, FLIC, CEL, etc.), shock-capture method and assembly method, fractional step method and operator splitting method, line method, spectrum method, random selection method, and finite element and boundary element method. These methods were more precise and could be adapted to a variety of practical problems. At the same time, the research of mathematical model and the theoretical analysis of discretization method had been developed deeply, and lots of analytical, discrete, and statistical hydrodynamic models had been established; the qualitative analysis theory of difference method, from previous analysis of compatibility, convergence, and stability, to the theoretical analysis of dissipation, dispersion, transmission, monotonicity, and conservation, and in addition, the analysis method had been developed from Fourier analysis to energy analysis and residual effect analysis. These theoretical studies had provided the scientific basis for the design, selection, and application of numerical methods.

In 1965, the American scientists Harlow and Welch put forward the idea of a staggered grid, in which the velocity component and pressure are stored on a grid with half step difference. In this way, the problem of the chessboard unreasonable pressure field when the velocity and pressure are stored on the same set of grids was effectively solved, and the original variable method (the method with velocity and pressure as the variables) for solving the N-S equation (the differential equation of incompressible viscous fluid motion) was established.

In 1966, Gentry, Martin, and Daly, Barakat and Clark, respectively, introduced the application of upwind scheme in the solution of compressible and unsteady laminar flow. The introduction of staggered network and upwind difference of convection term laid a foundation for the numerical solution of flow and convective heat transfer problems. In 1967, Patankar and Spalding published the P-S method for solving parabolic flow.

In 1972, Patankar and Spalding put forward the SIMPLE algorithm, which uses separate solution technology to solve the coupling problem of velocity and pressure. The basic idea is: first solve a velocity component, taking others as constants, and then solve other variables one by one; at any level of iterative solution of flow field, the velocity field must meet the mass conservation equation, which is a very important principle to ensure the convergence of iterative calculation of flow field.

In 1974, the American scholars Thompson, Thames, and Martin proposed the method of generating body fitted coordinates (TTM) using differential equations. It provides a new way for the finite difference method and the finite volume method to deal with the irregular boundary problems. The irregular region in the physical plane (two-dimensional problem) is transformed to the regular region in the calculation plane by transformation, so that the calculation is completed in the calculation plane, and then the results are transferred to the physical plane. After the TTM method was put forward, the branch of grid generation technology was gradually formed in CFD field.

Especially through numerical simulation and computer experiments, some new physical phenomena had been found. For example, in 1965, Zabusky and Kruskal revealed the conservation and particle properties of solitary waves in KdV equation through numerical simulation. Since then, the existence of solitons had been found in many different fields through numerical experiments; in 1968, Campbell and Mueller found the phenomenon of tilt induced separation of subsonic velocity in computer experiments, which was later confirmed by wind tunnel experiments. These findings make people no longer regard numerical calculation as an auxiliary means of theoretical and experimental research, but a subject direction independent of theory and experiment. It could help people seek objective laws of complex flow problems through numerical simulation.

Leonard published the famous quick format in 1979. This is a discrete scheme with third-order accuracy, and its stability is better than that of the central difference scheme. At present, the quick format has been widely used in CFD Research and application.

With the further development of the computer industry, CFD develops from 2 to 3D, from regular area to irregular area, from orthogonal coordinate system to non-orthogonal coordinate system. Therefore, in order to overcome some weaknesses of the staggered grid introduced by the chessboard pressure field, in 1982, Rhee and Chou proposed the non-staggered grid method. This method draws on the successful experience of staggered grid and arranges all the solution variables on the same set of the grid. It is widely used in the calculation of non-orthogonal curvilinear coordinate system. In order to deal with the coupling correlation algorithm of velocity and pressure in the calculation of incompressible flow field, SIMPLER and SIMPLEC algorithms have been put forward successively.

The rapid development of computational fluid dynamics, in addition to the solid material conditions brought by the development of computer hardware industry, is mainly due to the limitations of theoretical analysis methods and experimental methods of fluid mechanics. Due to the complexity of the problem and the high cost, neither an analytical solution nor experimental determination can be made, so the CFD method just makes up for the defects of theoretical and experimental research on complex problems, and it is highly valued by people with its advantages of low cost and simulation of complex physics (as shown in Figs. 4.5, 4.6, 4.7, 4.8 and 4.9). In recent years, a large number of CFD software has been tested, which has greatly broadened the scope of practical application. A numerical simulation of a phenomenon carried out on a computer under given parameters is equivalent to a numerical experiment. There have been examples in history where a new phenomenon was first discovered by CFD numerical simulation and then confirmed by experiments.

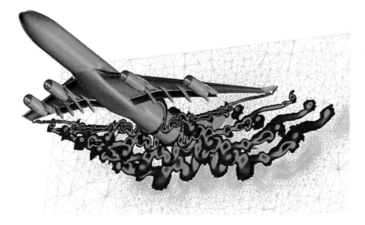

Fig. 4.5 Numerical simulation of vortex field around a large aircraft (from DLR)

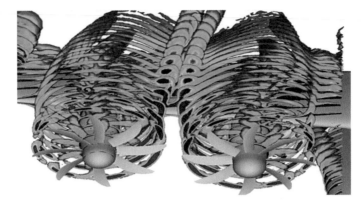

Fig. 4.6 Numerical simulation of slipstream vortices behind propeller (from DLR)

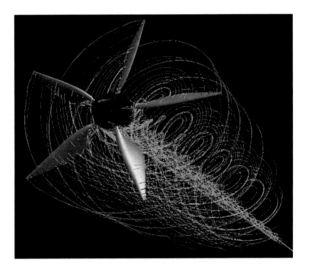

Fig. 4.7 Numerical simulation of slipstream behind propeller (from BeiHang University)

4.3   Application of Computational Fluid Dynamics

4.3.1   Numerical Solution of Low Velocity Flow

For low-speed flow (flow with constant density), there are two main types of flow in fluid mechanics, one is inviscid flow, and the other is viscous flow. For the flow without considering viscosity, classical hydrodynamics focuses on the incompressible and irrotational flow (potential flow) under the influence of potential force. The solution to this kind of problem is mainly based on the

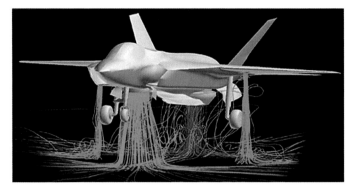

Fig. 4.8 Numerical simulation of F35b vertical takeoff and landing jet flow field (NASA)

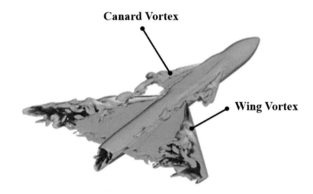

Fig. 4.9 Numerical simulation of dynamic vortex interaction on a pitching canard configuration (from BeiHang University)

velocity potential function or flow function as the unknown function, which satisfies the Laplace equation or Poisson equation. In classical hydrodynamics, many plane problems are solved by complex variable function or conformal mapping. However, it is an effective way to directly solve the approximate solutions of the original Euler equations or N-S equations for the flow around objects with complex geometry by various numerical methods, and it has been widely used in practice, as shown in Figs. 4.10, 4.11, 4.12 and 4.13.

(1) **Iteration method**

This is a method of solving simultaneous equations by gradual approximation, and it is also the main numerical solution of elliptic differential equations. The program of this method is simple, and the storage and calculation are small. Generally, a group of initial values is assumed first, and then the new values on each node are calculated. Taking the

Fig. 4.10 The flow field around the car

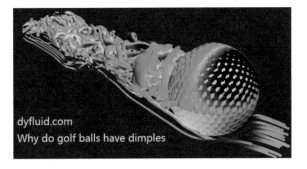

Fig. 4.11 The flow field around the golf ball

Fig. 4.12 Flow-field around Multi-segment airfoil

five point scheme as an example, the new values on the nodes are the average of the initial values of the adjacent four points. After the new value is calculated, the old value is retained to calculate the new value of other points. This kind of simple iteration converges very slowly and is rarely used now. However, if it is improved slightly, the old value will

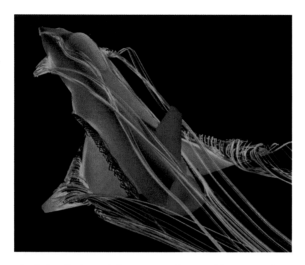

Fig. 4.13 Numerical simulation of flow around canard layout at low speed and high angle of attack (from DLR)

be washed out with the new value, and a relaxation factor will be introduced to accelerate convergence, and the weighted average of the new value and the old value will be used, which is the successive over relaxation method developed in the 1950s. Successive over relaxation method (SOR method) is one of the common iterative methods for solving linear equations. It is obtained by Gauss–Seidel iterative method through linear acceleration. Adjusting the relaxation factor, the convergence of the iteration can be greatly accelerated.

(2) **Time correlation method**

This is a method to solve the steady problems using the unsteady equations. It is often used to solve the N-S and Euler equations. Although the unsteady partial differential equations are used, the solved problem is not unsteady. According to the given initial conditions and time-varying constraints, the unsteady problem is to study the evolution of flow with time. This kind of unsteady behavior is closely related to the given initial value. However, in principle, the initial value of time-dependent method is chosen randomly, only the boundary conditions of the steady-state problem must be satisfied. In the solution process, the change in flow with time does not represent the real physical process. When the time is long enough, the unknown function value is gradually independent of time, and then it tends to the steady solution. So the time correlation method is actually an iterative method. The time variable is only used to record the number of iterations.

(3) **Alternating direction implicit method**

Alternative direction implicit method (ADI method) is one of the finite difference methods. It is mainly used to solve parabolic or elliptic partial differential equations, especially for two-dimensional and higher dimensional heat conduction and diffusion equations. Traditionally, the Crank–Nicolson method is used to solve the heat conduction equation, which is time-consuming. The advantage of ADI is that in each iteration step, the equation solved has a simpler structure, so it is easier to solve. The application of hydrodynamics is usually two-dimensional and three-dimensional. Due to the requirement of stability, the time step is limited by the dimension. The higher the dimension, the smaller the time step required and the larger the calculation workload. In the mid-1950s, the American scientist J. Douglas Faires et al. proposed the alternating direction implicit method to speed up the calculation. For example, in two-dimensional unsteady equations, the first step is to use the implicit difference for the derivative of X, while the derivative of Y-direction is the previous value. In the second step, implicit difference is used for the derivative of Y, and the derivative of X-direction is the value of the first step. The advantage of this method is that it has good stability and enough second-order accuracy. The difference equation is tridiagonal matrix equation, which is easy to solve. SIMPLE algorithm is a pressure modified semi implicit iterative method. In 1972, S.V. Patankar, the American scholar, and D.B. Spalding, the British hydrologist, proposed the algorithm. Now it is widely used in computational fluid dynamics and computational heat transfer in the world. This algorithm soon became the main method to calculate the incompressible flow field. Later, this algorithm and various subsequent improvements have been successfully extended to the compressible flow field. It has become a numerical method that can calculate the flow of any velocity.

(4) **Finite basic solution**

It is a numerical method for solving potential flow. In the design of low-speed aircraft in the aviation industry, potential theory is used to calculate various aerodynamic parameters, that is, to solve the two-dimensional or three-dimensional Laplace equation. In classical hydrodynamics, it is very successful to solve the Laplace equation by superposition of basic solutions. The main point of this method is to replace the influence of wing and fuselage on the flow field with the distribution of source, sink, and dipole. Their strength is determined by boundary conditions, and the results need to be solved by integral equations. It can be solved in some simple cases, but it is difficult in general cases. The appearance of

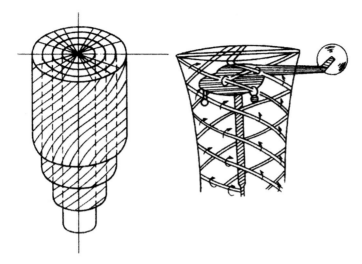

Fig. 4.14 Flow field calculated by vortex panel method

high-speed computer makes a breakthrough in the numerical solution of this integral equation. Its main idea is to discretize the integral equation, which represents the sum of singular points such as source and sink in space. For example, if the airfoil and fuselage surface is divided into many small elements, the singular point strength on each element is averaged and the sum of the singular point strength is the total effect of the flow field. Therefore, it uses the sum of finite terms instead of integral, and the final solution is a set of algebraic equations. Because the basic solutions are all functions with singularities, this method is also called the finite singularities method or the scale method. In the aerodynamic calculation of aviation, the panel method is most commonly used (as shown in Fig. 4.14). In this method, the characteristic surfaces such as the surface of the object or the arc surface of the wing are discretized. After the mesh is generated, a plane or surface is used to replace the original surface of the object, which is called the surface element. The singular points such as source, vortex, dipole, and their combination are arranged on the surface element to solve the aerodynamic problem.

4.3.2 Numerical Solution of Transonic Flow

For the steady potential flow, the governing equation of subsonic flow is the elliptic partial differential equation, while the governing equation of supersonic flow is the hyperbolic partial differential equation. The flow field of transonic flow is a mixture of subsonic and supersonic regions. The boundary between the subsonic region and the supersonic region is the sonic line. Before the solution, the position of the sonic line is unknown, so it is necessary to solve the mixed partial differential equation, which brings difficulties to the theoretical analysis and numerical calculation of transonic flow. Any small disturbance in the airflow usually propagates at the local sound velocity. In the transonic flow, most of the flow velocity is close to the sound velocity, which is similar to the propagation velocity of the above disturbance, so the disturbance is mainly concentrated in the direction almost perpendicular to the direction of the incoming flow. Therefore, in the wind tunnel experiment, the disturbance from the surface of the model will directly reflect the model from the wall of the wind tunnel, or even reflect back and forth many times. This kind of serious wall interference makes the experimental study of transonic flow very difficult. In the numerical calculation of transonic flow, solving the shape and position of the sonic line is an important subject. The closer the Mach number of the incoming flow is to 1, the larger the area of the flow near the speed of sound is. The small difference in velocity in the flow field will cause great changes in the position and shape of the sonic line. The change in the sonic line has a direct impact on the calculation format of the flow field. If there is a slight deviation in the calculation of the sonic line and the ultimate calculation results will be directly affected, which brings many difficulties to the numerical calculation of transonic flow. In 1971, the American scientists E. M. Muman and J. D. Cole first used the mixed difference scheme, and successfully solved the steady small disturbance velocity potential equation using the relaxation method. The mixed difference scheme is to use the central difference scheme in the subsonic region, and the conditions on all the adjacent nodes will affect the calculation points. In the supersonic region, the upwind scheme is used, because the upstream upwind node is just the dependent region of the hyperbolic wave equation. As shown in Fig. 4.15, the numerical simulation of transonic flow around airfoil is carried out.

4.3.3 Numerical Solution of Supersonic Flow

In supersonic flow, the main problem is how to deal with the shock wave. In the flow field with the shock wave, the characteristic scale of flow structure in different regions is very different. The characteristic scale of inviscid shock wave thickness is zero, while the characteristic scale of the flow field is limited. The flow parameters change discontinuously when passing through the shock wave, which brings great difficulty to the calculation of simulated shock wave.

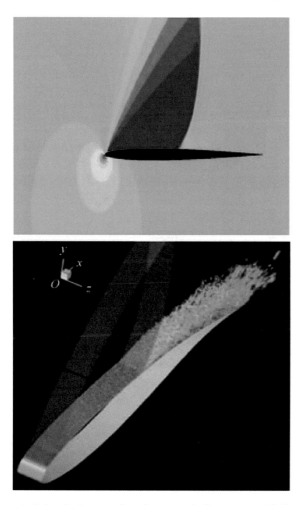

Fig. 4.15 Numerical simulation results of transonic flow over airfoil

There are two methods to deal with the shock wave in the supersonic flow field. One is Shock-Capturing, the other is Shock-Fitting.

Shock-Fitting method regards shock wave as an unknown moving boundary, separates shock wave according to shock discontinuity condition Rankine–Hugonun shock relation, and accurately calculates the position of the shock wave. It solves differential equation with the approximate method in the smooth area and Rankine–Hugonun relation at the discontinuity. The method has the advantages of high precision, accurate shock position, and clear physical image. However, the calculation is very complicated, so it is only suitable for those flows with simple shock wave motion and clear flow pattern.

The Shock-Capture method does not need any special treatment for the shock itself, but directly or indirectly introduces the "viscosity" term in the calculation formula, and automatically calculates the position and strength of the shock to "capture" the shock. There are so-called artificial viscosity and format viscosity. The artificial viscosity method was first proposed by the American scientists J. von Neumann and R. D. Richtmall in 1950. It is an automatic shock wave approximation method based on the physical theory of real viscous fluid. In this method, a viscous term is added artificially in the shock layer to make the shock discontinuity become a smooth transition region. In recent years, it has been widely used in supersonic flow. The scheme viscosity is a kind of difference scheme which indirectly introduces the viscosity term Lax scheme. The Shock-Fitting method treats the shock wave as a discontinuity and satisfies the condition of shock jump before and after the shock wave. But it is very difficult to realize in common coordinates. Generally, the coordinate transformation is used to make the shock position (unknown at this time) coincide with a coordinate axis, and then the shock is regarded as the inner boundary. This kind of processing is more accurate, but also very troublesome and inconvenient. The best way is to combine the Shock-Capturing method with the Shock-Fitting method. For example, the Shock-Fitting method is used for the shock wave in the outer of the flow field, and the Shock-Capture method is used for the shock wave in the inner of the flow field. The successive development of shock wave capture formats are TVD (Total Variation Diminishing) format (Harten 1983); NND (Non-Oscillatory Containing No Free Parameter and Dissipative Scheme) format (Zhang Hanxin 1984); ENO (Essential Non-Oscillatory Scheme) format (Harten et al. 1987); WENO (Weighted Essential Non Oscillatory Scheme) format (Liu et al. 1994), as shown in Figs. 4.16 and 4.17.

4.4 Commercial Software for Computational Fluid Dynamics

CFD commercial software generally includes a variety of physical models, such as steady and unsteady flows, laminar and turbulent flows, incompressible and compressible flows, heat transfer and chemical reaction, etc. For each kind of physical problem, there are suitable numerical solutions for users. Users can choose explicit or implicit difference schemes in order to achieve the best calculation speed, stability, and accuracy. CFD software can easily exchange numerical values and adopt unified pre-processing and post-processing tools, which saves researchers' repeated and inefficient work

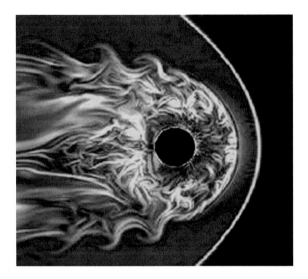

Fig. 4.16 Supersonic flow around a cylinder

Fig. 4.17 Supersonic flight of space shuttle

in computer methods, programming, pre-processing, and other aspects, and enables them to devote their main energy to the study of physical problems.

In 1979, the British scholar Spalding first released the PHOENICS 1.0 calculation software, in which PHOENICS is the abbreviation of Parabolic, Hyperbolic or Elliptic Numerical Integration Code Series (which means a series of programs for numerical integration of parabolic, hyperbolic and elliptic equations). In 1981, CHAM Company in the UK formally put PHOENICS software into the market, creating a precedent for CFD commercial software market. With the rapid development of computer group, parallel algorithm, direct numerical simulation (DNS) and large eddy simulation (LES) have been greatly promoted. A variety of commercial general software for calculating heat transfer and flow problems have been put into the market, such as PHOENICS (1981), FLUENT (1983), FIDAP (1983), STAR-CD (1987), FLOW-3D (1991, now called CFX), etc. In addition to the finite element method adopted by FIDAP, the finite volume method is used for other products. FIDAP later merged with FLUENT and became a part of the FLUENT software family. In 1989, Professor Patankar, the famous scholar, introduced a series of compact software to calculate the flow, heat transfer, and combustion processes. At present, there are at least 50 kinds of commercial software for the flow and heat transfer problems around the world, which play an important role in the application of CFD technology. The following describes the CFD commercial software widely used in the world.

(1) CFX

The software adopts the finite volume method, the pieced structured grid, and the non-orthogonal curvilinear coordinate (body fitted coordinate) system for discretization. The layout of variables adopts the same position grid method. The discrete schemes of convection terms include first-order upwind, mixed scheme, QUICK, CONDIF, MUSCI, and higher order upwind scheme. The coupling relationship between pressure and velocity is based on a simple algorithm, which includes line iteration, algebraic multiple grids, ICCG, strong implicit method and block implicit method (BIM). The software can calculate incompressible and compressible flows, coupled heat transfer, multiphase flow, chemical reaction, gas combustion, etc.

(2) FIDAP

FIDAP is the abbreviation of Fluid Dynamics Analysis Package, which was launched by Fluid Dynamics International Inc. in 1983. It is the first CFD software using finite element method (FEM) in the world. It can accept the grid generated by famous grid generation software such as I-DEAS, PATRAN, ANSYS, and ICEMCFD. The software can calculate the flow problems of compressible and incompressible flow, laminar and turbulent flows, single-phase and two-phase flows, Newtonian fluid and non-Newtonian fluid.

(3) FLUENT

This software was introduced by fluent Inc. in 1983. It is the second software based on the finite volume method put on the market after Phoenix software. It includes structured and unstructured grid versions. In the version of the structured grid, there are pre-processing software for body fitted coordinates, which can also be available to the grid generated by famous grid generation software such as I-DEAS, PATRAN, ANSYS, and ICEMCFD. The velocity and pressure are coupled by SIMPLEC algorithm on the same position grid. Different schemes of convection term include first-order upwind, central difference quick scheme, etc. The software can calculate compressible and incompressible flows, evaporation with particles, combustion process, chemical reaction process of multi-component medium, etc.

(4) PHOENICS

This is the first commercial CFD software put on the market in the world, which can be regarded as the ancestor of CFD commercial software. Some basic algorithms used in this software, such as simple method and mixed format, are just put forward by the founder of this software, D.B. Spalding and his collaborator, S.V. Patankar, which have a great impact on the later commercial software development. In recent years, PHOENICS software has made great improvements in function and method, including the technology of patch multigrid and fine grid embedding, the same position grid and unstructured grid. For the turbulence model, the general zero equation, low Reynolds k-epsilon model, and RNG k-epsilon model have been developed. This software can be used to calculate a large number of practical problems, including urban pollution prediction, flow in impeller, and pipe flow.

(5) STAR-CD

STAR-CD is the abbreviation of Simulation of Turbine Flow in Ambient Region, and the CD after the hyphen is the abbreviation of the developer Computational Dynamics Ltd. It is the general software based on the finite volume method. In the aspect of mesh generation, unstructured mesh is used. The shape of the element can be hexahedron, tetrahedron, prism with triangle section, pyramid shaped cone, and other polyhedron with six shapes. This software can be used to calculate steady and unsteady flow, Newtonian and non-Newtonian flow, flow in porous media, subsonic and supersonic flow, and it is widely used in the automobile industry.

4.5 Numerical Simulation of Flow Field for a Large Axial Flow Fan

4.5.1 Problem Description

The numerical calculation process and results of a large axial flow fan with a diameter of 12.35 m are presented in this example. The incompressible N-S equations are used to calculate the full turbulent RANS equations, and the k-epsilon model is used to calculate the turbulent closure equations. The numerical calculation is carried out in FLUENT; the implicit solver based on pressure is selected; the second-order MUSCL scheme is used for convection term; the second-order central difference scheme is used for diffuse term and the simple algorithm is used for velocity and pressure coupling. In the iterative calculation, the discrete nonlinear momentum equations, pressure correction equations, energy equations, turbulent kinetic energy equations, and turbulent energy dissipation rate equations will be solved successively. In the application of the finite volume method, the computational domain can be divided into any polyhedron, which makes the finite volume method very convenient for mesh generation. The finite volume can be divided into structural grid or unstructured grid. Structural grid is easy to construct high-precision discrete equations; unstructured grid is easy to generate, has strong applicability. The combination of these two types of grid can also be used in the simulation.

4.5.2 The Physical Model

The structure of the fan system is shown in Fig. 4.18. It is composed of front fairing, rotor, and rear fairing, in which the fan diameter is 12.35 m,

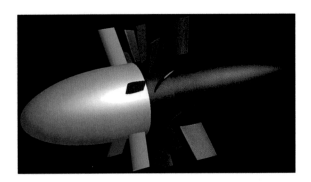

Fig. 4.18 The structure of the fan system

Fig. 4.19 Hub and blade of the fan

the diameter ratio of hub to propeller is 0.50, and the diameter of hub is 6.175 m (as shown in Fig. 4.19). It is a low-noise axial flow fan composed of blades and anti-twist guide vanes. The design adopts the theory of arbitrary circulation ($\alpha = 0.85$), and the modified technology of propeller and fan. The number of fan blades is 12 (GOE797 airfoil in the root area, GOE796 airfoil in the tip area, blade span is 3.0875 m). The number of anti-twist guide vanes is 7 (C4 airfoil). The number of the support plate of fan head is 5 (NACA0012 airfoil), the length of fan head cover is 7.9 m, the length of tail cover is 11.8 m, the equivalent diffusion angle of tail cover is 8°, the length of column section is 4.846 m, and the total length of fan system is 24.564 m. The fan has a design flow of 4800 m³/s, a pressure increase of 2051pa, and a design speed of 200 rpm. The distance between the model and the inlet of the computational domain is 50 m and the distance between the model and the outlet of the computational domain is also 50 m.

4.5.3 Mesh Generation and Boundary Conditions

The sliding grid technique is used to simulate the rotation of the fan blades relative to the deflector, fairing, and tunnel wall. In order to simulate the curved surface shape of the fan and deflector conveniently, unstructured grid was used to divide the flow field, and the whole flow field included 10.18 million grids. The grid distribution on the surface of fan, deflector, and fairing is shown in Fig. 4.20, and the grid distribution on the surface of the tunnel wall is shown in Fig. 4.21. As shown in Fig. 4.22, the inlet boundary

Fig. 4.20 Surface grid of upstream guide vane, blade, and downstream anti-twist guide vane

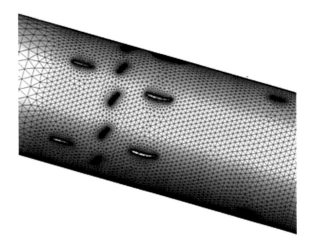

Fig. 4.21 Mesh on the tunnel wall

of the calculation area is the velocity inlet, and the outlet boundary is the pressure outlet. The boundary condition of the fluid domain is adopted for the rotor rotation, and the rotation axis and rotation speed are given.

4.5.4 Results

In this calculation, the numerical simulation is carried out for the rotating speeds of 80, 120, 160, and 200 rpm, respectively. The relationship curve between fan shaft power and rotating speed is shown in Fig. 4.23, in which the error of the CFD result is 1–2% compared with the measured result. Figure 4.24 shows the total pressure distribution curve of different sections along the axial direction at the design speed. Figure 4.25 shows the contour of the total pressure distribution along the axial direction at the design speed.

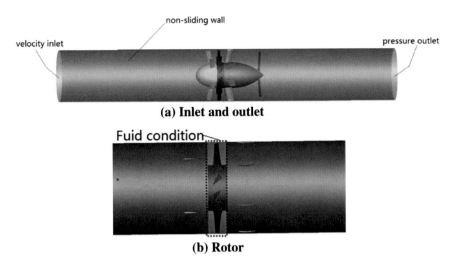

(a) Inlet and outlet

(b) Rotor

Fig. 4.22 Boundary conditions of the computational domain

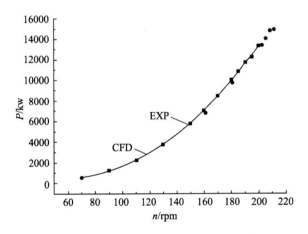

Fig. 4.23 The comparison between CFD result and EXP result

Figure 4.26 shows the surface pressure distribution of the fan fairing at the design speed. Figure 4.27 shows the surface pressure distribution on the blade at the design speed. Figure 4.28 shows the contour of the axial speed distribution of six sections along the axial direction at the upstream and downstream of the fan system. Figure 4.29 gives the streamline around the fan system.

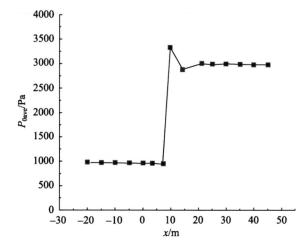

Fig. 4.24 The total pressure distribution curve at the design speed

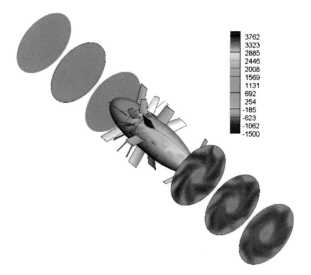

Fig. 4.25 The contour of the total pressure distribution at the design speed

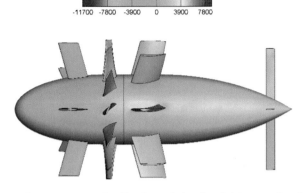

Fig. 4.26   The surface pressure distribution of the fan fairing at the design speed

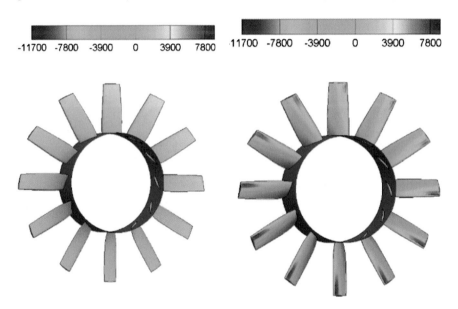

Fig. 4.27   The surface pressure distribution of the blade at the design speed

4.6   Numerical Simulation of Flow-Field in a Large Lowspeed Closed-Circuit Aeroacoustics Wind Tunnel

4.6.1   Problem Description

The numerical simulation process and results of a large low-speed closed-circuit aeroacoustics wind tunnel are presented in this example. The incompressible N-S equations are used to calculate the full turbulent RANS

Fig. 4.28 The contour of the axial speed distribution of six sections along the axial direction

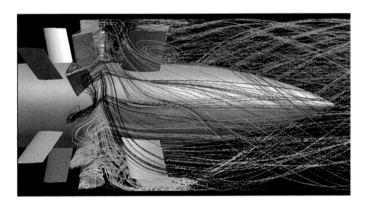

Fig. 4.29 The streamline around the fan system the axial

equations, and the k-epsilon model is used to calculate the turbulent closure equations. The numerical calculation is carried out in FLUENT; the implicit solver based on pressure is selected; the second-order MUSCL scheme is used for convection term; the second-order central difference scheme is used for the diffuse term and the SIMPLE algorithm is used for velocity and pressure coupling. In the iterative calculation, the discrete nonlinear momentum equations, pressure correction equations, energy equations, turbulent kinetic energy equations, and turbulent energy dissipation rate equations will be solved successively. In the application of the finite volume method, the computational domain can be divided into any polyhedron, which makes the finite volume method very convenient for mesh generation. The finite volume can be divided into structural grid or unstructured grid. Structural grid is easy to construct high-precision discrete equations; unstructured grid is easy to generate, has strong applicability. The combination of these two types of grid can be also used in the simulation. Because the simulation condition is low-speed flow, the numerical simulation employs the incompressible flow to calculate. The whole wind tunnel computational domain is closed, the whole flow is static at the outset, and the boundary condition is that the fan blades rotate at the design speed.

The honeycomb and damping network are simulated by porous media. The so-called porous media refer to the pore space composed of porous solid matrix and filled with single-phase or multiphase media. The solid matrix covers the volume space occupied by the porous media, and the pore space is interconnected. The momentum equation of porous media is to add an additional momentum source term into the flow momentum equation. The source term consists of two parts, one is the viscous loss term (Darcy, the first item on the right side of the below formula), the other is the internal loss term (the second item on the right side of the below formula).

$$S_i = -\left(\sum_{j=1}^{3} D_{ij} \mu V_j + \sum_{j=1}^{3} C_{ij} \frac{1}{2} \rho |V| V_j \right)$$

where S_i is the momentum source term of i direction (x, y, or z), D_{ij} and C_{ij} are the given coefficient matrix, V_i is the velocity component, ρ is fluid density, and μ is dynamic viscosity coefficient. In porous media, momentum loss contributes to pressure gradient, and pressure drop is proportional to fluid velocity (or velocity matrix).

4.6.2 The Physical Model

The example is a large low-speed closed-circuit aeroacoustics wind tunnel. The wind tunnel has a closed test section (section size is 4 m wide, 3 m high, and 10.5 m long). The maximum velocity in the test section is 100 m/s, turbulence intensity is 0.05%, and contraction ratio is 9. The wind tunnel is composed of stable section (including 1-layer honeycomb and 8-layer damping net), contraction section, test section, first diffusion section, first corner section, second corner section, fan front transition section, power section, fan rear transition section, second diffusion section, third corner section, and fourth corner section. The four corner deflectors are 81 pieces in total and all of them are noise reduced. The power system adopts fan anti-twist deflector system, with motor power of 3500 kW and fan design speed of 350 rpm, which is composed of 16 blades, 7 anti-twist deflectors, 5 front cover support pieces, 3 rear cover support pieces, fairing, and motor. The fan system is designed with high efficiency and low noise aerodynamic optimization. The total length of the wind tunnel is 96 m, the total width is 38 m, and the total height is 9 m. The three-dimensional layout of the wind tunnel is shown in Fig. 4.30.

The diameter of the fan system in this wind tunnel is $D = 7.5$ m, the ratio of propeller to hub is 0.6 (too large ratio of propeller to hub will cause too long tail cone of fairing or too large tail diffusion loss, too small ratio of propeller to hub will cause low axial speed of fan passage, decreased efficiency, and increased noise), the number of fan blades is 16 (GOE797 airfoil in blade root area, GOE796 airfoil in blade tip area), the number of anti-twist guide vanes is 7 (C4 airfoil), and the number of the fan front cover support plates is 5 (NACA0012 airfoil) and the number of the fantail cover support plates is 3 (NACA0012 airfoil). The total length of the fan system is 21 m, the chord

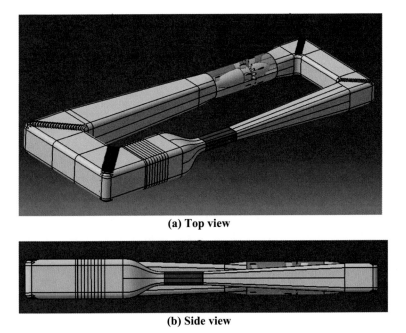

(a) Top view

(b) Side view

Fig. 4.30 The three-dimensional model of the wind tunnel

length of the fan blade is 0.586–0.838 m, the chord length of anti-twist guide vane is 1.838–1.915 m, the chord length of the front cone support plate is 2.0 m, and the chord length of the tail cover support plate is 1.5 m. Under the design working condition of fan, the wind speed in the experimental section is 100 m/s, the volume flow is 1200 m^3/s, the pressure rise is 2450 Pa, the design rotation speed of fan is 310 rpm, corresponding to the linear speed of propeller tip 119.7 m/s, which is within the limit of 150 m/s of propeller tip speed. The efficiency of the fan system is 86.2%, the thrust of the fan system is 64.3 kN, the torque is 105.1 kN*m, and the power is 3412 kW, which also determines the power and type selection of the drive motor. In order to reduce the noise of the coupling and bearing between the motor and the hub, the direct connection is adopted according to DNW wind tunnel (the fan hub is directly installed on the motor shaft). As shown in Fig. 4.31, the three-dimensional mode of fan system is given.

4.6.3 Mesh Generation and Boundary Conditions

As the fan section is a rotating machine, the grid quality should be good; while there are many corner deflectors in other working sections of the wind tunnel, especially in the corner section, it is difficult to generate mesh for the model. Therefore, in the actual simulation, the meshes of the fan segment and the rest of the working segment of the model are generated separately. The special software NUMECA is used to generate the structural mesh of the fan segment. The unstructured mesh of the rest section of the wind tunnel is generated by employing ICEM. After the meshes of the two sections are generated, combine them to generate the whole meshes of the model, and carry out the next numerical simulation (as shown in Figs. 4.32, 4.33, 4.34 and 4.35). The number of fan section structural mesh is 3 million (Fig. 4.32); the number of the unstructured mesh of the rest section is 20 million (Fig. 4.34).

4.6.4 Results

In the simulation, the wind speed of the test section is 20, 40, 60, 80, 100 m/s, respectively, the corresponding rotation speed of the fan is 62, 124, 186, 248, 310 rpm, the temperature (static temperature) is 288.15 k, and the air density is 1.225 kg/m^3. Firstly, the simulation results of the design conditions are analyzed. The wind speed of the test section is 100 m/s, the temperature (static temperature) is 288.15 k, the density is 1.225 kg/m^3, and

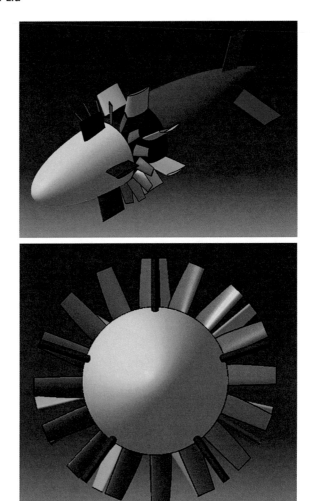

Fig. 4.31 The three-dimensional model of the fan system

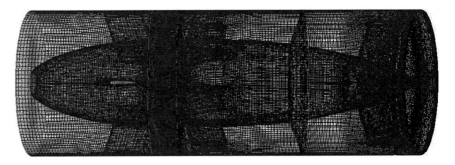

Fig. 4.32 The structural mesh of the fan system

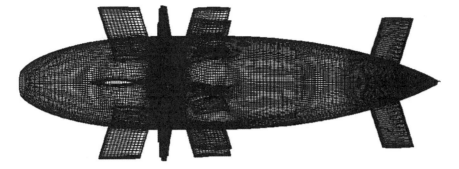

Fig. 4.33 The surface mesh of the fan

Fig. 4.34 The unstructured mesh of the model (without the fan section)

Fig. 4.35 The mesh of the model

the design speed is 310 rpm. Figure 4.36 shows the three-dimensional stream-line of the whole wind tunnel, Fig. 4.37 shows the top view of the streamline of the wind tunnel flow field, and Fig. 4.38 shows the velocity distribution curve of each section along the wind tunnel.

Figure 4.39 shows the static pressure distribution contour on the symmetrical plane of the wind tunnel and Fig. 4.40 shows the velocity distribution contour on the symmetrical plane of the wind tunnel. It can be seen from Fig. 4.40 that when the fan rotates at the design speed, the wind speed in the test section of the wind tunnel is about 102 m/s. The flow is directed to the inner wall of the wind tunnel at the corner deflector, and after passing through the honeycomb and damping net, the velocity is uniformly distributed along the section, which meets the design requirements. From Fig. 4.39, it can be seen that the fan pressurizes 1540 Pa, the total pressure

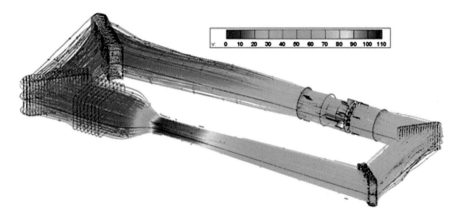

Fig. 4.36 The three-dimensional streamline of the whole wind tunnel

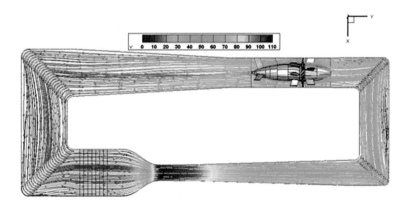

Fig. 4.37 The top view of the streamline of the whole wind tunnel

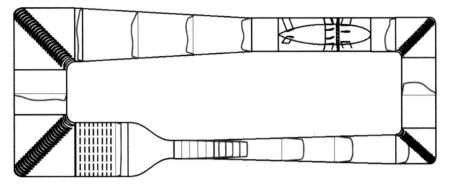

Fig. 4.38 The velocity distribution curve of each section along the wind tunnel (V_{design} = 100 m/s)

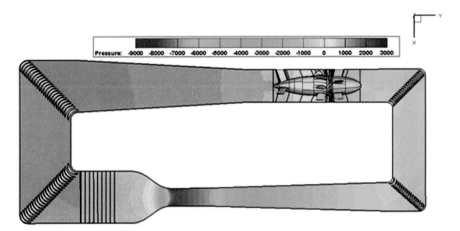

Fig. 4.39 The static pressure distribution contour on the symmetrical plane of the wind tunnel

loss after the flow passing through the honeycomb and damping network is 360 Pa, the static pressure of the contraction section decreases due to the acceleration of the airflow, and the static pressure of the diffusion section increases due to the deceleration of the airflow. Figure 4.41 shows the total pressure contour on the symmetrical plane of the wind tunnel, from which we can see the total pressure loss along the wind tunnel (especially the loss through the honeycomb, damping network, and corner deflector), and the total pressure increase through the fan. Figure 4.42 shows the distribution curve of total pressure along the wind tunnel.

Figure 4.43 shows the distribution of the limit streamline on the surface of the fairing and the hub and Fig. 4.44 shows the three-dimensional streamline distribution of the fan flow field. According to the two figures, the streamline

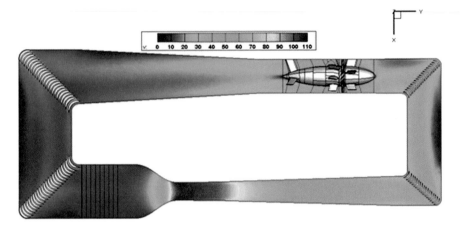

Fig. 4.40 The velocity distribution contour on the symmetrical plane of the wind tunnel

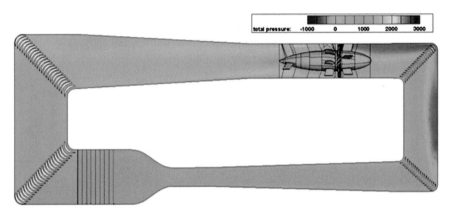

Fig. 4.41 The total pressure distribution contour on the symmetrical plane of the wind tunnel

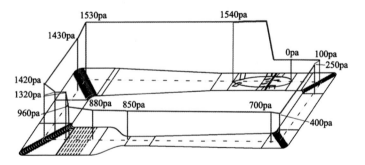

Fig. 4.42 The total pressure distribution curve along the wind tunnel

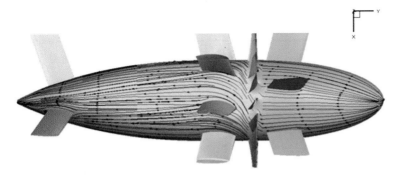

Fig. 4.43 The limit streamline on the surface of the fairing and the hub

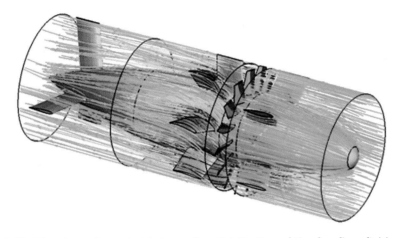

Fig. 4.44 The three-dimensional streamline distribution of the fan flow field

is twisted when it is accelerated by the blade, but it basically moves along the axis after passing through the deflector, so the flow field quality of the fan meets the design requirements.

Figures 4.45 and 4.46, respectively, show the curve of volume flow and total pressure increment with the rotation speed of the fan. It can be seen that the total pressure increment meets the requirements of wind tunnel characteristic curve under different wind speeds.

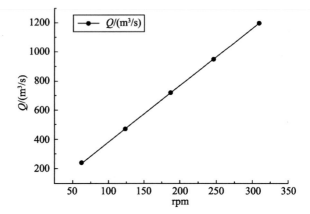

Fig. 4.45 The volume flow curve with the rotation speed of the fan

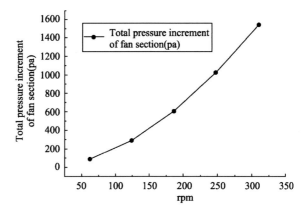

Fig. 4.46 The total pressure increment curve with the rotation speed of the fan

5

Experimental Fluid Mechanics

5.1 Classical Fluid Mechanics Experiment

Experimental fluid mechanics, theoretical fluid mechanics, and computational fluid mechanics constitute the three branches of fluid mechanics. Experimental fluid mechanics includes experimental aerodynamics and experimental fluid dynamics, involving nearly all fields of fluid mechanics. Experimental fluid mechanics plays a key role in the development of fluid mechanics theory. On the one hand, physical phenomena can be observed carefully and repeatedly through experiments, so as to reveal the flow mechanism and discover the flow law, on the other hand, an important basis for the establishment of physical models and theories can be provided by direct experimental measurement of physical quantities.

In 250 BC, Archimedes proposed the famous buoyancy theorem of fluid mechanics through a large number of experimental studies. Galileo (1564–1642), an Italian physicist, first introduced experimental methods into mechanical studies and then demonstrated the resistance to motion of objects in air in 1632. In 1643, E. Torricelli (1608–1647), an Italian scientist, completed the atmospheric pressure measurement experiment, and proved the basic law of constant orifice outflow, and proposed that the orifice outflow speed is proportional to the square root of the head value. In 1653, Blaise Pascal (1623–1662, as shown in Fig. 5.1), a French scientist, proposed the hydrostatic pressure transfer theorem on the basis of experiments and made a hydraulic press (as shown in Fig. 5.2). In 1686, Newton, a British scientist, completed the law of internal friction of fluid, and in 1687, he used

© Science Press 2021
P. Liu, *A General Theory of Fluid Mechanics*,
https://doi.org/10.1007/978-981-33-6660-2_5

Fig. 5.1 Blaise Pascal (1623–1662, French scientist)

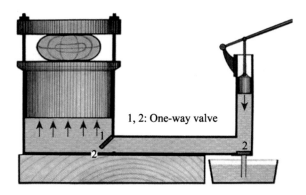

Fig. 5.2 Principle of hydraulic press (shdhyq.blog.163.com)

a pendulum and a vertical falling ball to conduct experiments on the resistance of flow around water and air. In 1732, H. Pitot, a French engineer, invented a device for measuring the total pressure in the fluid, namely Pitot tube. Later, in 1905, Prandtl developed it into a device for simultaneously measuring the total pressure and static pressure of the fluid and established Prandtl anemometer, also known as Pitot velocimeter (as shown in Fig. 5.3). In 1915, G. I. Taylor, a British mechanist, designed a multi-tube manometer for measuring fluid pressure distribution (as shown in Fig. 5.4). In 1799, G.B. Venturi (1746–1822), an Italian physicist, found that the pressure at the

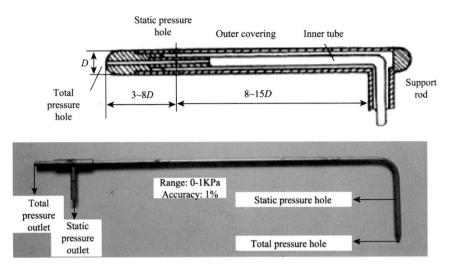

Fig. 5.3 Pitot velocimeter

Fig. 5.4 Multi-tube manometer

Fig. 5.5 Venturi pipe

minimum cross-section decreased sharply through the experiment on a variable cross-section pipeline, and proposed a shrink-expanded type pipeline for measuring the fluid flow in the pipeline, namely the Venturi pipe (as shown in Fig. 5.5). In 1872, William Froude, a British scholar, gave a method for calculating the friction resistance of ships in Torquay, England. He presided over the construction of a towed pool with a size of 85 m × 14 m × 4 m for ship experiments, which accurately measured the resistance of ship models and plates, and divided the ship resistance into three parts: surface resistance, traveling wave resistance and vortex resistance. In 1839, Hagen, a German scholar, completed the experiment on the flow characteristics in a circular tube. In 1880, Reynolds, a British scholar, conducted a flow transition experiment of circular tube, and proposed the concepts of laminar flow and turbulence in 1883. In 1904, Prandtl proposed the famous boundary layer theory based on a large number of experimental observations.

The famous Karman vortex street refers to the periodic vortex structure that exists behind the flow around the cylinder (as shown in Fig. 5.6). In 1911, Von Karman worked as an assistant professor in the Prandtl laboratory at the University of Göttingen. At that time, Dr. Karl Hiemenz, a Ph.D. student, developed a water tank and performed a circular flow test to check the splitting point of the boundary layer. To do this, it is necessary to know how the pressure intensity around the cylinder in the steady flow of water is distributed. When Hiemenz conducted the experiment, he unexpectedly found that the water flow in the tank oscillated constantly and violently. At this time, Von Karmen thought that if the water flow was always swinging, there must be an internal objective reason for the phenomenon. Von Karman used a rough calculation method to calculate the stability of the vortex system. He assumed that only one vortex could move freely, and all other vortices were stationary. Then he let the vortex move slightly to see what happened. Von Karman concluded that if the vortex was arranged

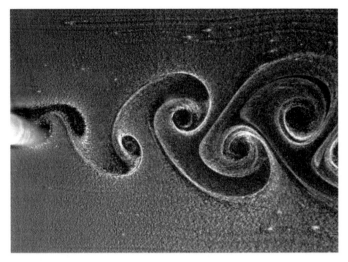

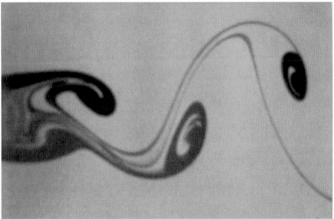

Fig. 5.6 Cylindrical flow test and Karman vortex street

symmetrically, it must be farther away from its original position; and for the antisymmetric arrangement, the same result is obtained, when there was a certain ratio between the rows and the neighboring vortices and when there was spacing between the rows and columns. When there was a certain ratio of the spacing of adjacent vortices, the vortex stayed in the vicinity of its original position and makes a slight circular path motion around the original position. Von Karman confirmed the existence of vortex street through theoretical analysis of Hiemenz's tank experiment. In later generations, due to the detailed and successful study of its mechanism by Von Karmen, it was named after Karmen's surname, known as Karmen Vortex Street. Von Karman wrote later in his book: "I do not claim to have discovered these vortices. Before

I was born, everyone knew that there was such a vortex. The first thing I saw was a picture in Bologna Church, Italy. The picture depicts St. Christopher holding her young Jesus across the river. The painter painted a staggered vortex behind Christopher's bare feet.

Another example is the well-known Rayleigh–Bénard phenomenon (as shown in Fig. 5.7), which is a widely studied thermal convection phenomenon and is considered to be one of several simple system models for studying turbulence problems. In 1900, for the first time, Bénard carried out experiments on a thin horizontal liquid layer with a free surface on its upper surface to study the convection caused by heating the bottom of the liquid layer. In this experiment, he observed the convective structure, and the structure given was a relatively regular hexagon. Bénard's experimental results attracted wide attention from researchers. To explain this phenomenon, Rayleigh first analyzed the problem in 1916 and proposed a dimensionless parameter to study and determine the stability of the heated fluid flow at the bottom, thus giving a theoretical method. This method was also used as the basis for modern research on heat convection. For more than a century since then, the Rayleigh–Bénard convection structure has attracted the interest of many researchers, especially in terms of basic structural features and their instability. In the Rayleigh–Bénard convection experiment, a transparent container was used, the distance between the upper and lower plates was H, and the bottom plate had good thermal conductivity; the inside was filled with liquid, water was used in most of the experiments. In this experiment, the temperatures of the upper and lower substrates were kept constant,

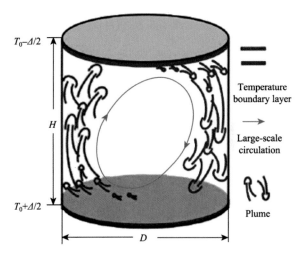

Fig. 5.7 Rayleigh–Bénard thermal convection

and the temperature of the upper substrate was lower than that of the lower substrate, and the temperature difference was constant at T. When T is relatively small, that is, the temperature difference between the upper and lower plates is relatively small, and the liquid in the system is in a non-circulating state. However, due to the thermal expansion and contraction effects of the liquid, the liquid close to the lower plate causes a decrease in density due to an increase in temperature, which is smaller than that of the liquid above. Due to the gravity of the liquid itself, there is a downward flow trend for the liquid with high density on the top of the container. As the temperature difference between the upper and lower plates continues to increase, the difference in liquid density between the upper and lower layers also increases. When the temperature difference reaches a critical value, the gravity of the liquid overcomes the viscous diffusion of the liquid, and the liquid that is initially in stability changes, causing the liquid to convectively flow within the container.

5.2   Similarity Principle

Fluid dynamic tests (aerodynamic and hydrodynamic tests) are generally divided into physical tests and model tests. Physical tests (such as aircraft flight tests and missile live-fire launch, various prototype observation tests) do not cause the distortion of model and environment simulation, and have always been the means to finally identify physical flow dynamics and observation flow fields, but the test costs are expensive and the test conditions are difficult to control. The model test is carried out under the condition of artificial control using the model which is similar to the real object geometry. In order to make the model test results applicable to the actual situation, it is necessary to make the flow around the model similar to the flow around the real object. In this way, their dimensionless hydrodynamic characteristics can be the same. This requires the same ratio of homogeneous forces acting on the volume element at all similar corresponding points. In fluid dynamic tests, the ratios of these dimensionless numbers are called similar parameters. There are many similar parameters, such as Mach number, Reynolds number, Froude number, and so on. In a model test, it is difficult to make all parameters exactly equal to the real object. However, for a specific model test, the functions of similar parameters can be divided into primary and secondary, and the similar parameters to be simulated should be determined according to the purposes, requirements, and other specific conditions of the test.

In order to establish the similarity relationship between the scaled model test and the real test, as early as 1851, Stokes, a British mechanic and mathematician, proposed the dynamic similarity theory based on the differential equations under the conditions of geometric similarity and motion similarity theory. Later in 1873, Helmholtz, a German scientist, further demonstrated this theory. In 1892 and 1904, Rayleigh, a British physicist, used the dimensional analysis method to propose two similar dynamic similarity parameters. In 1914, E. Buckingham (American physicist, 1867–1940, as shown in Fig. 5.8) proposed a famous π theorem based on the dimensional analysis method, which laid a solid theoretical foundation for the similarity test. In 1870, Froude conducted a ship model test according to the similarity criterion of water flow gravity. In 1885, the British physicist Renault conducted tidal model tests on estuaries based on gravity similarity criterion. In 1898, H. Engels established the world's first hydraulic laboratory in Germany and conducted a river channel model test.

The world's first recognized wind tunnel was built in 1871 by F.H. Wenham (British engineer, 1824–1908) and the resistance to the relative motion of objects and air was measured. The wind tunnel is a wooden box with openings at both ends, with a section of 45.7 cm × 45.7 cm and a length of 3.05 m. The wind tunnel uses a steam engine to drive the fan to rotate at a high speed, generating a flow rate of 65 km per hour, which is used to design a reasonable aircraft. People are surprised by the wind tunnel experimental

Fig. 5.8 E. Buckingham (1867–1940, American physicist)

results that the high-speed airflow can produce lifts that are several times the weight of the wing when passing over the wing. This finding confirmed people's belief in making airplanes. The use of wind tunnel tests to study the shape of the aircraft marks the transition from adventure to science in the creative activities of humanity's long-awaited flight. Later, in September 1901, American inventors Wilbur Wright and Orville Wright brothers (as shown in Fig. 5.9) designed a small wind tunnel to accurately measure the lift force generated by airflow on the wing and the forces that made the aircraft turn when the control panel deflected and obtained the most important data for aircraft design. The wind tunnel test section is 40.6 cm × 40.6 cm in size, 1.8 m in length, and 12 m/s in velocity. In 1902, Jukowsky built a 2-foot-diameter wind tunnel at Moscow University.

The similarity theory is a law and condition to demonstrate that prototype and model physical phenomena remain similar. It is a theory that studies the relationship between personality and commonality, or special and general relations, and internal contradictions and external conditions. In the model test, only when the model and the prototype remain similar, can the corresponding results of the prototype be derived from the model test results. Similarity theory is mainly applied to guide model test to determine the similarity grade between "model" and "prototype". Similarity theory has not only become the basis of physical model experiments, but also become one of the guiding theories in the field of computer "simulation" and other fields.

Fig. 5.9 Wright brothers (Left: Orville Wright (1871–1948, American inventor); Right: Wilbur Wright (1867–1912, American inventor))

Similarity theory, from the phenomenon of occurrence and development of internal regularity (mathematical equations) and the external conditions (definite condition), based on the main premise that the mathematical equations to inherent homogeneity on the dimension and the correctness of the mathematical equations is not affected by the influence of the measurement system of units, comes to a conclusion through the mathematical deduction methods such as linear transformation. Similarity theory is characterized by a combination of high abstraction and application. The similarity theory provides guidance for simulation test, determines the reduction or enlargement of the model scale, increases or decreases the parameters, and changes the medium properties, etc. The purpose is to find out the internal regularity of the model with the lowest cost and the shortest operating cycle. Although the similarity theory itself is a more rigorous mathematical logic system, once it is applied in practice, it cannot be very accurate in many cases, because the problems dealt with by similarity theory are usually extremely complex. Three theorems in similarity theory are based on the following:

(1) Definition of similar phenomena;
(2) The relationship between the physical quantities involved in any physical phenomenon is subject to various objective laws, and they cannot be changed arbitrarily;
(3) The size of each physical quantity involved in the physical phenomenon exists objectively, regardless of the measurement unit used.

If the prototype and the model are proportional to each physical quantity corresponding to each point and time, the two systems are similar. The similarity number (called similarity scale, similarity coefficient, etc.) is the ratio of the physical quantity of the prototype to the physical quantity of the corresponding model. There are mainly geometric similarities, similar motions, and similar dynamics. Traditionally, length, time, and mass are taken as basic physical quantities, and the similarity number relationship between prototypes and models is called the similarity index. If the two are similar, the similarity indicator is 1. The dimensionless quantity group derived from similar indicators is called the similarity criterion.

If two flows are similar, they are similar as a single-valued condition, and the inertial force acting on the two systems corresponds to the proportion of other forces. In a fluid mechanics problem, if the forces acting on a particle satisfy the dynamic similarity, the ratios between the following forces must be equalized, including the ratio of inertial force to pressure (or pressure difference), the ratio of inertial force to gravity, the ratio of inertial force to viscous force, the ratio of inertial force to elastic force, the ratio of inertial force to surface tension, the ratio of convective inertial force to non-stationary inertial force, and the following six dimensionless numbers are introduced:

(1) Euler number (Eu) represents the pressure coefficient of the surface pressure distribution of the object, as well as the lift coefficient and drag coefficient. Physically, the Euler number represents the ratio of inertial force to differential pressure force.
(2) Froude number (Fr) represents the ratio of the flow inertial force to the gravity and represents the ratio of the water flow velocity to the microwave velocity of gravity wave.
(3) Reynolds number (Re) represents the ratio of flow inertial force to viscous force.
(4) Mach number (Ma) represents the ratio of inertial force to elastic force, which is a measure of compressibility of gas. It is usually used to express the ratio of flight speed to acoustic speed of aircraft.
(5) Weber number (We) represents the ratio of inertial force to surface tension.
(6) Strouhal number (St) represents the ratio of convective inertial force to unsteady inertial force.

On the premise of geometric similarity, the decisive criterion of flow phenomenon similarity is only the Reynolds number criterion, so the dynamic similarity of the model experiment must obey the Reynolds number similarity criterion.

Similarity First Theorem: Two similar systems have the same single-valued condition and the same value of similarity criterion.

Similarity Second Theorem: if any physical phenomenon is expressed by the functional relation of n physical quantities, and these physical quantities contain m basic dimensions, then (n-m) similarity criteria can be obtained.

Similarity Third Theorem: for any phenomenon with the same characteristics, when the single-value conditions (geometric properties of the system, physical properties of the medium, initial conditions and boundary conditions, etc.) are similar to each other, and the similarity criteria composed of physical quantities of the single-value conditions are equal in numerical value, then these phenomena must be similar.

These three theorems form the core of similarity theory. The third similarity theorem defines what conditions a model satisfies to make physical phenomena similar. It is the law that the model test must follow.

5.3 Application of Similarity Theory

1. *Derivation of similarity criteria*

A method of deriving similarity criteria from physical equations is called similarity transformation method. The specific steps of the similarity transformation method to derive the similarity criteria are as follows: list physical equations; list similar transformations of each physical quantity, and substitute them into the physical equations; obtain the similarity index composed of similar numbers, and set it equal to 1; substitute similar transformations into similar indicators, and similar criteria can be obtained.

Take the length dimension L, the time dimension T, and the mass dimension M as the basic dimensions, and the other physical quantity dimensions as the derived dimensions, and the corresponding dimension expression is

$$[q] = L^x M^y T^z$$

where, x, y, and z are dimension indexes, which can be determined using physical theorems or definitions (only the power product of the basic quantity can be used in the dimension expression, but the exponential, logarithm, trigonometric function, and addition and subtraction operations cannot be used). If all the dimension indexes in a dimensional expression of a physical quantity are zero, the physical quantity is a dimensionless quantity; otherwise, it is a dimensional quantity. Compared with pure numbers, dimensionless quantities have specific physical meanings and quantitative properties. The value of the dimension quantity varies with the unit, and the value of the dimensionless quantity does not vary with the unit.

Based on two similar flows, the principles that must be described for the same physical equation can be expressed as a system of dimensionless equations for the N–S equations that characterize the incompressible flow.

For dimensional incompressible fluid N–S group (mass force only gravity) is

$$\frac{\partial u}{\partial t} + u\frac{\partial u}{\partial x} + v\frac{\partial u}{\partial y} + w\frac{\partial u}{\partial z} = g_x - \frac{1}{\rho}\frac{\partial p}{\partial x} + v\left(\frac{\partial^2 u}{\partial x^2} + \frac{\partial^2 u}{\partial y^2} + \frac{\partial^2 u}{\partial z^2}\right)$$

$$\frac{\partial v}{\partial t} + u\frac{\partial v}{\partial x} + v\frac{\partial v}{\partial y} + w\frac{\partial v}{\partial z} = g_y - \frac{1}{\rho}\frac{\partial p}{\partial y} + v\left(\frac{\partial^2 v}{\partial x^2} + \frac{\partial^2 v}{\partial y^2} + \frac{\partial^2 v}{\partial z^2}\right)$$

$$\frac{\partial w}{\partial t} + u\frac{\partial w}{\partial x} + v\frac{\partial w}{\partial y} + w\frac{\partial w}{\partial z} = g_z - \frac{1}{\rho}\frac{\partial p}{\partial z} + v\left(\frac{\partial^2 w}{\partial x^2} + \frac{\partial^2 w}{\partial y^2} + \frac{\partial^2 w}{\partial z^2}\right)$$

$$\frac{\partial u}{\partial x} + \frac{\partial v}{\partial y} + \frac{\partial w}{\partial z} = 0$$

If it is to become a dimensionless form, a dimensionless transformation is performed on the quantities in the system of equations. It is illustrated by the x-direction equation (the other two directional component equations are similar) and the continuous equation. Introduce dimensionless variables into the equations, there is

$$t^* = \frac{t}{T}, x^* = \frac{x}{L}, u^* = \frac{u}{V_0}, p^* = \frac{p}{p_0}, \ldots$$

where L, T, V_0, p_0 are characteristic length, time, speed, and pressure, respectively.

The dimensionless continuous equation becomes

$$\left(\frac{\partial u^*}{\partial x^*} + \frac{\partial v^*}{\partial y^*} + \frac{\partial w^*}{\partial z^*} \right) = 0$$

The equation in the x-direction is

$$\frac{V_0}{T} \frac{\partial u^*}{\partial t^*} + \frac{V_0^2}{L} \left(u^* \frac{\partial u^*}{\partial x^*} + v^* \frac{\partial u^*}{\partial y^*} + w^* \frac{\partial u^*}{\partial z^*} \right)$$
$$= g - \frac{p_0}{\rho L} \frac{\partial p^*}{\partial x^*} + v \frac{V_0}{L^2} \left(\frac{\partial^2 u^*}{\partial x^{*2}} + \frac{\partial^2 u^*}{\partial y^{*2}} + \frac{\partial^2 u^*}{\partial z^{*2}} \right)$$

finished into an infinite formation

$$Sh \frac{\partial u^*}{\partial t^*} + u^* \frac{\partial u^*}{\partial x^*} + v^* \frac{\partial u^*}{\partial y^*} + w^* \frac{\partial u^*}{\partial z^*}$$
$$= \frac{1}{Fr^2} - Eu \frac{\partial p^*}{\partial x^*} + \frac{1}{Re} \left(\frac{\partial^2 u^*}{\partial x^{*2}} + \frac{\partial^2 u^*}{\partial y^{*2}} + \frac{\partial^2 u^*}{\partial z^{*2}} \right)$$

where Sh is the Strohal dimensionless number, i.e.

$$Sh = \frac{L}{V_0 T}$$

Fr is the Froude dimensionless number, i.e.

$$Fr = \frac{V_0}{\sqrt{gL}}$$

Eu is the Euler number, i.e.

$$Eu = \frac{p_0}{\rho V_0^2}$$

Re is the Reynolds number, i.e.

$$Re = \frac{V_0 L}{\nu}$$

Obviously, if two flows are similar, they must be described by the same dimensionless physical equations. For a similar phenomenon between prototype and model, the similarity scale of basic physical quantities is

$$\lambda_L = \frac{L_p}{L_m}, \lambda_T = \frac{T_p}{T_m}, \lambda_M = \frac{M_p}{M_m}$$

In the equations, the subscript *p* represents the prototype, and the subscript *m* represents the model, where λ_L is the length scale, λ_T is the time scale, and λ_M is the mass scale. According to the dimensionless equations, the similarity criteria between the prototype and the model are as follows:

The Strohal number similarity criterion is

$$Sh_p = Sh_m, \frac{L_p}{V_p T_p} = \frac{L_m}{V_m T_m}, \frac{\lambda_L}{\lambda_V \lambda_T} = 1$$

Froude number similarity criterion is

$$Fr_p = Fr_m, \frac{V_p}{\sqrt{g_p L_p}} = \frac{V_m}{\sqrt{g_m L_m}}, \frac{\lambda_V}{\sqrt{\lambda_g \lambda_L}} = 1$$

Euler number similarity criterion is

$$Eu_p = Eu_m, \frac{p_p}{\rho_p V_p^2} = \frac{p_m}{\rho_m V_m^2}, \frac{\lambda_p}{\lambda_\rho \lambda_V^2} = 1$$

Reynolds number similarity criterion is

$$Re_p = Re_m, \frac{V_p L_p}{\nu_p} = \frac{V_m L_m}{\nu_m}, \frac{\lambda_V \lambda_L}{\lambda_\nu} = 1$$

For compressible flows, in addition to the similarity criteria described above, Mach number similarity criterion, which characterizes compressibility, will be derived, i.e.

$$Ma_p = Ma_m, \frac{V_p}{a_p} = \frac{V_m}{a_m}, \frac{\lambda_V}{\lambda_a} = 1$$

where a is the sound wave speed.

2. *Application of similarity criteria*

Theoretically, in the fluid mechanics model test, the single-value conditions between the prototype flow field and the model flow field are similar, and the number of similarity criteria with the same name is equal. Guaranteeing the similarity of single-valued conditions including geometric similarity is an important premise of model experiment, which is related to factors such as model design, model attitude, parameter distribution in flow field, model support form, and other factors. In the fluid mechanics experiment, there are more than a dozen similarity criteria, and five are often used. According to the similarity theory, the corresponding similarity criteria should be simulated simultaneously in the experiment to achieve "completely similar". However, this complete similarity is actually difficult to achieve, and generally can only be "partially similar." Fortunately, the physical meanings of various similarity criteria are different. In a specific case, not all similarity criteria are equally important. In other words, for a particular experiment, some similarity criteria must be simulated, and some similarity criteria can be ignored. To do this correctly, the physical meaning of each similarity criterion and its impact on similar flows should be understood. For example, to simulate a periodic non-stationary flow of an incompressible fluid, except for geometry and motion, the dynamic similarity must satisfy the Strouhal number similarity criterion. In order to simulate the resistance problem of the pipeline flow, the Reynolds number similarity criterion must be satisfied. For the flow problem of gravity, the Froude number similarity criterion must be satisfied. For compressible flow, the Mach number similarity criterion is needed.

Set a propeller diameter $D_p = 4$ m, propeller speed $n = 1075$ rpm, standard takeoff speed $V_p = 258$ km/h, static takeoff pulls force $T = 4300$ kg, try to design the propeller wind tunnel scaled model according to different similarity criteria.

Determination of length scale λ_L. Set the model propeller diameter D_m, which is determined according to $\frac{\pi D_m^2}{4}/A = 0.2 - 0.3$, where A is the cross-sectional area of the wind tunnel test section. For wind tunnel cross-sectional area $A = 1.52 = 2.25$ m^2, the diameter of the propeller of the scaled model is $D_m = 0.8$ m. Prototype diameter $D_p = 4.0$ m. Length scale is $\lambda_L = 5$.

(1) Design the model according to the Ma number and Sh number criteria

The design set $\lambda_L = 5.0$, $D_m = D_p/\lambda_L = 4.0/5.0 = 0.8$ m, according to Mach number similarity criterion, obtain that

$$\frac{V_p}{a_p} = \frac{V_m}{a_m}, a_p \approx a_m, \lambda_V = 1$$

Obtained by the Strouhal number similarity criterion

$$\frac{D_p f_p}{V_p} = \frac{D_m f_m}{V_m}, \frac{V_m}{D_m n_m} = \frac{V_p}{D_p n_p}, \frac{\lambda_V}{\lambda_L \lambda_n} = 1, \lambda_n = \frac{1}{\lambda_L} = 1/5$$

Accordingly, the similar parameters of the model obtained and prototype are given in Table 5.1.

(2) Design the model according to the Re number and Sh number criteria

The design set $\lambda_L = 5.0$,according to Reynolds number similarity criterion, obtain that

$$\frac{\lambda_V \lambda_L}{\lambda_\nu} = 1, \lambda_\nu \approx 1, \lambda_V = 1/\lambda_L = 1/5$$

Obtained by the Strouhal number similarity criterion

$$\frac{\lambda_V}{\lambda_L \lambda_n} = 1, \lambda_n = \frac{\lambda_V}{\lambda_L} = \frac{1}{\lambda_L^2} = 1/25$$

Table 5.1   Related variables prototype and model values (Ma criterion)

Name	Prototype value	Model value	Scale
Propeller diameter D(m)	4.0	0.800	5
Propeller rotate speed n (rpm)	1075	5375	1/5
Standard takeoff speed V_0(km/h)	258.0	258.0	1.0

Table 5.2 Related variables prototype and model values (Re criterion)

Name	Prototype value	Model value	Scale
Propeller diameter D(m)	4.0	0.800	5
Propeller rotate speed n (rpm)	1075	60100	1/25
Standard takeoff speed V_0(km/h)	258.0	1290	1/5

Table 5.3 Related variables prototype and model values (St criterion)

Name	Prototype value	Model value	Scale
Propeller diameter D(m)	4.0	0.800	5
Propeller rotate speed n (rpm)	1075	2150	0.5
Standard takeoff speed V_0(km/h)	258.0	103.2	2.5

Accordingly, the similar parameters of the model obtained and prototype are given in Table 5.2.

(3) Design the model according to the *Sh* number criterion

The design set $\lambda_L = 5.0$, wind tunnel model propeller speed is limited, set

$$\lambda_n = 1/2, n_m = 2n_p$$

Obtained by the Strouhal number similarity criterion

$$\frac{\lambda_V}{\lambda_L \lambda_n} = 1, \lambda_V = \lambda_L \lambda_n = 5/2 = 2.5$$

Accordingly, the similar parameters of the model obtained and prototype are given in Table 5.3.

5.4 Flow Visualization Measurement Technique

Fluid mechanics experiments generally include flow visualization and flow measurement methods. The main task of the flow visualization experiment is to visualize the fluid flow process, while the main task of the flow measurement experiment is to obtain the quantitative information of the fluid flow process, which complement each other and are an integral part of the experimental fluid dynamics. Through various flow visualization and measurement experiments, people can understand complex flow phenomena, explore physical mechanisms and motion laws, and provide scientific basis for establishing new concepts and mathematical models. As the famous master of fluid mechanics in the world, German scientist Prandtl said, "I only came up with mathematical equations after I believed I had a deep understanding of the nature of physics. The use of the equation is to say the size of the quantity, which is not intuitive, and it also proves whether the conclusion is correct." Flow visualization and measurement technology itself is also the primary means of solving practical engineering problems. It can be said that every theoretical breakthrough in the development of fluid mechanics and its application in engineering almost starts from the observation of the flow phenomenon. Such as the Reynolds transition experiment in 1880, the Mach shock phenomenon test in 1888, the concept of the boundary layer proposed by Prandtl in 1904, and the analysis of the vortex street around the cylinder in 1912. All are based on the results of flow visualization and measurement. The in-depth analysis of the flow phenomenon is the key to establish and verify new concepts and discover new laws.

It should be pointed out that in recent decades, due to the urgent need for engineering practice and the rapid development of modern optics, laser technology, computer technology, electronic technology, and information processing technology, it has brought vigor and vitality to the flow visualization and measurement technology. There is considerable progress, especially in the ability to display the spatial flow and flow internal structure, as well as the quantitative extraction and analysis of flow information. Three-dimensional, unsteady complex flow quantitative display and measurement information can be obtained simultaneously. The flow visualization and measurement methods that have appeared so far are often divided into two categories, conventional and computer-aided. The former is called the conventional method, and the latter is called the method of combining flow visualization and measurement with computer image processing. Among the conventional methods, the wall surface tracking method, the wire method, the tracer method, and the optical method are included.

1. *Conventional flow visualization technique*

(1) Wall surface tracking method

The method is to coat a thin layer of material on the surface and produce a certain visible flow pattern on the surface when it interacts with the fluid. It is used to qualitatively or quantitatively display the flow characteristics of the surface, such as laminar flow, turbulence and transition position, separation

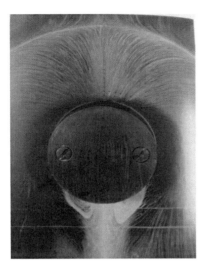

Fig. 5.10 Two-dimensional cylindrical flow surface oil flow spectrum at *Ma* = 5 (Quoted from Li Suxun's "shock and boundary layer dominated complex flow")

point, separation area, etc. Surface oil flow (as shown in Fig. 5.10) and fluorescent oil flow are the common methods for flow spectrum identification by means of topological criteria; sublimation techniques are often used to determine the boundary layer transition, separation flow, near-wall flow structure, and wall mass exchange. The thermal deep layer is used to show surface heat exchange.

(2) Wire method

In this method, one end of the filament, the wool, and the nylon thread are attached to the object surface, and the other end is swung in the airflow to display the flow state of the object surface. It is possible to display the laminar flow, the position of the turbulent flow and the separation zone, and the position of the vortex core. A mesh array can also be used to display the spatial flow field. In the early 1980s, a fluorescent microfilament method was developed. The fluorescent microfilament is a nylon monofilament dyed with a fluorescent substance and antistatically treated, and has a diameter of only 0.01 to 0.04 mm. It is attached to the surface of the model, and under the irradiation of ultraviolet light, the surface flow state of the model can be clearly displayed (as shown in Fig. 5.11).

Fig. 5.11 Wind tunnel test of automobile with a wire method (Quoted from WeChat public number: AutoWindTunnel)

(3) Tracer method

The tracer substance is directly injected into the flow field as a contrast medium, or a chemical reaction is used to generate a tracer substance in the fluid, or an electronic control is used to generate a tracer substance that moves along with the fluid to make the flow visible. The smoke trace (including tobacco) method (as shown in Fig. 5.12), the dyeing line method, the hydrogen bubble method, the helium bubble method, the laser-fluorescence method, the steam screen method, etc., all belong to the tracer method. Due to the injection of particles into the fluid, there is a problem of particle

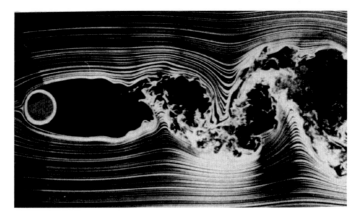

Fig. 5.12 Circular flow with a smoke line method (Quoted from An Album of Fluid Motion)

followability, so the display effect is better when it is used for a steady flow and the display error is larger when it is used for an unsteady flow.

(4) Optical method

The conventional optical flow visualization method mainly refers to a shadow method, a schlieren method (including color schlieren, as shown in Fig. 5.13), an interference method, and the like, for a flow field having a density change. The image is formed by the refraction effect of light or by the relative phase shift of different rays and shows physical phenomena such as shock wave, vortex, boundary layer transition, and shock boundary layer interference. The change in density field can also be given quantitatively by the interference method. The optical method does not interfere with the flow field, but has greater limitations. Images such as shadows, schliers, and interferometers are the result of the accumulation of light across the flow path. Although the density distribution of the flow field can theoretically be obtained from image reduction, the resolution is not high and is limited to two-dimensional flow field. Some methods based on the principle of interference can directly reflect the change in density and give quantitative results. These have attracted people's attention. Therefore, various types of interferometry have been developed, such as laser holographic interference, heterodyne interference, laser speckle interference, etc. Optical methods are mostly used in high-speed flow field measurements.

In addition, the simultaneous implementation of flow visualization and measurement and the use of computer graphics and data processing techniques have matured. In the experiment, based on the flow display and image

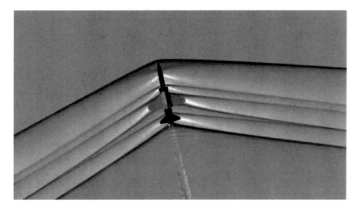

Fig. 5.13 Schlieren photograph of the supersonic flight of the aircraft (from NASA)

equipment, the computer image processing system is used to complete the image display and data processing, and then the change in color display parameters is used to provide rich flow field information and high-quality images.

2. *Combined flow visualization and measurement technique*

In view of the increasing complexity of the flow phenomenon, people urgently need to rely on laser technology, computer technology, information processing technology, etc., thus promoting the joint technique of flow visualization and measurement. For example, laser Doppler velocimetry (LDV), particle image velocimetry (PIV), laser-induced fluorescence (LIF), computed tomography (CT), and optical surface pressure measurement (ospmt) developed in recent decades are all visualization and measurement techniques. The combination technique of quantity, which has the function of both qualitative display and quantitative measurement, has greatly promoted the research progress of complex flow.

(1) Doppler effect of waves

The Doppler effect of waves is a phenomenon in which the observed frequency differs from the source frequency due to the relative motion between the wave source and the observer. This was discovered by Christian Doppler in 1842 (Austrian physicist, 1803–1853, as shown in Fig. 5.14). A famous example is: When a whistling train passes through the observer (as shown in Fig. 5.15), he will find that when the train is approaching from a distance, the sound of the whistle is changed from low to high; When the train moves from near to far, the sound will change from high to low, which is called Doppler effect. The reason is determined by the frequency of sound wave vibration. If the frequency is high, the tone will sound high; if the frequency is low, the tone will sound low. It can be seen that the wavelength of the sound wave emitted by the whistle of a train moving at a constant speed changes as it propagates. The result is that for trains coming from afar, the wavelength of sound waves becomes shorter as if the waves are compressed. On the contrary, when the train moves far away, the wavelength of the sound wave becomes larger, as if the wave has been stretched. The frequency of the wave received by the observer relative to the sound wave of the moving object

Fig. 5.14 Christian Doppler (1803–1853, Austrian physicist)

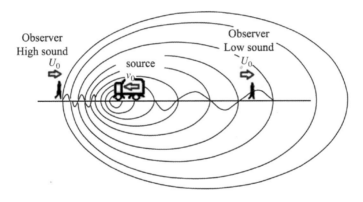

Fig. 5.15 Doppler effect

is

$$f = \frac{a \pm U_0}{a \mp V_0} f_0$$

where f_0 is the natural frequency of the wave source, a is the acoustic velocity of the stagnant air, U_0 is the speed of the observer relative to the air, and V_0 is the velocity of the wave source relative to the air. When the observer moves toward the source, a positive sign is taken in front of U_0; when the observer moves away from the source (i.e., along the source), a negative sign is taken in front of U_0. When the wave source moves toward the observer, a negative sign is taken in front of V_0; when the wave source moves away from the observer, a positive sign is taken in front of V_0. It can be seen from the above formula that when the observer and the sound source are close to each other, $f > f_0$; when the observer and the sound source are distant from each other, $f < f_0$.

(2) Laser velocimetry technique

The spectral technique of laser velocimetry relies on the absorption line frequency of the measured flow field medium component or the Doppler frequency shift in the fluorescence emission or scattering spectrum. Since this method directly acquires the velocity from the molecular motion, so as to avoid the disadvantages caused on the particle, velocity is not only real and of high precision, but is suitable for high-speed flow measurement (as shown in Fig. 5.16). Based on a variety of optical velocimetry techniques, such as laser-induced Doppler integrated velocimetry (DGV), filtered Rayleigh scattering (FRS), coherent Raman spectroscopy (including coherent anti-Stoke Raman spectroscopy (CARS),), inverse Raman spectroscopy (IRS), and stimulated Raman gain spectroscopy (SRGS) have been developed.

Fig. 5.16 Laser velocimetry technique

(3) Particle imaging velocimetry technique

Particle imaging velocimetry technique is developed under computer technology and graphics processing technique, and is an optical fluid measurement technique (as shown in Fig. 5.17). Since the 1950s, laser velocimetry technique has been used and gradually formed on this basis. The biggest advantage of this technique is that it breaks through the limitations of spatial single-point measurement technique such as laser Doppler velocimetry, which not only has the accuracy and resolution of single-point measurement technique, but also can obtain the overall structure and transient image of planar or spatial flow field display. The information about the entire flow field can be recorded at the same time, and the average speed, the pulsation speed, etc., can be, respectively, given, which is a non-contact measurement method. The basic principle is: the particles are placed in the flow field during the experiment, and the laser beam emitted by the pulsed laser forms a modulated laser irradiation flow field through a series of optical elements, and the images of the particle field at different times are recorded by multiple exposures. By measuring the displacement ΔL of each particle at Δt time interval, the velocity of the particle can be calculated. Under the condition

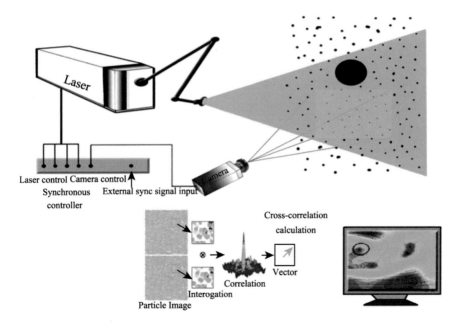

Fig. 5.17 Particle imaging velocimetry technique

that the followability of the particle satisfies the requirement, the velocity of the particle can represent the velocity of the fluid.

(4) Laser-induced fluorescence flow visualization and measurement technique

Laser-induced fluorescence flow visualization and measurement technique (as shown in Fig. 5.18) is a photoluminescence flow visualization and measurement technique developed in the 1980s that enables qualitative display of flow structures and quantitative measurement of flow parameters. Photoluminescence visualization and measurement techniques dissolve or mix certain substances (such as iodine, sodium or fluorescent dyes, etc.) into the fluid. The molecules of these substances absorb photons and are excited by light at a specific wavelength of light. During the experiment, the pulsed laser light is used to illuminate, and the excited light can not only display the flow structure, but also measure the velocity using the Doppler frequency shift effect of the absorption and emission lines. The intensity of the light is also a function of the gas flow density and temperature in the excited zone, so the parameters such as density, temperature, velocity, pressure, and concentration of the flow field can be measured while displaying the flow structure.

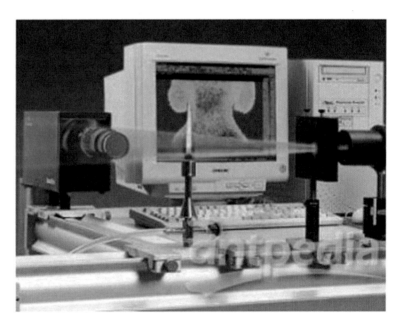

Fig. 5.18 Laser-induced fluorescence flow visualization and measurement technique

The method is a non-contact transient flow field measurement technique, which is suitable for high-speed flow and large-speed gradient flow. In this case, the Doppler velocity measurement method based on particle scattering is no longer applicable because the particles cannot completely follow the fluid, and the molecular luminescence-based LIF technology does not have this limitation because LIF does not have particle followability problems.

(5) Optical tomography

Chromatography began in 1967, and began to be used for medical diagnostics and nondestructive testing of materials. CT was used for flow visualization in the mid-1980s. It is an advanced computer-aided flow visualization and measurement technique (as shown in Fig. 5.19). The tomography technique reconstructs a three-dimensional image from a "projection" obtained by observing a multi-directional convective field, and is adapted to the structural measurement and analysis of the three-dimensional flow field. Optical tomography is the most commonly used method in flow visualization, such as shadow, schlieren, interferometry, etc., which can be used to obtain

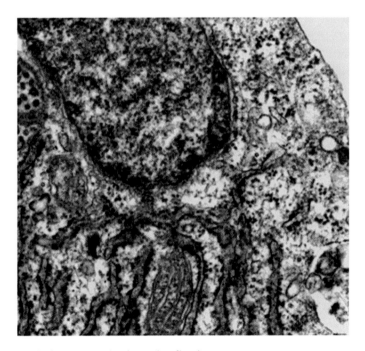

Fig. 5.19 Optical tomography (vascular flow)

multi-directional "projection", but it is more important to combine the interferometer with computer tomography technology to obtain quantitative 3D flow visualization. The three-dimensional density field and concentration field can be obtained by conventional interferometric tomography, and the three-dimensional temperature field can also be obtained by holographic interference tomography. Images of different sections can be obtained at different instants or at the same instant using multi-directional interferometry. The former can only be used for constant flow, and the latter can also be used for unsteady flow.

(6) Optical surface pressure measurement technique

The basic principle of optical surface pressure measurement is to apply a pressure sensitive paint (PSP) to the surface of the model. Under ultraviolet light or other light of a given wavelength, the coating emits fluorescence of visible wavelength. Its brightness is inversely proportional to the absolute pressure of the air or any oxygen containing gas acting on the surface of the coating. The image of the surface of the model is recorded with a CCD camera and processed by a computer to give the pressure distribution on the surface of the model (as shown in Fig. 5.20). Optical surface pressure measurement requires sufficient brightness and uniformity of illumination. The brightness of the surface coating of the model is calibrated with absolute pressure and stored in a computer. The image of the surface of each model is recorded in

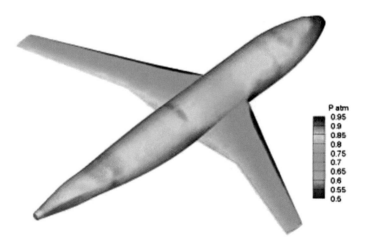

Fig. 5.20 F4 surface pressure (pressure sensitive coating PSP) wind tunnel test (from DLR)

two states of "no blowing" and "blowing" in the wind tunnel and the pressure distribution of the surface of the model is obtained by comparing the computer with the calibration pressure data.

5.5 Flow Velocimetry Technique

1. *Traditional measurement of time-average speed*

For the time-average velocity of any point in the flow field, Pitot tube velocimeter is often used. Pitot tube (as shown in Fig. 5.3) is a simple device for measuring velocity based on Bernoulli equation of steady, incompressible, and inviscid fluid. Since the test hole is facing the flow direction of the fluid, it is not suitable for the flow of solid particles in the fluid. Pitot tube is a single-point contact speed measurement, which has a great impact on the measured flow field, and the actual flow rate measurement has certain limitations. The porous probe is a device for measuring the magnitude and direction of the velocity, and mainly includes three-, five-, and seven-hole probes. Due to the small range of flow angles measured by the three-hole probe, five- or seven-hole probes are currently used. The five or seven-hole probe solves the problem that the Pitot tube cannot measure the speed direction and can measure the speed and pressure. The five-hole probe is inserted into the flow field measuring point position, and the pressure measured by each hole is appropriately corrected to obtain the velocity magnitude and direction of the measured point of the flow field. When measuring the velocity with the seven-hole probe, the pressure values of the seven holes are measured, and the velocity, direction, total pressure, static pressure, and other parameters of the measuring point can be obtained. The five or seven-hole probe has high measurement accuracy, good reliability, simple structure, easy to damage the probe, easy maintenance, and low cost. However, the five or seven-hole probe is limited by its own measurement angle, and the measurement calibration and calibration data processing are cumbersome.

2. *Measurement of instantaneous speed*

For the instantaneous velocity field, the velocity values include the time-average velocity and the pulsation velocity, and in particular, the velocity pulsation with time is the basic data of the surface flow field characteristics. A commonly used measurement method in wind tunnels is a hot wire anemometer by means of a hot wire (usually a platinum wire or tungsten wire having a diameter of 1 to 2 microns) or a hot film (usually a 0.1 micron thick platinum film or Nickel film) changes in convective heat transfer to measure speed. This instrument is particularly suitable for studying the turbulent structure of the flow field. Another instrument for measuring instantaneous velocity is a laser Doppler velocimeter, which is measured based on the principle that the Doppler shift value of the light in the airflow is proportional to the velocity of the airflow. In such measurements, it is necessary to utilize scattering particles that are naturally present or artificially added to the gas stream, and the size and concentration of the particles must be limited to a certain range. The following is described:

(1) Hot wire Anemometer

Hot wire Anemometer (HWA for short), invented in the 1920s, is a single-point contact device for measuring the instantaneous velocity of flow field, as shown in Fig. 5.21. The basic principle is to use a thin metal wire (fine wire with a diameter of $d = 0.5 \sim 5$ μm, length $L > 300d$ of platinum wire or tungsten wire), and then heat the wire after passing it to make the temperature higher than the temperature of the fluid. Therefore, the wire is called a "hot wire". When the fluid flows through the wire in the vertical direction, a part of the heat of the wire is taken away by convection heat transfer, causing

2005   3 31

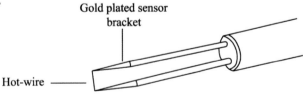

Gold plated sensor
bracket

Hot-wire ———

Fig. 5.21 Hot wire Anemometer

the temperature of the wire to drop. According to the theory of forced convection, heat exchange in an infinitely long cylinder, the relationship between the heat taken away by the hot wire and the velocity U of the fluid can be derived. This relationship is called the formula of L.V. King (1914), which is

$$I^2 R = (T_s - T_0)(A + B\sqrt{U})$$

where R and I are, respectively, the resistance of the hot wire and the current intensity flowing, Ts is the operating temperature of the hot wire, T0 is the ambient temperature of the fluid to be measured, and A and B are physical constants related to the fluid and the hot wire. The hot wire anemometer can measure the velocity in real time, dynamically and continuously without the addition of tracer particles. It has a high-speed response frequency and can be used to measure opaque fluid and study the frequency response range of all turbulence fluctuations.

The length of the hot wire is generally in the range of 0.5 to 2 mm, and the diameter is in the range of 1 to 10 μm. The material is platinum, tungsten or platinum-rhodium alloy. If a thin metal film (thickness less than 0.1 micron) is used instead of the metal wire, it is a hot film anemometer, which functions similarly to the hot wire, but is mostly used to measure the liquid flow rate. In addition to the ordinary single-wire type, the hot line can also be a combined two-wire or three-wire type to measure the velocity component in all directions. The electric signal output from the hot wire is amplified, compensated and digitized and input into the computer, which can improve the measurement accuracy, automatically complete the data post-processing process, and expand the speed measurement function, such as simultaneously completing measurement on the instantaneous value and the time-average value, the combined speed and the sub-speed, and the turbulence and other turbulence parameters (as shown in Fig. 5.22). Compared with the Pitot tube, the hot wire anemometer has the advantages of a small probe volume and small interference to flow field, fast response, measuring unsteady flow rate, measuring flow with very low speed (such as low as 0.3 m/s).

The hot wire anemometer converts the flow rate signal into an electrical signal, using the formula of L.V. King (as shown in Fig. 5.21) and can also be used to measure fluid temperature or density. It has two working modes: ① constant electric current. The electric current through the hot wire remains unchanged. When the temperature changes, the heat line resistance changes, so the voltage at both ends changes, thereby measuring the flow rate; ② constant temperature type. The temperature of the hot wire remains the same, such as maintaining 150 °C, the flow velocity can be measured according to

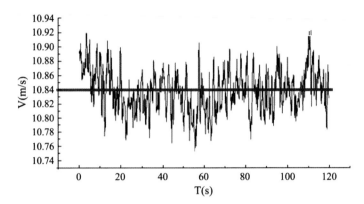

Fig. 5.22 Instantaneous speed process

the electric current applied. The constant temperature type is more widely used than the constant electric current type.

(2) Particle Image Velocimetry

The particle image velocimeter is called PIV for short. This is a transient, multi-point, contactless laser, fluid mechanism velocity measurement technique developed in the late 1970s. In recent decades, it has been continuously improved and developed. The characteristics of PIV technique are beyond the limitations of single-point speed measurement technology (such as LDA). It can record the velocity distribution information at a large number of spatial points in the same transient state and can provide rich information about flow field structures and flow characteristics (as shown in Fig. 5.23). The principle

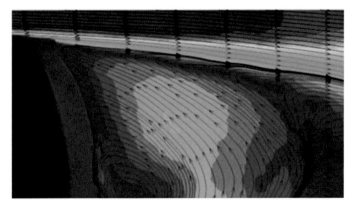

Fig. 5.23 PIV flow field information

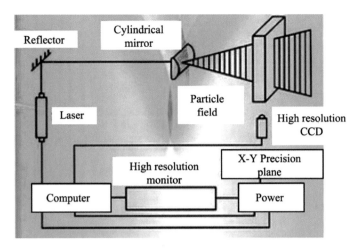

Fig. 5.24 PIV speed measurement principle

of PIV is to spread a certain number of tracer particles into the flow field (as shown in Fig. 5.24) and use a laser light source to illuminate a certain flow plane in the measured flow field, while using two consecutive exposures perpendicular to the plane camera. The image of the particles is recorded on a negative or CCD camera. Using image processing technique, instantaneous flow field velocity distribution information can be obtained. In addition to spreading the tracer particles to the flow field, all measurement devices are not involved in the flow field (as shown in Fig. 5.24) and have high measurement accuracy. Due to the above advantages of PIV technique, it has become one of the main equipments for measuring velocity in the fluid mechanics laboratory. PIV velocity measurement technique has many classifications, but regardless of the form of PIV, its velocity measurement relies on the tracer particles scattered in the flow field. PIV is to indirectly measure the transient velocity distribution of the flow field by measuring the displacement of the tracer particles in a known short time interval. If the tracer particles have a sufficiently high flow following, the motion of the tracer particles can truly reflect the motion state of the flow field (as shown in Fig. 5.25). Therefore, the tracer particles are very important in the PIV velocimetry. In PIV velocity measurement technique, high-quality tracer particle requirements are: (1) more important than the experimental fluid; (2) small enough scale; (3) shape as circular and as uniform as possible; (4) high-light scattering efficiency. Hollow microbeads or metal oxide particles are usually used in liquid experiments, smoke or dust particles are used in air experiments (nanoparticles are used for supersonic measurement), and fluorescent particles are used in microchannel experiments.

Fig. 5.25 PIV instantaneous flow field

(3) Laser Doppler Velocimetry

The laser Doppler velocimeter is a high-precision laser fluid velocity measurement technique established in the 1970s with the development of laser technology. Laser Doppler Velocimetry (LDV) is a Doppler signal that measures the tracer particles passing through the laser probe and then derives the velocity based on the relationship between velocity and Doppler frequency. Because of the laser Doppler interferometric velocity measurement, there is no interference to the flow field, and the speed measurement range is wide. Since the Doppler frequency and the speed are linear, it has no relationship with the temperature and pressure of the point, and it is currently the speed measurement instrument with the world's highest accuracy. The measuring principle is that the transmitting source emits a certain frequency of the transmitting wave. When there is a relative speed between the transmitting source and the detector, there is a certain frequency shift between the receiving frequency and the transmitting frequency, and the frequency shift is relative between the two. The speed caused by the frequency shifts the reaction speed. The commonly used speed measuring optical system is a double beam type based on an interference fringe model. Since the laser has good coherence, the focusing lens concentrates the incident light at an angle θ, and the light forms a light-dark interphase interference fringe at the focus point. When the tracer particles in the fluid pass through the vertical stripe region, a list of light waves are sequentially scattered, its light intensity changes with time, called Doppler signals. The frequency of change in light intensity is called the Doppler frequency. Set U as the tracer particle velocity, and the relationship between Doppler shift and particle velocity is

$$f_d = \left(\frac{2n}{\lambda} \sin \frac{\theta}{2}\right) U$$

where f_d is the Doppler frequency shift, λ is the emission wavelength, θ is the incident angle, and n is the refractive index of the fluid.

The LDV system is functionally divided into: the optical path section (as shown in Fig. 5.26) and the signal processing section. Optical path section: He-Ni lasers or Ar ion lasers are used because they can provide high-power lasers of three wavelengths of 514.5, 488, and 476.5 nm. The new generation of LDA systems uses solid-state lasers that significantly reduce the operator's experience. The optical splitter with frequency shifting device divides the laser into two beams of equal intensity. After passing through the single-mode polarization-maintaining fiber and the fiber coupler, the laser is sent to the laser emitting probe, and the laser is adjusted at the same point in the waist part to ensure the minimum measurement volume, which is the measurement body, namely the optical probe. The receiving probe sends the received Doppler signal to the photomultiplier tube and converts it into an electrical signal and processes the concurrency. After the Doppler signal analyzer analyzes the processing and records it to the computer, the supporting system software can perform data processing. In the case where appropriate tracer particles are present in the flow field, the three-direction velocity of the flow can be measured simultaneously, and even after upgrading to the PDA system, the particle diameter of the spherical transparent particles can be measured.

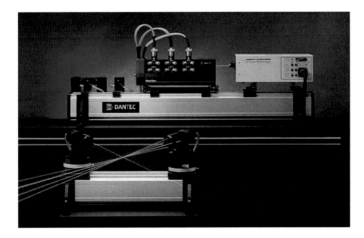

Fig. 5.26 LDA velocimetry principle

With the application of narrow linewidth single-mode solid-state lasers in laser Doppler, argon-ion gas lasers have been eliminated. LDA laser emission optical systems such as DopplerLite and FlowExplorer DPSS have been developed. The new DPSS laser not only provides an unregulated integrated LDA transmitter probe, but also a reliable fiber optic transmitter probe. Among them, the integrated LDA transmitting probe has a built-in device such as a laser and a Bragg unit, and the adjustable components such as the fiber coupler are eliminated, thereby greatly reducing the adjustment difficulty and directly using without any optical coupling adjustment. However, the disadvantage is that due to the built-in laser, the temperature and humidity requirements and vibration requirements of the probe are relatively high. Due to the photoelectric integration, the probe cannot be dustproof and moistureproof. Fiber optic probes based on DPSS lasers inherit all the advantages of traditional fiber laser Doppler. The watertight probe has no strict requirements on the cleanliness of the use environment, can withstand certain vibrations, has high adaptability to the temperature and humidity of the environment, and can even be configured with the internal gas purging system of the probe. The disadvantage is that when the ambient temperature and humidity change greatly, it is necessary to periodically check the fiber coupling condition to prevent the fiber coupling efficiency from decreasing.

The latest laser Doppler signal processor, whether in the time benchmark, the highest data rate, the highest processing Doppler frequency, bandwidth, transit time, and other core indicators, or memory, sampling bits, and other auxiliary indicators, has been greatly improved. There are significant optimizations in power management, cooling, and stability, and almost all aspects that can affect the final data quality are improved. It not only provides laser power adjustment, signal optimization, and other functions, but also achieves data quality monitoring through real-time signal oscilloscope function, through the vector graphics and a variety of combined 3D graphics to obtain intuitive data display results. In addition, a special coordinate frame system is set up to ensure stability and accuracy during the movement. For single-point measurement systems (as shown in Fig. 5.27), measurement time can be significantly reduced and data quality can be improved.

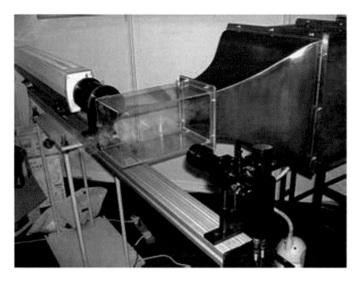

Fig. 5.27 LDA velocimetry device

5.6 Experimental Measurement Method for Dynamic Forces

(1) Pressure measurement method

Pressure measurement in the wind tunnel was first measured by a liquid pressure gauge such as a U-tube pressure gauge. The most commonly used measure of total air pressure and static pressure is the Pitot tube. In the supersonic flow, a positive shock wave is generated before the Pitot tube, so only the total post-wave pressure can be measured. Nowadays, pressure sensors have been widely used to measure pressure. There are many types of pressure sensors, which can be divided into resistance strain type, capacitive type, inductive type, diaphragm type, solid state piezoresistive type, and piezoelectric type according to the conversion principle. When a very low pressure is encountered in a hypersonic wind tunnel, a diaphragm or solid piezoresistive sensor is often used. Piezoelectric sensors are mainly used for pulsed wind tunnels or for measuring transient pressures. When measuring multi-point pressure, a pressure measuring system consisting of a pressure sensor and a pressure scanning valve or an electronic scanning pressure measuring system (as shown in Fig. 5.28) is widely used.

Fig. 5.28 PSI 9816 electronic pressure scan valve system

(2) Measurement method of temperature and heat flow

For high-speed flowing airflow, the total temperature and static temperature of the airflow need to be determined. When the total temperature of the airflow is lower than 2000 K, the total temperature probe can be used to measure the total temperature, and the static temperature of the airflow is generally calculated based on the total temperature and the Mach number.

The temperature measuring element of the total temperature probe can be used with a thermistor at a lower temperature, and a copper-constantan, nickel-chromium-nickel aluminum thermocouple at a higher temperature. Platinum-platinum rhodium thermocouples can measure temperature up to 2000 K. In addition to thermocouple measurement, the temperature sensitive coating or phase change coating can also be used as the sensitive element. The measurement range is from room temperature to hundreds of K. The measurement accuracy is slightly low, but the information is large. Infrared photographic technique can be used to measure surface temperature distribution from room temperature to 1200 K. This method does not interfere with the flow field and has a large amount of information. For the temperature of the airflow and the surface temperature in the range of 1200 to 4000 K, it is usually measured by an optical measuring instrument such as a radiant pyrometer, a photoelectric pyrometer, or a colorimeter pyrometer. Modern coherent anti-Stokes-Mann spectroscopy (CARS) has a high spectral radiation conversion rate and has attracted widespread attention. The low-density wind tunnel airflow itself does not radiate the spectrum, so electron beam probes are commonly used to excite the gas to ionize and radiate the spectrum, and then the vibration and rotation temperatures are obtained by spectral analysis. For wind tunnel heat transfer experiments, the rate of change in temperature over time can generally be used to measure heat flow. Different types of wind tunnels often use different types of calorimeters. Thin-wall calorimeters are commonly used in conventional hypersonic wind tunnels, zero-point calorimeters are commonly used in arc wind tunnels, and thin-film resistance calorimeters are commonly used in shock tunnels (as shown in Fig. 5.29).

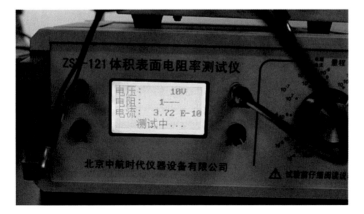

Fig. 5.29 Thin-film resistance calorimeter

(3) Measurement method of density

Optical methods are generally used to measure the density of the gas stream. The shadow method, the schlieren method, the interference method, etc., can only be used for qualitative measurement of density, and the laser hologram can quantitatively measure the three-dimensional density distribution of the wind tunnel flow field. In the holography method, an interference pattern formed by two columns of interfering light waves (signal wave and reference wave) is recorded by a photographic method, and the light wave of the "flow field" can be reproduced by appropriately illuminating the hologram afterwards. The holographic method records all the information of the flow field (including phase difference, direction difference, and optical path difference), so the reproduced light wave can be quantitatively measured afterwards (as shown in Fig. 5.30).

(4) Measurement methods of force and moment

The aerodynamic forces and aerodynamic moments acting on the model can usually be measured directly using a wind tunnel balance (as shown in Fig. 5.31) and by measuring the motion of the free flying tunnel model. Free-flight force measurement is the calculation of the trajectory of free flight

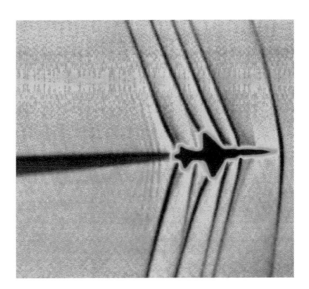

Fig. 5.30 Schlieren technique

Fig. 5.31 Six-component force balance

in the wind tunnel and the determination of acceleration by means of a high-speed camera. The aerodynamic forces and aerodynamic moments are calculated according to Newton's second law. In order to make the aircraft have good stability and maneuverability, it is also necessary to know the derivative of the angular change rate of aerodynamic forces and aerodynamic moments acting on the model, that is, the dynamic derivative. In a wind tunnel, the dynamic derivative can be measured by a balance that is freely vibrating or forced to vibrate, or it can be reversed by a free-flight test.

(5) Surface friction measurement method

It is often required to measure the surface shear stress when studying viscous flow. In addition to the indirect method, the friction balance can be directly measured. The friction balance often uses a differential transformer or a strain sensor, a thermal film, and a Prius tube.

5.7 Test Error Analysis

Measurement and analysis are indispensable means and ways for human beings to understand the laws governing the development of things. Through experimentation and measurement, people can obtain qualitative concepts and quantitative relationships with things, and rise to rationality through processing analysis, and discover the general regularity of things development. Many new discoveries and breakthroughs in science are done on the basis of experiments. An important purpose of the test is to obtain the quantitative relationship between the physical quantities through measurement, so it is necessary to carefully examine the accuracy of the physical quantity measurement, which inevitably leads to the measurement error and its analysis.

Due to the limitations of the test equipment (wind tunnel, water tunnel, etc.) methods, test environment, measuring instruments, measurement procedures, etc., there is always a certain difference between the measured value and the real value of any physical quantity. Absolute error, relative error, or significant figures are often used to illustrate the accuracy of an approximation. In order to assess the accuracy of the test data, it is necessary to analyze the source of the error and its effects. Therefore, it is determined which factors are the main aspects affecting the accuracy of the test, thereby further improving the test plan in the test, reducing the difference between the test observation value and the real value, and improving the accuracy of the test.

1. *Concept of error*

(1) True value and mean value

The real value is a certain value that is objectively present in the measured physical quantity, also called a theoretical value or a defined value. Usually, the real value is unmeasurable. If the number of measurements is infinite in the test, the probability of occurrence of positive and negative errors is equal according to the law of distribution of errors. After careful elimination of systematic errors, the measured values are averaged to obtain values that are very close to the real value. But, in fact, the number of test measurements is always limited. The average value obtained with a finite measurement can only be approximated to the real value. The commonly used average values are as follows: arithmetic mean, geometric mean, root mean square average, and log average. It should be noted that the logarithmic mean of the variables is always less than the arithmetic mean. The purpose of taking various averages is to find the one closest to the real value from a set of measured values. In wind tunnel tests, the distribution of data is mostly normal, so the arithmetic mean is usually used.

(2) Classification of errors

According to the nature of the error and the reasons for its occurrence, it is generally divided into three categories:

(1) Systematic error: Systematic error refers to the error caused by factors that are not detected or confirmed in the measurement and test, and

these factors affect the result always shifting in one direction, and the size and symbol are exactly the same in the same set of test measurements. When the test conditions are determined, the systematic error obtains an objective constant value. When the test conditions are changed, the variation of the systematic error can be found. The cause of the system error: the measuring instrument is not good, such as the scale is not accurate, the instrument zero point is not corrected or the standard table itself has deviation; Changes in the surrounding environment, such as temperature, pressure, humidity, etc., deviate from the calibration value; Test personnel's habits and biases, such as errors caused by high or low readings. For the shortcomings of the instrument, the influence of changes in external conditions, and the individual's bias, the system error can be basically cleared after being corrected separately.

(2) Random error: In the observation of all the measured values that have eliminated the system error, the measured data are still different in the last digit or the last two digits, and their absolute value and symbol change, sometimes large or small, sometimes positive or negative, with no definite law. Such errors are called random errors or accidental errors. The cause of the random error is unknown and cannot be controlled and compensated. However, if a certain value is measured with enough precision many times, it will be found that the random error completely obeys the statistical law, and the size of the error or the probability of positive and negative occurrence can be determined. Therefore, as the number of measurements increases, the arithmetic mean of the random errors approaches zero, so the arithmetic mean of multiple measurements will be closer to the real value.

(3) Negligence error: Negligence error is an error that is obviously inconsistent with the facts. It is often caused by carelessness, excessive fatigue, and incorrect operation of the tester. Such errors can be found without rules. As long as the sense of responsibility, vigilance, and careful operation are strengthened, the negligence error can be avoided.

2. *Precision, Correctness, and Accuracy*

The amount that reflects how close the measurement is to the true value is called accuracy. It corresponds to the error size, and the higher the accuracy of the measurement, the smaller the measurement error. "Accuracy" should include two meanings of precision and correctness.

(1) Precision: The degree of reproducibility of the measured values in the measurement is called precision. It reflects the degree of influence of random errors, and high precision means that the random error is small.
(2) Correctness: The degree of deviation between the measured value and the real value, called correctness. It reflects the influence accuracy of the system error, and high correctness means that the system error is small.
(3) Accuracy: It reflects the degree of influence of all systematic errors and random errors in the measurement.

In a set of measurements, the correctness of high precision is not necessarily high; the precision with high correctness is not necessarily high; but the accuracy is high, and the precision and correctness are high.

In order to illustrate the difference between precision and correctness, the following target shooting example can be used to illustrate. As shown in Fig. 5.32a, the precision and correctness are good, so the accuracy is high; Fig. 5.32b shows that the precision is good, but the correctness is not high; Fig. 5.32c shows that the precision and correctness are not good. In the actual measurement, there is no clear real value like the bull's-eye, but it is trying to determine the unknown real value.

3. *Error Expression Method*

When measuring with any measuring tool or instrument, there is always an error, and the measurement result cannot be exactly equal to the real value of the object measured, but only its approximation. The quality of measurement is based on the measurement accuracy, and the measurement accuracy is estimated according to the measurement error. The smaller the error of the measurement result is, the more accurate the measurement is considered.

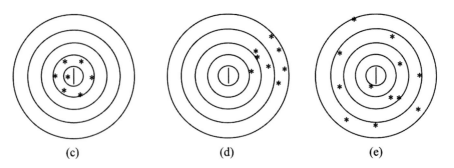

(c)   (d)   (e)

Fig. 5.32 The relationship between precision and accuracy

The expression methods of error are: (1) absolute error represents the difference between the measured value and the real value, which is usually called error; (2) relative error is to measure the accuracy of a certain measured value, which is generally expressed by relative error. That is to say, the percentage of the ratio of absolute error to the actual value to be measured is the relative error; (3) the reference error is the ratio of the absolute error of the instrument to the measuring range; (4) the arithmetic mean error is the mean value of the error of each measuring point; (5) the standard error is the root mean square error. The standard error is not a specific error, and its size indicates the dispersion degree of each observation value of equal precision measurement set to its arithmetic mean value under certain conditions.

4. *Accuracy of Measuring Instrument*

The accuracy level of the measuring instrument is indicated by the maximum reference error (also known as the allowable error), which is equal to the percentage of the ratio of the maximum absolute error of the instrument to the range of the instrument. Generally, standard instruments are used to calibrate lower level instruments. Therefore, the maximum absolute error is the maximum absolute error between the calibrated instrument and the standard instrument. The accuracy grade of the measuring instrument is uniformly stipulated by the state. Remove the percentage sign from the allowable error, and the remaining figures are called the accuracy grade of the instrument. The accuracy level of the instrument is usually indicated on the panel of the instrument by the number in the circle. For example, the allowable error of a certain pressure gauge is 1.5%, and the accuracy grade of the electrical instrument of this pressure gauge is 1.5, which is usually referred to as level 1.5 instrument for short.

5. *Error properties*

(1) Normal distribution of errors

If the system error and the negligence error are not included in the measurement series, the magnitude of the random error is found in a large number of tests as follows: (1) The error with a small absolute value has more chances than the error with a larger absolute value, that is, the probability of error is related to the size of the error. This is the unimodality of the error; (2)

The number of positive or negative errors with equal absolute values is equivalent, that is, the probability of error is the same, which is the symmetry of the error; (3) the probability of occurrence of a large positive or negative error is very small, that is, a large error generally does not appear, this is the boundedness of the error; (4) As the number of measurements increases, the arithmetic mean of the random error approaches zero. This is called the low compensation of the error. According to the above error characteristics, the probability distribution of random errors appears to satisfy the Gaussian positive distribution function (as shown in Fig. 5.33). This Gaussian error distribution function is called the error equation. Among them, the smaller the standard error σ is, the higher the measurement accuracy is, and the higher the peak of the distribution curve is; the narrower the σ is, the wider the distribution curve is. It can be seen that the smaller the σ is, the larger the proportion of the small error is, and the higher the measurement accuracy is. On the contrary, the larger the proportion of the large error is, the lower the measurement accuracy is.

(2) Optimal value of measurement set

In the case of the same measurement accuracy, a series of observation values are composed of measurement sets. When using different methods to calculate the average value, the error values obtained are different, and the probability of error occurrence is also different. If an appropriate calculation method is selected, the error is minimized and the probability is the largest, and the average value calculated therefrom is the optimum value. According

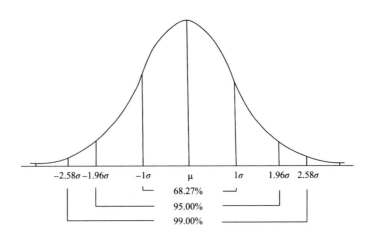

Fig. 5.33 Error normal distribution curve

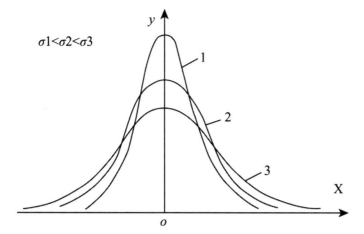

Fig. 5.34 Error distribution curve with different σ

to the Gaussian distribution law, the maximum probability can be achieved only when the sum of square errors of each point is the smallest (as shown in Fig. 5.34). This is the least squares value. It can be seen that for a group of observations with the same accuracy, the value obtained by arithmetic average is the best value of the group of observations.

(3) Analysis of indirect measurement error

The above discussion is mainly the error analysis of direct measurement, but in many cases, indirect measurement variables are often involved. The so-called indirect measurement is the quantity given by a certain functional relationship by the directly measured quantity, which is called the physical quantity of indirect measurement. If the velocity is determined by the measured displacement of the particle, the velocity is the physical quantity measured indirectly. Therefore, the indirect measurement is a function of each measurement obtained by direct measurement. The measurement error is a function of the error of each measurement value. Indirect measurement error is also called function error. The general expression is that the indirect measurement is a multivariate function of direct measurement, i.e.: $y = f(x_1, x_2, x_3, \ldots, x_n)$, where Y is the indirect measurement and $(x_1, x_2, x_3, \ldots, x_n)$ is the direct measurement. Expanded by the Taylor series, the maximum absolute error value of y is obtained:

$$\Delta y = \left| \frac{\partial f}{\partial x_1} \Delta x_1 + \frac{\partial f}{\partial x_2} \Delta x_2 + \cdots + \frac{\partial f}{\partial x_n} \Delta x_n \right|$$

where $\frac{\partial f}{\partial x_i}$ is the error transfer coefficient and Δx_i is the error of the direct measured value. If the direct error satisfies the normal distribution, the standard deviation of the indirect error is:

$$\sigma_y = \sqrt{\left(\frac{\partial f}{\partial x_1}\right)^2 \sigma_{x_1}^2 + \left(\frac{\partial f}{\partial x_2}\right)^2 \sigma_{x_2}^2 + \left(\frac{\partial f}{\partial x_3}\right)^2 \sigma_{x_3}^2 + \cdots + \left(\frac{\partial f}{\partial x_n}\right)^2 \sigma_{x_n}^2}$$

6

Wind and Water Tunnel Equipment

6.1 Development of Wind Tunnel Equipment

The wind tunnel facility is the most widely used pipe-type device for testing aerodynamics and the development of aircraft. Through a tubular device, airflow can be artificially generated and controlled to simulate the flow around an aircraft or an object, and the force of airflow on the object can be measured. Almost no aircrafts and missiles in the development process are without wind tunnel test. Moreover, with the development of aviation and aerospace technology, the requirements for wind tunnel test are increasingly high. With the continuous improvement of computer and numerical simulation technology, the measurement technology and precision of wind tunnel are also greatly improved. At the same time, the phenomenon of wind tunnel test simulation is becoming more and more complicated. At the beginning of the twentieth century in America, when the Wright brothers were developing the first manned aircraft with power in the world, the time of wind tunnel test was only more than 20 h. In the 1970s, the wind tunnel test time to develop an advanced aircraft reached tens of thousands of hours. In addition, wind tunnel tests are widely used in many sectors of the national economy, such as studying the appearance of vehicles with least resistance, wind load of tall buildings, wind-induced vibration of wind turbines and Bridges, and environmental air pollution.

The world's first known wind tunnel was built in 1869–1871 by the Englishman F. Wenham, who measured the drag force of objects moving relative to air. It is a wooden box with openings at both ends, with a cross section of 45.7 cm × 45.7 cm and a length of 3.05 m. Before they success-fully made the world's first powered flight, the American Wright brothers

© Science Press 2021
P. Liu, *A General Theory of Fluid Mechanics*,
https://doi.org/10.1007/978-981-33-6660-2_6

Fig. 6.1 Wind tunnel constructed by the Wright brothers (The test section size is 40.6 cm × 40.6 cm, 1.8 m long, and the free-stream speed is 12 m/s)

built a wind tunnel (as shown in Fig. 6.1) in 1900, with a cross section of 40.6 cm × 40.6 cm, a length of 1.8 m, and a velocity of 40–56.3 km/h. In 1901, the Wright brothers built a wind tunnel with a wind speed of 12 m/s to test their plane. The development of wind tunnels was first at low speed. In 1907, Prandtl built a wind tunnel with a tunnel pipe of equal section, which consumed a lot of energy. In 1917, Prandtl changed the uniform section tunnel body into a variable section type, similar to the current single-loop wind tunnel. In 1914, the French engineer Eiffel (1832–1923) built an open wind tunnel without loops. After World War I, due to the use of metal structure materials, monoplane appeared. The power of the engine increased greatly and the speed of the aircraft improved greatly. By that time, in the case of propeller efficiency and drag, a series of basic aerodynamic problems were proposed. The construction of conventional low-speed wind tunnel in developed countries such as European countries and America mainly focused on the 1920s to the 1950s, and mainly resolved the problems of aerodynamic power and efficiency of aircraft and propellers, including the aerodynamic layout of aircraft. On May 27, 1931, the world's first full-size wind tunnel in Hampton, Virginia, near Langley research center came into use (as shown in Fig. 6.2). The test section is 18.28 m (60 feet) wide and 9.144 m (30 ft) high. It was used in various aerodynamic tests such as fighters in World War II, the capsules, the submarines, and modern jets. In 1980, Ames research center in NASA converted an old low-speed wind tunnel, forming the largest full-size low-speed wind tunnel in the United States with an experimental section size of 24.4 m (width) × 12.2 m (height). A new test section of 36.6 m ×

Fig. 6.2 The first full-size wind tunnel in the world built in America

24.4 m was added to the wind tunnel, and the fan motor power was increased from 25 to 100 megawatts. This kind of large wind tunnel can provide flow field conditions for real aircraft and full-size scaled model to study aerodynamic forces of aircraft components. The royal aeronautical research institute (RAE)'s 3.5 m × 2.6 m low-speed wind tunnel built in 1944 basically met the needs of the model development at that time.

In 1932, Swiss scientist G. Ackeret built a continuous supersonic wind tunnel with $Ma = 2$ in the test section to solve the aerodynamic problems of shells and the general law of supersonic flow. To solve the problem of insufficient driving power, the wind tunnel was operated below atmospheric pressure. In the 1950s, due to the emergence of high-thrust jet engines, the development of aircraft passed the "sound barrier" and entered the developing period of low supersonic. During this period, countries with developed aviation built a number of supersonic wind tunnels to meet the needs of aircraft type development. For example, NASA Ames research center built the world's largest supersonic wind tunnel in 1956, with a section size of 4.88 m × 4.88 m and Mach number $Ma = 0.8-4.0$. The T-106 wind tunnel built in Russia in 1952 is a sub- and transonic wind tunnel with a diameter of 2.48 m. The T-109 wind tunnel, built in 1953, is an impulse sub-, trans-, and supersonic wind tunnel with a section size of 2.25 m × 2.25 m. In 1957, a supersonic wind tunnel with a cross-sectional diameter of 2.5 m was built by RAE in the UK. In 1961, the S2-MA wind tunnel in the French aerospace lab (ONERA) was a sub- and transonic wind tunnel with a cross-sectional size of 1.94 m × 1.75 m.

Since the 1960s, with the emergence of third-generation fighter jets and large passenger aircrafts, the flow around aircraft has become increasingly complex and the Reynolds number (Re) effect has become a major problem. The development of high-Reynolds number wind tunnels, including low-temperature wind tunnels and pressurized wind tunnels, was facilitated. Typical examples of the former were the US national transonic equipment (NTF) and the European transonic wind tunnel (ETW). The examples of the latter were the F1 wind tunnel of ONERA in France and the 5-m supercharged wind tunnel of RAE in the UK.

Since the 1970s, as the problem of revision in Reynolds number and aerodynamic noise came out, the typical production wind tunnel was built and put into use successively, such as the LLF wind tunnel (aeroacoustic wind tunnel) at German-Dutch Wind Tunnels (DNW) built in 1979, the American NTF transonic wind tunnel (low-temperature transonic wind tunnel) built in 1982, and the European ETW wind tunnel (low-temperature transonic wind tunnel) built in 1993.

China has more than 50 low-speed wind tunnels and more than 30 trans- and supersonic wind tunnels. All the low-speed wind tunnels and trans-supersonic wind tunnels used for aircraft development were built after the founding of the People's Republic of China in 1949. From the 1950s to the 1960s, the construction of 3-m low-speed wind tunnels and 0.6-m trans- and supersonic wind tunnels came to a climax. The low-speed wind tunnels built during this period were as follows: the 3.5 m × 2.5 m low-speed wind tunnel (FL-8) at the AVIC Aerodynamic Research Institute and 3 m × 3 m low-speed wind tunnel (FD-09) at the China Academy of Aerospace Aerodynamics (CAAA). The transonic supersonic wind tunnels were as follows: 0.6 m × 0.6 m transonic supersonic wind tunnel (FL-1) constructed by CAAA; 0.76 m × 0.53 m sub- and transonic wind tunnel (FD-08) and 0.6 m × 0.6 m transonic supersonic wind tunnel (FD-06) at CAAA; 0.64 m × 0.52 m sub-and transonic wind tunnel (FL-7) at AVIC Aerodynamic Research Institute; 0.6 m × 0.6 m transonic supersonic wind tunnel (NH-1) at Nanjing University of Aeronautics and Astronautics. Since the end of 1960s, with the establishment and development of the China Aerodynamics Research and Development Center (CARDC), the wind tunnel construction in China reached the second climax. By the early 1980s, the production low-speed wind tunnels CARDC built were as follows: FL-12 (4 m × 3 m) and FL-13 (8 m × 6 m, 12 m × 16 m). The supersonic wind tunnels were FL-21 (0.6 m × 0.6 m), FL-23 (0.6 m × 0.6 m), and FL-24 (1.2 m × 1.2 m). In addition, there were 2.5-m diameter low-speed wind tunnel (DFD-03) built in AVIC Aerospace Life-support Industries, LTD.; the 0.6-m diameter transonic supersonic wind tunnel (CG-01) in Xi'an Institute of Modern Control Technology; and the low-speed wind tunnel with tandem test section NH-2 (3.0 m × 2.5 m, 5.1 m × 4.25 m) in Nanjing University of Aeronautics and Astronautics.

With the development of aviation and space flight vehicle in China, the wind tunnel test demands stronger simulating ability (Re, Ma, etc.) and a wider range of experimental ability (two-dimensional airfoil test, propeller performance test, etc.). Therefore, the low-speed wind tunnel NF-3 built in Northwestern Polytechnical University in the 1990s has three test sections including the 2D (3 m × 1.6 m), 3D (3.5 m × 3.5 m), and propeller (2.2 m × 2.2 m). The FL-2 transonic supersonic wind tunnel built by CAAA is with the test section size of 1.2 m × 1.2 m, which can be pressurized to 0.8 mpa. The designed maximum testing Reynolds number is 1.6×10^7. The FL-26 transonic wind tunnel completed by CARDC in 1999 is with the test section size of 2.4 m × 2.4 m, which can be pressurized to 0.45 mpa, and the maximum testing Re is 1.2×10^7.

At the beginning of the twenty-first century, AVIC Aerodynamics Research Institute built a low-speed supercharged wind tunnel (FL-9) with a section size of 4.5 m × 3.5 m, a trans- and supersonic wind tunnel with a section size of 1.6 m × 1.5 m (FL-3), and a sub-hypersonic wind tunnel with a section size of 1.2 m × 1.2 m (FL-60). FL-9 low-speed supercharged wind tunnel can be supercharged to 0.4 mpa, the highest testing Reynold number can be reached to 8.6×10^6, with belly brace, tail brace, half mode, and other supporting forms, which can meet the requirements of aircraft development for low-speed and high-Re testing conditions. FL-3 trans- and supersonic wind tunnel is a dual-purpose three-sound-speed wind tunnel with a test section $Ma = 0.3-1.35$ (2.25), which can be used for the measurement of force, pressure, an inlet with a large angle of attack, and other test technologies. The CAAA has built a trans- and supersonic wind tunnel (FD-12) with a section size of 1.2 m × 1.2 m, which can be pressurized to 0.8 mpa. The designed maximum testing Re is 1.6×10^7. The CARDC has built a 2 m × 2 m supersonic wind tunnel, which is a blow-down, impulse, and inject-type supersonic wind tunnel with a fully flexible nozzle. The range of the Mach number is 1.5−4.25, and the test Reynolds number range is $7.72 \times 10^6 \text{\textasciitilde} 74.16 \times 10^6$ (characteristic length is 1 m). The CARDC and Xi'an Institute of Modern Control Technology have built vertical wind tunnels with diameters of 5 m and 4.5 m, respectively, which are capable of studying and testing aircraft at stall/tail rotation and characteristics of vertical movement of various aircraft. NF-6 wind tunnel built by Northwestern Polytechnical University is the first and only continuous supercharged high-speed wind tunnel that has been already in operation in China. AVIC helicopter design and research institute has built a helicopter wind tunnel with a section size of 8 m × 6 m, improving the testing and researching level of helicopter wind tunnel further.

In addition, with the completion and put into use of 8 m × 6 m low-speed wind tunnel and 4-m dynamic wind tunnel, and the construction of large modern wind tunnels such as 5.5 m × 4 m aeroacoustic wind tunnel, 3 m × 2 m icing wind tunnel, and 2.4 m × 2.4 m continuous transonic wind tunnel, the ability of wind tunnel test will enter the ranks of the world's advanced level and will meet the demands of China's aviation and spacecraft development for wind tunnel test equipment.

6.2 Wind Tunnel Type

As a special kind of pipe, wind tunnel needs the power device of the fan system to produce adjustable airflow, so that the flow field around the model in the test section can simulate or partially simulate the original flow field. Since F. H. Wenham (English) built the world's first low-speed blow-down wind tunnel in 1871, the American Wright brothers built another low-speed blow-down wind tunnel in 1901 and tested their aircraft. It took 149 years for the world to build wind tunnels. In order to meet the requirements of various aerodynamic tests, a large number of different types of wind tunnel equipment were built in the middle of the twentieth century. Wind tunnels are classified according to the Mach number (or speed) of the airflow in the test section, including low-speed wind tunnels, subsonic wind tunnels, transonic wind tunnels, supersonic wind tunnels, and hypersonic wind tunnels.

Wind tunnels are widely used to study the basic laws of aerodynamics, to verify and develop relevant theories, and directly serve for the development of various aircraft. Through wind tunnel experiments, the aerodynamic layout of aircraft can be determined and its aerodynamic performance can be evaluated. Modern aircraft designs depend heavily on wind tunnels. For example, in the 1950s, about 10,000 h of wind tunnel tests were conducted on the U.S. b-52 bomber, and about 100,000 h were conducted on the first space shuttle in the 1980s.

The wind tunnel tests are necessary for the design of a new aircraft. The wind tunnel requires flow with various velocities, densities, and even temperatures to simulate the real flight state of various aircraft. The flow velocity in the wind tunnel is characterized by the Mach number (Ma) in the test section. The wind tunnel can be classified according to the Ma range of the test section. Wind tunnel with the test section $Ma \leq 0.3$ is a low-speed wind tunnel (as shown in Figs. 6.3, 6.4, 6.5, 6.6, 6.7, 6.8, 6.9, 6.10, 6.11 and 6.12 for various low-speed dc and reflux wind tunnels). In this wind tunnel, the air density in the airflow is almost constant and is treated as a constant. The wind tunnel is called a subsonic wind tunnel when $0.3 < Ma \leq 0.8$. The wind tunnel is called a transonic wind tunnel when $0.8 < Ma < 1.2$. The wind tunnel is called a super-sonic wind tunnel when $1.2 < Ma \leq 5.0$. The wind tunnel with the test section $Ma \geq 5$ is called a hypersonic wind tunnel. In addition, wind tunnel types are also divided by application, structural type, and type of test section.

Due to the controllability and repeatability of the airflow in the wind tunnel test section, wind tunnels are also widely used to test automobile aerodynamics and wind engineering such as the wind load and vibration of structures, building ventilation, air pollution, wind power generation, environmental wind field, flow conditions in complex terrain, and the effectiveness of wind protection facilities. All of these problems can be solved by using the principle of geometric similarity, placing the terrain and ground

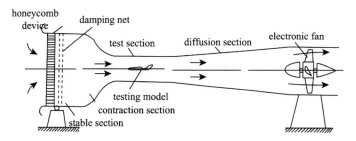

Fig. 6.3 Typical low-speed blow-down wind tunnel

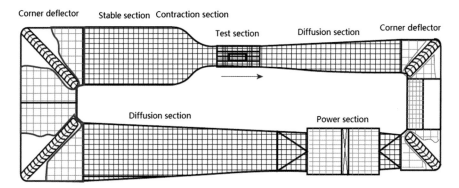

Fig. 6.4 Low-speed loop wind tunnel

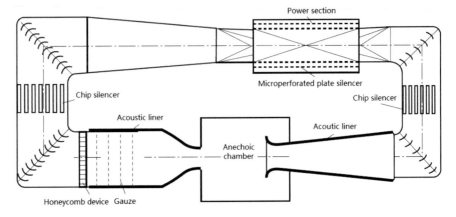

Fig. 6.5 Aeroacoustic wind tunnel

Fig. 6.6 Blow-down low-speed wind tunnel with a diameter of 0.13 m (Beihang University)

objects in the wind tunnel with the scaled model, and then measuring the size and distribution of the wind force on the model with the instrument.

The total number of wind tunnels in the world has reached more than 1,000. The largest low-speed wind tunnel is the national full-size facility (NFSF) at NASA's Ames Center. The large transonic wind tunnel with the highest Reynolds number is the national transonic facility (NTF) of Langley Center in the United States. It is a low-temperature wind tunnel with a section size of 2.5 m × 2.5 m in the test section. Liquid nitrogen injection technology is used to decrease the gas temperature in the test section so that the Reynolds number in the test section reaches or approaches the true value of the aircraft. The development trend of the wind tunnel is to increase the simulation capability of wind tunnel further and improve the flow field

Fig. 6.7 Low-speed blow-down wind tunnel with a diameter of 1 m (Beihang University)

Fig. 6.8 Low-speed blow-down wind tunnel with a diameter of 0.5 m (Beihang University)

quality, eliminate the tunnel wall interference under transonic velocity, and develop a self-correcting wind tunnel.

The main advantages of wind tunnel test equipment are as follows: (1) The test conditions (including airflow state and model state) are easy to control; (2) Flow parameters (wind direction and wind speed) can change

Fig. 6.9 Low-speed blow-down wind tunnel with 2 m × 2 m (China Institute of Water Resources and Hydropower)

Fig. 6.10 Low-speed blow-down wind tunnel with 1 m × 1 m (China Institute of Water Resources and Hydropower)

independently; (3) The model is static and the measurement is convenient and accurate; (4) Test is generally not affected by changes in the atmosphere. The disadvantage is that it is difficult to meet all similarity criteria, and there is cave wall and model support interference, which can be overcome by data correction.

Fig. 6.11 Blow-down low-noise wind tunnel driven by a centrifugal fan (Beihang University)

Fig. 6.12 Low-speed and low-noise loop wind tunnel with 0.2 m × 0.2 m (Beihang University)

6.3   Low-Speed Wind Tunnel

A wind tunnel in which the compressibility of air is negligible is called low-speed wind tunnel. In terms of the size of the test section (referring to the equivalent diameter of the cross-sectional area of the test section), there are micro low-speed wind tunnels of tens of millimeters, small low-speed wind tunnels of 1−1.5 m mainly used for teaching and researching, medium low-speed wind tunnels of 2−4 m (most aviation wind tunnels of 3 m), and large low-speed wind tunnels of over 8 m. Large wind tunnels in which real aircraft or full-size models can be put into a test section are called full-size wind tunnels. The power of the fan motors that drive the airflow in a low-speed wind tunnel is only a few hundred watts. In addition to low-speed aerodynamic tests of aircraft, low-speed wind tunnels can also be used for low-speed aerodynamic tests of recoverable satellites, spacecraft, and the final stages of the shuttle re-entry process. In the non-aerospace field, low-speed wind tunnels are also needed for the study of vehicle air resistance, wind load and wind vibration of buildings and structures, wind energy development and research, as well as the study of air pollution, cold and heat, rain and snow, light, dust storms, and storms.

1. **Low-speed wind tunnel**

   There are two basic forms of low-speed wind tunnel: blow-down and loop. According to the different structures of the test section, it can be divided into open-jet and closed-section. The boundary of the closed-section wind tunnel is the solid wall and the boundary of the open-jet is the free boundary. Most low-speed wind tunnels in the world are loop wind tunnels. The difference between the loop and blow-down wind tunnels is the extra reflux channel. The main function of the reflux channel is to make the airflow in the wind tunnel free from the interference of the outside atmosphere, so that the temperature can be controlled and noise pollution can be reduced. The blow-down wind tunnel is vulnerable to the interference of the outside atmosphere (referring to the large direct flow wind tunnel with the inlet and outlet outside, which will be affected by a gust, rain, snow, and foreign matters if no measures are taken), and the noise will pollute the environment. The pressure in the test section is lower than the pressure in the atmosphere outside the tunnel. These are the downsides of straight wind tunnels. The advantages of a blow-down wind tunnel are that the large transverse flow changes generated during the vertical short-takeoff and landing aircraft test will not lead to backflow, and the engine can run in clean air without the problems of air temperature rise and cooling in the wind tunnel. Blow-down wind tunnels are mostly closed-section. If the blow-down low-speed wind tunnel is open-jet, an airtight chamber must be placed outside the open test section. Otherwise, since the fan is located behind the test section, air will flow directly into the test section, resulting in transverse flow in the test section.

The loop wind tunnel is divided into a circular loop and an ordinary single loop. A loop wind tunnel is usually a pressure wind tunnel. This wind tunnel requires the total pressure of the gas in the wind tunnel to be 0.125−25 times the atmospheric pressure before the wind tunnel works. The Reynolds number can be increased when the pressure is high, and the Mach number can be increased when the pressure is low. The cave of this kind of wind tunnel can withstand the pressure difference between inside and outside. Common single-loop wind tunnels, including open and closed test sections, are familiar. Most low-speed wind tunnels in the world are horizontal, that is, the center line of the wind tunnel is on the horizontal surface. If the center line of the test section of the wind tunnel is in the vertical direction, the wind tunnel is vertical. Aircraft tail-spin test, manned spacecraft re-entry module-parachute system dynamic stability test, etc. need to be carried out in vertical wind tunnel.

In the field of aeronautics and astronautics, low-speed wind tunnels are classified according to their purposes, mainly as follows:

(1) Two-dimensional wind tunnel (binary wind tunnel). The cross section of the test section of this wind tunnel is rectangular, and the ratio of the length on both sides is 2.5−4. The two-dimensional wind tunnel is mainly used to study the aerodynamic characteristics of airfoils.

(2) Three-dimensional wind tunnel (ternary wind tunnel), namely the general wind tunnel. In this kind of wind tunnel, a full-machine model or semi-model test can be carried out, mainly used to measure the aerodynamic force, pressure distribution, and velocity field acting on the model, which is the most widely applied wind tunnel.

(3) Low-turbulence wind tunnel. The turbulence of the airflow in the test section of the wind tunnel is very low and can be adjusted. Generally, the turbulence in the cleared atmosphere is about 0.01−0.03%. The turbulence in a normal medium-large low-speed reflux wind tunnel can reach 0.1−0.5% or more, while the turbulence in a low-turbulence wind tunnel should be less than 0.05%. Low-turbulence wind tunnels are mainly used to study the flow phenomena that are greatly affected by turbulence, such as boundary layer transition, separation flow, vortex rupture, and other experimental studies. Such wind tunnels can be two-dimensional or three-dimensional, characterized

by a large contraction ratio and a multi-layer damping net installed in the stable section.

(4) Variable density wind tunnel, in which the density of airflow can be artificially changed to obtain different experimental Reynolds numbers. One way to change the density of the air is to use the gas denser than the air, such as nitrogen, as a wind tunnel medium. Another method is to change the total pressure of the airflow, that is, to increase the pressure of the wind tunnel.

(5) Tailspin wind tunnel is a special wind tunnel that studies the development and modification of aircraft tailspin by free flight. Most of these wind tunnels are vertical wind tunnels. The test section is set vertically. The airflow direction in the test section is from down to up. The airflow velocity at the center of the test section is about 5−10% lower than that at the edge, that is, the velocity distribution on the cross section of the test section is in a dish shape, so that the tailspin model is kept near the center of the test section.

(6) Rafale wind tunnel, also known as a gust wind tunnel, is a kind of special wind tunnel that produces artificial airflow simulating gust and studies the ability of aircraft to adapt to natural gust in flight through the model test. During the experiment, the model launcher throws the model roughly horizontally and makes it pass through the vertical airflow. The motion trajectory of the model is photographed with a high-speed camera. The cross-sectional area of the vertical flow is large and the velocity is not large (about less than 15 m/s). A gust wind tunnel is relatively rare, the general wind tunnel after modification can also carry out gust tests.

(7) Free-flying wind tunnel is a special wind tunnel that allows the model to fly freely in the airflow of the test section. The low-speed free-flying wind tunnel is usually blow-down, which is characterized by the fact that the axis of the wind tunnel and the velocity of the airflow can be adjusted rapidly to simulate various dynamic flight states. Models can be powered or not. The deflection angle of the control surface of the model is controlled by remote control outside the tunnel. High-speed camera can be used to record the model during the experiment.

(8) Icing wind tunnel is a special wind tunnel that studies the phenomenon of icing on the surface of aircraft body and ground structures in flight and the methods to prevent or eliminate it. It is characterized by a cooling device in front of the stability section and a sprayer in the stability section, so that the icing conditions of the prototype flow field can be simulated in the test section.

(9) Vertical short-takeoff and landing wind tunnel is a wind tunnel to study the aerodynamic performance of vertical short-takeoff and landing aircraft. Vertical short-takeoff and landing aircraft (especially aircraft with full-wingspan high-efficiency high-lift device) has a strong downdraft, and the interaction between the downdraft and the boundary layer of the wind tunnel wall will cause the model flow to be different from the real situation, so the data obtained from such experiments in the general low-speed wind tunnel are not accurate. The test section of the vertical short-takeoff and landing wind tunnel is larger than the normal wind tunnel to reduce the interference between the jet flow and the wall of the wind tunnel. This type of wind tunnel does not require high speed, with a maximum speed of about 40 m/s. In the wind tunnel, the influence of the ground is simulated, and the belt type moving floor with the same speed as the wind speed is used to eliminate the influence of the floor boundary layer.

2. **Main components of the low-speed wind tunnel (as shown in Fig. 6.13)**

   Although there are many types of low-speed wind tunnels, the components and how they work are basically the same. Here, the names, functions, and basic principles of the main components of the low-speed wind tunnel are introduced.

   (1) The test section is the place where the prototype flow field is simulated in the wind tunnel for the model aerodynamic test, and it is an important part of the wind tunnel. In order to simulate the prototype flow field, it is necessary to meet the requirement that the test

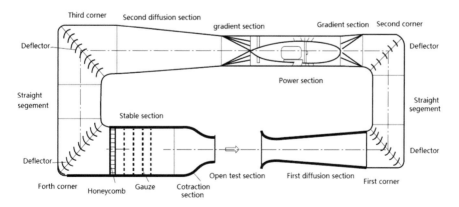

Fig. 6.13 Components of a low-speed wind tunnel

Reynolds number, the size of the test section, and the size of the flow velocity reach a certain value. In addition, the airflow in the test section should be stable, the velocity and direction should be evenly distributed in space, and the turbulence, noise intensity, and static pressure gradient of the incoming flow should be low. These characteristics of the flow in the test section are called flow field quality. The flow field quality index has been stipulated in China. As seen in the military standard, in the model area of the test section, the flow direction $|\Delta\alpha|$ and $|\Delta\beta|$ shall not be greater than 0.5 and the absolute value of the average flow deviation angle shall not be greater than 0.1. Along the axial static pressure gradient in the center line of the test section, the length of the model area L shall be within the range of $L \cdot \left|\frac{dc_p}{dx}\right| \leq 0.005 \leq 0.1\%$ (Cp is the static pressure strength coefficient) and the turbulence degree in the center of the model area shall not be greater than 0.1%. The cross-sectional shapes of the low-speed wind tunnel test section are rectangular, square, round, oval, and octagonal. Most of the existing wind tunnels are rectangular with angle cutting. The length of the closed section is generally 1.5–2.5 times of the equivalent diameter of the cross-sectional area, and the length of the open section is generally 1–1.5 times. Generally, the wind tunnel has a diffusion angle along the axial direction (along the direction of incoming flow), or the area of each section is gradually reduced along the axial cutting angle, so that the cross-sectional area is gradually expanded along the axial direction. The absolute value of the negative static pressure gradient caused by axial thickening of the wall boundary layer is then reduced, so as to meet the requirements of flow field quality.

(2) Diffuser section includes the first and second diffusion sections. In a low-speed wind tunnel, the diffuser section is a section of the pipe expanding along the direction of the airflow, so it is also called the diffusion section. Its function is to slow down the airflow in the test section and change the kinetic energy into pressure energy, so as to reduce the energy loss of the airflow in the wind tunnel and reduce the power required for the operation of the wind tunnel. The total power lost by the airflow in the pipe is generally proportional to the cubic velocity. The diffuser decelerates the flow from the test section and reduces the power loss of the whole wind tunnel. However, the diffuser can also cause the energy loss of the flow, including friction loss and diffuser loss. The so-called diffusion loss refers to the loss caused by the airflow avoiding the thickening of the boundary layer

under the action of an adverse pressure gradient. If the expansion angle of the diffuser is larger, the friction loss can be reduced and the diffuser loss can be increased. On the contrary, when the expansion angle decreases, the friction loss will increase and the diffuser loss will decrease. A large number of experiments have proved that the optimal expansion angle of the three-dimensional circular section is 5–6°

(3) The corner and the deflector are to solve the separation problem caused by the airflow turning in the loop wind tunnel. A loop wind tunnel usually has four right-angle corners behind the airflow. The airflow from the test section passes through the first, second, third, and fourth corners at one time and bends 360° in total. When the air flows through the corner, it is easy to generate separation and vortex, forming the pulsation and non-uniform region in the flow. This is because, when the airflow enters the corner, the streamline bends and centrifugal inertial force occurs. The flow rate decreases and the pressure increases along the direction away from the center of curvature. The velocity increases and the pressure decreases along the direction of the center of curvature. So, if the flow rate on the inside of the corner increases and the pressure decreases, there is a contraction effect. If the flow rate on the outside decreases and the pressure increases, there is a diffuser effect. That leads to the pressure on the outside of the corner higher than the pressure on the inside. After the airflow flows through the corner, there is a diffuser effect inside the corner and a contraction effect outside. The diffusion results in the separation of the inner and outer walls of the airflow and the formation of many small-scale vortexes. This effect changes with the distance between the inside and outside. For this reason, the wind tunnel corners are equipped with deflectors with the function equivalent to dividing a large corner into several small corners. For each small corner, the distance between the inside and outside is significantly reduced, so the flow separation and vortex are significantly weakened. The function of the flow deflector is to reduce the generation of separation when the air flows through the corner and reduce the intensity of the secondary flow vortex, so as to reduce the energy loss of the airflow and to improve the flow field performance after the air flows through the corner. For example, the performance of the flow field behind the second corner is improved to reduce the wind tunnel energy loss. For example, the flow field performance behind the fourth corner is also improved to improve the flow field quality in the test section. The shape of the cross section of the deflector is a circular arc, arc plus straight line, and wing profile.

(4) The stable section is a pipe section of sufficient length with a constant cross section. It is characterized by a large enough cross-sectional area, low airflow speed, and rectifier device in the stable section. The function of the stability section is to stabilize the turbulent and non-uniform airflow from the upstream, so that the vortex attenuation, velocity size, and direction distribution are more uniform.

(5) The rectifier device refers to the honeycomb device and the rectifier net (damping net). The honeycomb device is composed of many square or hexagonal cells, such as honeycomb. The rectifier net is a small mesh wire diameter of the small metal net, can have a layer or several layers. The honeycomb device can guide the airflow and reduce the scale of the large eddy and the transverse turbulence of the airflow. The large-scale vortexes can be divided into smaller scale vortexes by the rectifier net, while the small-scale vortexes can be attenuated within a sufficient length of the stable section behind the rectifier network, so that the turbulence of the airflow, especially the axial turbulence, can be significantly reduced. In addition, the energy loss of the airflow through the rectifier device is related to the size of the incoming flow velocity. If the distribution of the incoming flow velocity is not uniform in the cross section, the airflow with high velocity loses more while the airflow with low velocity loses less. Therefore, the rectifier device can make the distribution of the airflow velocity become even. The turbulence, flow direction, and velocity distribution uniformity of the airflow can be improved obviously by the well-designed stable section, honeycomb, and rectifier net system.

(6) Contraction section: In a low-speed wind tunnel, the contraction section is located between the stable section and the test section, which is a smooth transition curved pipe, with the cross-sectional area gradually decreasing along the flow direction. If the contract period of the cross-sectional area of the entrance section is A1, cross-sectional area of exit is A_0, $\eta = A_1/A_0$ is shrinkage ratio. The function of the contraction section is to make the airflow from the stable section accelerate uniformly and improve the flow field quality (airflow uniformity, turbulivity, etc.) of the test section. The design of the contraction section should meet the following requirements: when the air flows through the contraction section, the flow rate increases monotonously to avoid the separation of the airflow on the

tunnel wall; the velocity distribution at the outlet of the contraction section should be uniform, the direction should be straight and stable. The length of the contraction section is appropriate. If the length is too long, the construction cost and the flow energy loss will be large. Whether the contraction segment can meet these requirements depends mainly on two aspects: contraction ratio and contraction curve. The general contraction ratio is between 6 and 12, and there are many design methods for the contraction curve, such as cubic curve, double cubic curve, and fifth curve (as shown in Figs. 6.14 and 6.15).

(7) Power section, which is the heart of a low-speed wind tunnel, generally adopts an axial flow fan. The main parts are as follows: power section shell (round section pipe); Blade; Drive motor; Fairing; Front guide piece or pre-twist piece; anti-twist vane or anti-twist vane. The motor can be mounted inside the fairing or outside the hole (driven by a long shaft fan). The function of the fan is to fuel the airflow in the wind tunnel to ensure that the airflow maintains a certain speed. Figures 6.14 to 6.16 show the single-stage axial flow fan system, and Figs. 6.17 and 6.18 show the two-stage axial flow fan system.

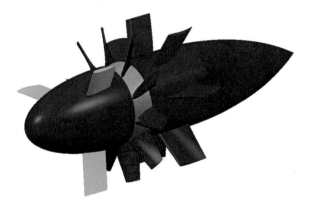

Fig. 6.14 Axial fan system for power segment (7 m in diameter)

Fig. 6.15 Axial fan system for power segment (12.35 m in diameter)

Fig. 6.16 Numerical simulation of the flow field of an axial flow fan with a diameter of 12.35 m

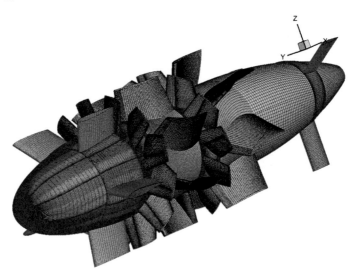

Fig. 6.17 Surface grid of 2-m diameter double-stage axial flow fan in the ice wind tunnel of 181 plant

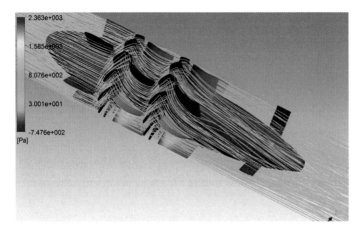

Fig. 6.18 Flow field of 2-m diameter double-stage axial flow fan in the ice wind tunnel of plant 181

6.4   Introduction to Typical Low-Speed Wind Tunnels

1. Aeroacoustic wind tunnel with low turbulence

Beihang University built a 1 m × 1 m low-speed loop wind tunnel (as shown in Fig. 6.19) in Shahe, Beijing in 2013. It is a low-turbulence aeroacoustic wind tunnel (referred to as "D5 aeroacoustic wind tunnel). Not only aircraft aerodynamic layout pre-study, steady and unsteady experiment, fluid mechanics and aerodynamics of fundamental research but also aerodynamic noise mechanism and air dynamic noise reduction

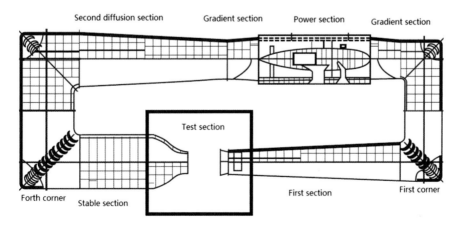

Fig. 6.19 Structure diagram of D5 general aeroacoustic wind tunnel

measures can be conducted in it. D5 wind tunnel has a good flow field quality and low background noise level, and its aerodynamic noise index meets the design requirements (as shown in Table 6.1). The D5 aeroacoustic wind tunnel is composed of a test section, diffusion section, corner section, power section, and stability section. The overall length, width, and height of the wind tunnel are 25.58 m, 9.2 m, and 3.0 m. The section of the test section is square, with a width of 1.0 m and a height of 1.0 m. The length of the closed test section is 2.5 m, the length of the open test section is 2.0 m, and the length of the air collector is 0.5 m. The test section is open and closed. There is a full anechoic chamber of 7 m × 6 m × 6 m surrounding the test section. When the wind tunnel runs in an open-jet form, it is a typical aeroacoustic wind tunnel. Both conventional aerodynamic and acoustic tests can be conducted in the D5 wind tunnel.

The center line of the wind tunnel test section is 2.0 m above the ground. The wind tunnel structure is made of pure steel, and the thickness of the tunnel wall is 6 mm. In order to improve the main frequency

Table 6.1 Main aerodynamic and acoustic design indexes of D5 wind tunnel

The serial number	Project
1	Test section dimensions: length 2.5 m, width 1 m, height 1 m, contraction ratio 9
2	Velocity of open test section: $V_0 = 1.5-80$ m/s Velocity of closed test section: $V_0 = 1.5-100$ m/s
3	Turbulence in the test section: $\varepsilon = \sqrt{\overline{u'^2}}\big/V_0 = 1 - 0.08\%$
4	Velocity instability of test section: <1%
5	Mean speed deviation of test section: $\Delta V < 1\%$
6	Deviation Angle of air flow at space point: $\Delta\alpha \leq \pm 0.5^0; \Delta\beta \leq \pm 0.5^0$ Mean flow deflection angle: $\Delta\alpha \leq \pm 0.1^0; \Delta\beta \leq \pm 0.1^0$
7	Axial static pressure gradient in the test section: $\left|\frac{dp}{dx}\right| < 10$ Pa/m
8	Microperforated plate silencer is used in the fan section of the wind tunnel
9	Noise index of test section: at the designed wind speed, the aerodynamic noise sound pressure level is less than 85 dBA at the wind tunnel diameter, one and a half times the distance from the wind tunnel center
10	Temperature control index: design wind speed $V_0 = 100$ m/s, $\Delta T_{max} < 5^0$ (running continuously for half an hour)
11	Power system: variable frequency ac motor, stepless speed control system
12	The measurement control system is controlled by computer (data acquisition, wind speed acquisition and control)

of the tunnel, longitudinal ribs and tunnel supports are arranged in many places. Rubber pads are added between the flanges at the joints of each section of the wind tunnel to eliminate and alleviate the vibration caused by gas flow and fan section. In order to adapt to the deformation caused by temperature, the two ends of the tunnel can be expanded freely. In order to reduce the vibration and noise, the foundation of fan section, test section, and cave section is poured separately. At the same time, vibration isolation material is used at the bottom of the fan section to reduce the sound and vibration transmission through the foundation. The motor is arranged in the fan fairing to avoid the long shaft connection passing through the corner deflector and affecting the airflow quality and uniformity in the fan room.

In addition to the requirements of flow field quality, the aeroacoustic wind tunnel also has acoustic field quality, including low background noise and non-reflection free field conditions, as well as sufficient space size to satisfy far-field acoustic measurements. In D5 wind tunnel, the background noise level of the wind tunnel is reduced by laying acoustic liner material on the tunnel wall, and no reflection condition and far-field sound measurement are achieved by building an anechoic chamber outside the test section. The background noise in the wind tunnel test section can usually be divided into two parts: one is the rotating noise and eddy noise generated by the axial flow fan in the wind tunnel during the operation process. The other is the regenerative noise generated by the airflow in the wind tunnel during the flow process as the secondary sound source. The background noise includes both broadband noise and discrete noise. Therefore, necessary noise reduction methods should be taken to acoustically treat the wind tunnel loop and achieve the goal of reducing background noise. The D5 wind tunnel adopts a noise reduction scheme with the combination of a single-channel resistive silencer and micro-perforated plate silencer to reduce the background noise (as shown in Fig. 6.20). A micro-perforated plate silencer is used in the fan power section to eliminate the low- and medium-frequency band noise generated by the fan, and a single-channel resistive silencer is used in the tunnel wall to eliminate the medium- and high-frequency band noise.

The background noise of the D5 wind tunnel is measured in the anechoic chamber during operation in the open test section. The measured position is located outside the airflow field, 1.5 times the wind tunnel diameter (1.5 m) from the central axis of the wind tunnel and 1 m from the downstream of the nozzle, flush with the central axis of the wind tunnel. In the test, the background noise value at the speed of 20−80 m/s

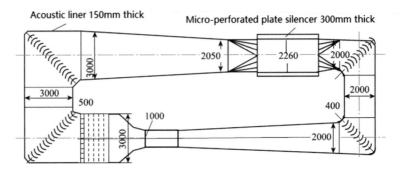

Fig. 6.20 D5 wind tunnel noise reduction processing

in the test section of the wind tunnel was measured and compared with the total sound pressure level of other acoustic wind tunnels abroad. Through analysis, it can be obtained that: in the case of no wind, the sound pressure level in the silencer room is 23.7 dB; at 80 m/s, the total sound pressure level in the muffling room is 87 dBA, and the sound pressure level corresponding to 8000 Hz is 67.7 dBA, which meet the design requirements (as shown in Fig. 6.21). At the same time, the acoustic characteristics of the D5 wind tunnel reach the background noise level of the acoustic wind tunnel of Hyundai motor in Korea, and the aeroacoustic test of the parts

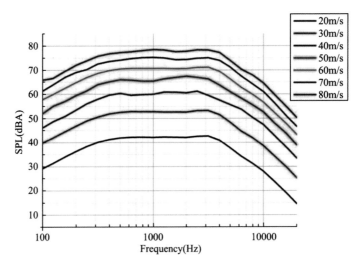

Fig. 6.21 Frequency spectrum of 1/3 octave of background noise at different wind speeds in D5 wind tunnel

can be carried out (as shown in Figs. 6.22, 6.23, 6.24, 6.25, 6.26, 6.27 and 6.28).

2. **4 × 3 m low-speed wind tunnel**

The FL-12 wind tunnel is a 4 m × 3 m single-loop closed low-speed wind tunnel located at the China center for aerodynamic research and development in Mianyang, Sichuan Province (as shown in Fig. 6.29). Wind tunnel design work began in 1965, and model tests began in 1971. After the completion of the wind tunnel, the measurement and control system

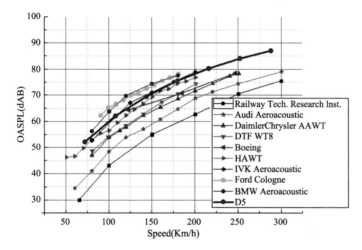

Fig. 6.22 Comparison of the background noise of different aeroacoustics wind tunnels

Fig. 6.23 D5 aeroacoustic wind tunnel

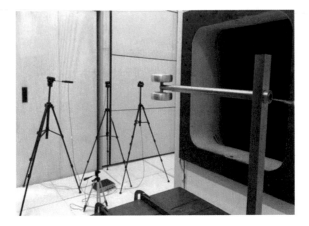

Fig. 6.24 D5 aeroacoustic wind tunnel landing gear noise test

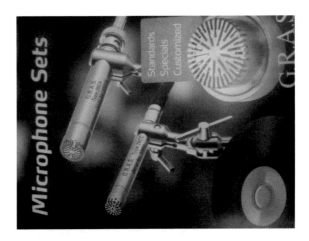

Fig. 6.25 D5 pneumatic microphone sensor for pneumatic noise measurement

has been reformed and updated many times, and the automatic computer control has been realized in the aspects of measurement, control, data acquisition, and processing. The measurement and control system consists of four parts: test management and database system, measurement system, pressure control system, and motion control system. In 2010, the wind tunnel power system was upgraded and replaced by the dc motor which had already been in service for a long time. The FL-12 wind tunnel is mainly composed of the test section, the first diffusion section, the first corner, the second corner, the fan section, the second diffusion section, the third corner, the fourth corner, the stability section, and the contraction section. The size of the test section is 4 m × 3 m × 8 m (width

PSI9816

Fig. 6.26 D5 aeroacoustic wind tunnel landing gear pressure test

× height × length), and the cross section is a chambered rectangle. The wind tunnel is equipped with a tower-type mechanical balance and a series of strain balances for force test. Equipped with pressure sensors and electronic scanning valve system with an appropriate range for pressure distribution measurement test; high pressure air source and high precision pressure flow control system for aircraft engine performance simulation and measurement; it is equipped with dynamic derivative test device, jet test device, inlet test device, shifting test rack device, ground effect test device and control system, launching test device, test device with propeller power, abdominal brace and tail brace support device, rotating balance test device, etc.

The fan section has a diameter of 5.98 m, 12 blades made of forged aluminum, and the drive power of ac variable frequency motor is 2050 kW. The motor is installed in the fairing, with 7 tail cone guide blades and 7 front support blades. Figures 6.30 and 6.31 show the wind tunnel test section.

3. **Pressurized wind tunnel**

FL-9 low-speed supercharged wind tunnel is a critical aviation infrastructure scheme proposed according to the needs of China's trans-century aviation industry development (as shown in Fig. 6.32). Its construction and use filled the gap of China's high-Re wind tunnel (the maximum

Fig. 6.27 D5 wind tunnel aerodynamic noise test of supercritical wing

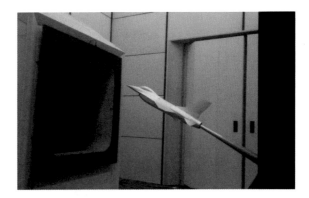

Fig. 6.28 Wind tunnel load test

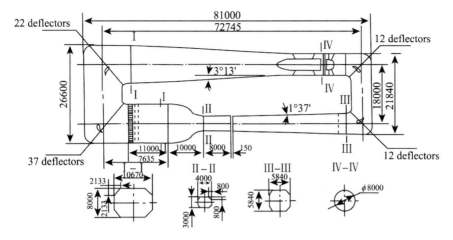

Fig. 6.29 FL-12 wind tunnel is a 4 m × 3 m single-loop closed low-speed wind tunnel

Fig. 6.30 Wind tunnel test section

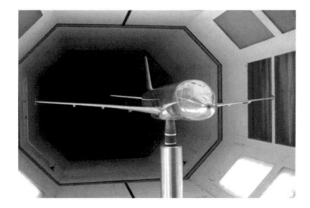

Fig. 6.31 FL-12 wind tunnel test

Fig. 6.32 FL-9 low-speed supercharged wind tunnel

Reynolds number in the test section reached 8.5×10^6), making China's low-speed wind tunnel test equipment and test technology level among the international advanced ranks, greatly increasing China's low-speed wind tunnel test capacity. FL-9 low-speed supercharged wind tunnel was built in 2006. After equipment debugging and auxiliary equipment construction, all flow field calibration contents were completed in 2008, and all flow field indexes met the national military standard requirements. The FL-9 low-speed pressurized wind tunnel is in the form of a closed single loop. The test section is 4.5 m × 3.5 m, and the pressure range is 0.1−0.4 mpa. The dimensions between the rectangular loop axes are 78 m × 18 m, and the maximum peripheral dimension is 86.2 m × 28 m. The height of the wind tunnel axis from the ground is 5 m. The volume of the wind tunnel is 13000 m³. The length of the test section is 10 m, among which the movable part is 8.5 m long. The supporting vehicle of the model serves as the lower wall of the test section, which can be moved out of the wind tunnel with the chamber. There are four model support vehicles in the whole wind tunnel, and each vehicle is equipped with a model support system and data acquisition system, which can make all preparations for the test in an independent debugging and preparation room. Curved doors are set at both ends of the test section to seal the high-pressure air in the wind tunnel when the model state is changed in the pressurization test, and a circular sealing door with a diameter of 1.8 m is set for the test personnel to enter and exit the test section. The pressure balance between the laboratory and the test section is achieved through the pressure regulating joint. The pressure in the laboratory is maintained by the circular sealing door and sealing ring.

The wind tunnel's power plant is made up of 16 carbon fiber blades, with three groups of four in front of the blades, a total of 12 pieces, including five pre-twist pieces. Behind the blade, there are 11 stop blades and 3 support blades. 9.5 MW ac variable frequency motor is located outside the wind tunnel, and the fan speed is controlled by a long shaft drive of about 10 m.

The inlet of the stable section of the wind tunnel is provided with a water circulation cooling system, which can effectively control the airflow temperature change in the test process and make the temperature rise in the test process very small. A honeycomb device is arranged in the middle of the contraction section. The honeycomb device has a hexagon with a side length of 15 mm and a thickness of 300 mm, which is made of aluminum alloy. After the honeycomb device is a three-layer damping network, and the distance from the outlet of the honeycomb device is 1 m, 2 m, and 3 m, respectively. The damping mesh is made of 0.6 mm steel wire with three mesh sizes of 10 meshes per square inch.

4. **Japan's high-speed railway wind tunnel with low noise**

   The RTRI large-scale low-noise wind tunnel was designed by the Japan institute of railway technology in 1994 and completed and put into operation in 1996 (as shown in Fig. 6.33). It is a single-loop, open and close wind tunnel with low turbulence and low noise, used for railway high-speed train aerodynamic and aeroacoustic test (as shown in Fig. 6.34), the open test section of 8 m × 3.0 m × 2.5 m, silent test section of section size 20 m × 5 m × 3 m, the corresponding maximum wind speed 111 m/s and 83 m/s, test section turbulence degree 0.2%, at 83 m/s (300 km/h) the background noise of test section 75 dBA. The wind tunnel is composed of a wind tunnel body, measurement and control system, power system, model support system, lifting system, etc. The axial flow fan in the power

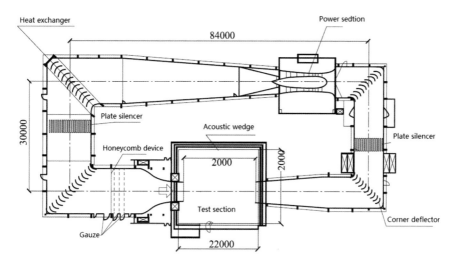

Fig. 6.33 Layout of RTRI large-scale low-noise wind tunnel

Fig. 6.34 RTRI large-scale low-noise wind tunnel test section

section has a diameter of 5 m, a design speed of 550 rpm, and a maximum power of 7000 KW; anechoic hall is 22 m long, 20 m wide, and 13 m high.

5. **DNW wind tunnel (aeroacoustic wind tunnel)**

DNW-LLF was built by Germany and the Netherlands in 1980 (as shown in Fig. 6.35). The wind tunnel test section has three closed sections. It is the biggest low-turbulence and low-noise wind tunnel in Europe. It can be used for aerodynamic and acoustic test of aircraft, helicopters, and other projects. It is also one of the few wind tunnels in the world that have the ability of aeroacoustic experimental study. It has three interchangeable test sections, the section sizes are as follows: 9.5 m × 9.5 m, 8 m × 6 m, 6 m × 6 m, respectively. The maximum wind speeds of the corresponding test sections are 62, 116, and 152 m/s. There is also an open-jet test section (as shown in Fig. 6.36) with a diameter of 8 m × 6 m, maximum wind speed of 80 m/s, and off-site noise index of 78 dBA in the test section. These four sections can be used interchangeably and can be quickly switched to another test after one is completed. DNW wind tunnel is composed of a wind tunnel body, measurement and control system, power system, model support system, and lifting system. Using this large aeroacoustic 8 × 6 m wind tunnel, airbus has completed aeroacoustic wind tunnel tests on A320, A340, A380, and other large passenger aircrafts. The DNW-LLF wind tunnel has attracted many customers, including the United States, due to its excellent aeroacoustic testing capability. The aeroacoustic tests of Chinese models such as the ARJ21 and C919 are also conducted in the wind tunnel.

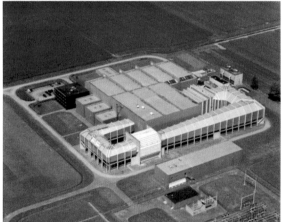

Fig. 6.35   DNW-LLF aeroacoustic wind tunnel in Germany and the Netherlands

Fig. 6.36   DNW-LLF aeroacoustic wind tunnel test section in Germany and the Netherlands

Fig. 6.37 Tongji University automobile wind tunnel test section

Fig. 6.38 Smoke flow test in automobile wind tunnel of Tongji University

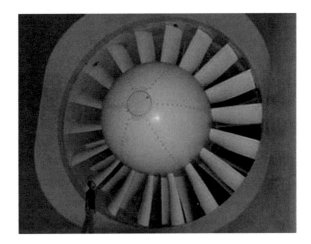

Fig. 6.39 Fan system in automobile wind tunnel of Tongji University

6. **Large aeroacoustic automobile wind tunnel**

Tongji University is guided by scientific and technological innovation and new energy vehicle development. It is also featured by innovative research mechanism and organizational structure. In 2009, Tongji University completed and put into operation an aeroacoustic vehicle wind tunnel. The section size of the acoustic wind tunnel is 6.3 m × 4.2 m, the maximum wind speed is 69.4 m/s, and the off-site noise index of the test section is 61 dBA under the wind speed of 44.4 m/s. This "wind tunnel" not only can do it for the automotive vehicle test but can also perform high-speed train simulation test, technology belonging to the international first-class level. There are more than 10 auto acoustic wind tunnels in the world. The wind tunnel of Tongji University is one of them. This wind tunnel laid a foundation for China's independent automobile brand creation (as shown in Figs. 6.37 to 6.40). As a key technology platform for public automobiles and rail vehicles, automobile wind tunnel will provide important basic services for China's automobile and rail vehicle industry, especially for the independent research and development of new energy vehicles, and also provide important technical support for China's automobile industry from "made in China" to "created in China". The test section, with a five-belt mobile ground system and a six-component test balance, will be the quietest wind tunnel of its kind in the world. The wind tunnel center is capable of carrying out a series of tests on all kinds of vehicles and parts, including cars, buses, SUVs and trucks, and rail vehicle models.

Fig. 6.40 Exterior of Tongji University's automobile wind tunnel

7. **Full-size wind tunnel**

To date, the world's largest low-speed wind tunnel is located at NASA Ames research center. It is a 24.4-m by 12.2-m full-size low-speed wind tunnel. A new test section of 36.6 m × 24.4 m has been added to the wind tunnel, and the fan motor power has been increased from 25 to 100 megawatts (as shown in Figs. 6.41, 6.42, 6.43 and 6.44).

Fig. 6.41   Full-size wind tunnel inlet

Fig. 6.42   Fan section of full-size wind tunnel

Fig. 6.43 Full-size wind tunnel exterior

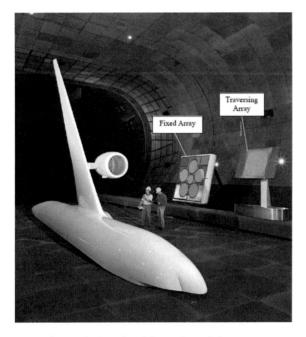

Fig. 6.44 Noise test of a civil aircraft with semi model

6.5 Supersonic Wind Tunnel

1. **General**

Supersonic wind tunnel refers to the wind tunnel where the flow Mach number of the test section is approximately within the range of 1.4–5.0. If the Mach number exceeds about 5, a heater is needed to improve the overall stability of the airflow to prevent air liquefaction in the test section, which belongs to the category of hypersonic wind tunnel. If the Mach number ranges from 0.8 to 1.4, the wall of the test section should be perforated or grooved (ventilation wall), which belongs to the category of transonic wind tunnel. Supersonic wind tunnel generally has no heater, and the test section is a solid wall (no holes, no grooves). Different from low-speed wind tunnel, two basic conditions must be met to obtain supersonic flow: contraction–expansion nozzle and the area ratio between throat section and nozzle outlet section must be changed to change the experimental Mach number; the ratio of pressure in the stable section to the pressure at the outlet of the diffuser section should be large enough and increase with the increase of the experimental Mach number. When the air expands to supersonic velocity, the temperature drops sharply and the moisture in the air condenses in the test section. In addition, a supersonic diffuser is usually used in supersonic wind tunnels, especially when the Mach number is high. These are all different from low-speed wind tunnels. A supersonic wind tunnel with the same test section size requires much more drive power than a low-speed wind tunnel. It is generally believed that the test section size equal to or less than 0.6 m or so belongs to a small supersonic wind tunnel, while the test section size of 1 m or so belongs to a medium supersonic wind tunnel. This division is not very strict, just a general reference. Supersonic wind tunnels first appeared in the 1920s. Many of the world's large supersonic wind tunnels were built in the 1950s. For example, the world's largest supersonic wind tunnel (the test section of 4.88 m × 4.88 m, $Ma = 1.5$–4.75) was built in the United States around 1956. The development trend of the supersonic wind tunnel is to improve the performance of the existing wind tunnel and improve the technical level of measurement and control. In order to study the drag reduction technology in supersonic flight, a supersonic wind tunnel with low turbulence and low noise needs to be built.

Supersonic wind tunnel is divided into continuous type and transient impact type. The continuous supersonic wind tunnel, like the general low-speed wind tunnel, works continuously, the experimental conditions are easy to control, the experiment is not limited by time, but the power equipment should be quite large. According to the different ways of generating pressure ratio, the transient supersonic wind tunnel can be divided into blowing type, inhaling type, ejecting type, blowing-suction type, and blowing-leading type. The working process of the transient supersonic wind tunnel is as follows: prior to the operation of the wind tunnel, the air is pumped into the gas cylinder. When the air pressure in the cylinder reaches a certain value, the wind tunnel can start to work. At this point, the globe valve is first opened, and then the quick valve is opened. The air from the cylinder passes through the pressure regulating valve, passing through the stable section, contraction section, nozzle section, test section, supersonic diffuser section, and subsonic diffuser section, and finally is discharged into the atmosphere.

2. **Typical wind tunnel**

   Located in Mianyang, China, a 2 m × 2 m supersonic wind tunnel was approved in 2003 (as shown in Fig. 6.45), started construction in 2007, and officially completed ventilation in 2010. It is the largest supersonic wind tunnel with the strongest simulation capability in China. Together with the 2.4-m ejector transonic wind tunnel, the wind tunnel has formed the sub-, trans-, and supersonic wind tunnel test capability of China's 2-m scale, which plays a very important role in the development of supersonic flying weapons in China. Wind tunnel type is downblow-ejector transient supersonic supercharged wind tunnel. The section size of the test section is 2 m × 2 m, with a length of 7.2 m. The Mach number of the test section of the wind tunnel is 1.5−4.0, and the root mean square deviation of distribution all meets the advanced index requirements of the national military standard (GJB 1179–91). The mean flow deflection angle of the model area in the test section meets the requirements of the national military standard (GJB 1179–91). The noise of the test section is better than that of the existing supersonic wind tunnel in China.

Fig. 6.45 2 m × 2 m supersonic wind tunnel

6.6 Transonic Wind Tunnel

1. General

The Mach number of the transonic wind tunnel is about $0.8-1.4$. The lower limit is 0.8 because the Mach number of congestion (blockage) in the tunnel wall (non-vented wall) of the test section is generally around 0.8. To make the Mach number more than 0.8, methods should be taken to solve the congestion problem. The Mach number is around 1.4, and the general supersonic nozzle and solid wall test section can be used, belonging to the category of supersonic wind tunnel. There are three differences between supersonic wind tunnel and transonic wind tunnel. First, the wall of the supersonic wind tunnel test section is the solid wall, the wall of the transonic wind tunnel test section is the breathable wall (hole or grooved wall). The four walls are the ventilation wall or the upper and lower two walls are the ventilation wall in transonic wind tunnel. Outside of the ventilation wall is a cavity called a chamber. The ventilation wall includes a perforated wall or slotted wall or a perforated and slotted wall. The ratio of the ventilation area of the wall to the total area of the wall is called the open-close ratio. Under certain conditions, part of the air in the test section flows into the compartment through the ventilation wall. Second, before the supersonic wind tunnel test section, there is a supersonic nozzle that first shrinks and then expands. During the wind tunnel operation, the Mach number at the nozzle throat is 1, and the Mach number at the nozzle outlet is greater than 1. The transonic wind tunnel is different. In front of the test section, the pipe is simply contracted, and the maximum Mach number at the outlet of the pipe is 1. Third, to get different Mach numbers in the supersonic wind tunnel, the area ratio of the nozzle should be changed. In a transonic wind tunnel, by changing the ratio of static pressure at the entrance of the test section to the static pressure at the chamber, different Mach numbers in the transonic range can be obtained by simply contracting the same pipe (i.e., the area ratio remains unchanged). In the development history of a transonic wind tunnel, the most important technical breakthrough is the adoption of a ventilation wall, whose main functions are as follows: (1) to prevent airflow congestion in the test section. In the real wall test section, when the incoming flow approaches the sound velocity, the phenomenon of congestion will occur. The model is installed in the test section and a minimum section is formed between the model and the cave wall. Even if there is no model in the test section, the section with the minimum effective cross-sectional area will be formed in the test section due to the existence of boundary layer of cave wall. When the Mach number of the incoming flow is approximately 0.8, the Mach number reaches 1 in the minimum section. Although the incoming flow is still in the subsonic velocity range, no matter how to improve the pressure ratio of the wind tunnel, the Mach number of the incoming flow cannot be further increased due to the limitation of the minimum section flow. In other words, there is no transonic flow in the solid wall section. The ventilation wall solves this problem. The ventilation wall is adopted in the test section and the air in the laboratory is removed appropriately. A part of the air in front of the model flows into the laboratory, and the part that could not pass due to the limitation of the minimum section could be discharged through the laboratory. In this way, the Mach number of the flow before the minimum section can exceed 0.8, thus establishing a transonic flow in the test section. As long as the indoor pressure is kept appropriately low, equal to the Mach number at the time of wind tunnel operation and the corresponding static pressure under the total pressure, then, once the airflow enters the test section, some air will pass through the ventilation wall and enter the chamber. This flow that passes through the ventilation wall continues until the pressure in the test section is equal to the pressure in the chamber. Finally, a uniform transonic flow with the Mach number in line with the requirements of wind tunnel operation is obtained. Thus, when the airflow flows downstream along the test section, using the ventilation wall to continuously reduce the airflow flow through the test section can play the same role as the geometric nozzle. In the process of the wind tunnel, how to keep the indoor air out? One method is pumping, in which a vacuum pump is used to remove air from the chamber, and the air extracted should flow back into the wind tunnel at the appropriate location for a recirculating continuous wind tunnel. The other method is the mainstream ejector type. Import cross-sectional area is slightly greater than the test section diffuser outlet cross-sectional area, forming the gap from the chamber to the diffuser. Under the influence of the mainstream of the ejector, the air in the chamber is drawn out through a slit into the diffuser. The size of the pressure in the chamber is related to the size of the aperture. The size of the aperture of imports can be adjusted by adjusting the diffuser wall plate. (2) To reduce or eliminate the interference of the reflected wave of the tunnel wall. When the Mach number of the test section is close to 1 or slightly greater than 1, the reflected shock wave generated from the model with a large wave angle meets the real wall and returns to the model. If the shock wave meets the free air boundary ("air wall"), it generates a reflection expansion wave that hits the model back. The reflected wave changes the pressure distribution on the surface of the model and destroys the normal flow around it, which is completely different from the flow around the prototype flow field. After the ventilation wall is adopted, there are both real wall and "air wall" on the ventilation wall. Some shock waves generated by the model meet the real wall and generate reflected shock waves, while waves generated by the other part meet the "air wall" and generate reflected expansion waves. These separately reflected shock waves and reflected expansion waves meet at a certain distance from the cave wall and cancel each other out. As long as the ventilation rate of the wall is appropriate, the interference of the wall reflection wave can be minimized. In particular, when the geometric parameters of the hole are properly selected, the "no reflection" degree can be achieved within a considerable Mach number range. (3) To reduce or eliminate the cave wall interference in subsonic velocity. The cave wall interference is very serious in subsonic velocity in transonic wind tunnel. The effect of wall interference in closed and open sections is the opposite. Therefore, the use of ventilation wall, as long as the open-close ratio of ventilation wall selection is appropriate, can reduce or nearly eliminate wall interference.

2. **Typical wind tunnel**

The world's first practical transonic wind tunnel was developed in 1947. At present, there are two development trends of transonic wind tunnel: one is to further increase the Reynolds number of transonic experiment; the other is to further reduce the interference of transonic tunnel wall. With the development of aerospace industry, the Reynolds number of modern large aircraft flying at transonic speed is as high as 6×10^7 while the experimental Reynolds number of general transonic wind tunnel can only reach 1/10 of the flight Reynolds number, and the maximum is not more than 1/6. The difference between the experimental Reynolds number and the flight Reynolds number is so large that many experimental results are unreliable. For example, because of the low Reynolds number in the experiment, the location and pressure distribution of shock waves on the wing surface in the wind tunnel experiment are far different from the flight situation. Due to the low experimental Reynolds number, it is difficult to carry out the experiment of large angle of attack and high maneuverability of aircraft, and to carry out the experimental research of advanced airfoils such as peak airfoils and supercritical airfoils. If the experimental Reynolds number is increased to more than or equal to 4×10^7 (the characteristic length is the average aerodynamic string), it is generally considered that the change in aerodynamic coefficient with Reynolds number is no longer significant. This experimental Reynolds number is close to the flying Reynolds number, known as a high-Reynolds number wind tunnel. The methods to increase the experimental Reynolds number are as follows: increasing the size of the wind tunnel; high density gas being used as the working medium of the wind tunnel; increasing the total pressure of airflow in the wind tunnel; reducing the total temperature of the airflow. Increasing the size of a wind tunnel, like a low-speed wind tunnel, is not realistic for a transonic wind tunnel because it is too expensive to build. It is theoretically feasible to change the working medium of the wind tunnel to increase the experimental Reynolds number, but there is no gas with the same specific heat capacity as air, higher density than air, and cheaper. Since 1966, pressure wind tunnel has been used to increase the experimental Reynolds number, which has achieved certain results. However, it has caused an increase in dynamic pressure, resulting in the strength of the model and the support. Meanwhile, the driving power of the wind tunnel has increased significantly, which has limited the further improvement of the Reynolds number. Since the 1970s, a new type of wind tunnel, low-temperature wind tunnel, has shown more and more advantages. A low-temperature wind tunnel is a wind tunnel where the temperature of the working medium is less than 173 K. As the temperature decreases, the viscosity coefficient μ and sound velocity a decrease, the density ρ increases, and the Reynolds number Re increases. With lower temperature, while the density increases, the sound velocity decreases, and wind speed decreases under the same Mach number, thus dynamic pressure can basically remain unchanged. The driving power of the wind tunnel will drop slightly, to avoid the general pressure due to the increase of Reynolds number in the wind tunnel model too much load and driving power too large. Compared with the conventional ambient temperature wind tunnel, by reducing the total temperature, the Reynolds number can be increased by as much as six times with the wind tunnel size unchanged and the gas flow pressure basically unchanged. If the total pressure of the airflow is appropriately increased while the total temperature of the airflow is reduced, that is, the combination of low-temperature wind tunnel and pressure wind tunnel, the experimental Reynolds number will be increased even more. For example, in transonic wind tunnel at NASA Langley Experiment Center National, the world's largest high-Reynolds number transonic wind tunnel was formal operating in 1983, the transverse dimension of the test section is 2.5 m × 2.5 m. The Mach number is within the testing range 0.2−1.2. The total pressure can be up to 9×10^5 Pa. The airflow temperature can be reduced to 100 K by using the method of liquid nitrogen injection. When the Mach number is 1, the Reynolds number based on the model chord length 0.25 m is 120×10^6, twice as high as the Reynolds number of cruise flight B747. Another example is the European Transonic Wind Tunnel ETW, scheduled for completion in 1993. The test section is 2.4 m × 2.0 m, with a minimum operating temperature of 90 K and a maximum experimental Reynolds number of 5×10^7.

(1) National transonic wind tunnel, Langley experimental center, NASA
    The National Transonic Facility (NTF), NASA Langley Laboratory, USA, is a single-loop, continuous, low-temperature pressurized wind tunnel driven by an axial flow compressor (or axial flow fan, axial fan), built in 1983 (as shown in Figs. 6.46, 6.47, 6.48, 6.49, 6.50 and 6.51). The wind tunnel test section parameters such as total temperature, total pressure, and fan speed can be independently controlled, so as to, respectively, study Ma number (compression), the Re number

Fig. 6.46 Outdoor scene of NTF

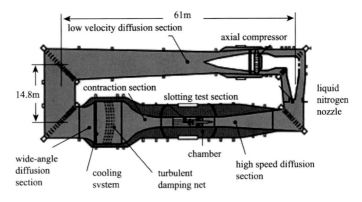

Fig. 6.47 US national continuous transonic low-temperature pressurized wind tunnel

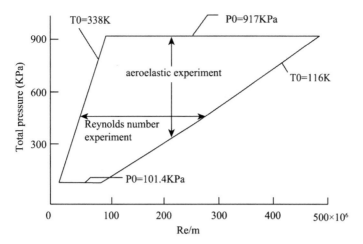

Fig. 6.48 Re and total pressure curve of NTF test section, US

Fig. 6.49 NTF test vent, US

(viscosity), and pneumatic elastic dynamic pressure effect, through total temperature and total pressure and the combination of fan speed. The range of Reynolds number per meter in the test section can be obtained as $6.6 \times 10^6 - 75.7 \times 10^6$/m, and Mach number range as $0.1 - 12$. The test section is 2.5 m wide, 2.5 m high, and 7.6 m long. In order to prevent the blocking of transonic flow, six grooves are

Fig. 6.50 Ac frequency converter motor outside NTF, US

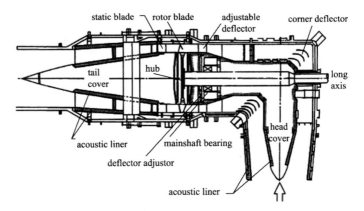

Fig. 6.51 US NTF axial flow compressor (To shorten the long shaft, the hood extends to the second corner.)

opened at the top and bottom of the test section. The liquid nitrogen injection device in the wind tunnel can make the wind tunnel test section run under a total temperature of 116 K, the total temperature of the test section is 116−338 K, and the total pressure in the test section is 101.4−917 kPa. The power system is located at the second corner section downstream wind tunnel by the hole of 101 MW of ac variable frequency motor and axial flow compressor. The diameter of the compressor is 6.1 m. The inside and outside diameters of the flow area are equivalent. The compressor adopts single stage, with

25 rotor blades, 26 static blades, and 24 import upgrade pieces. The rotor blades are made of fabricated fiberglass-reinforced plastic. The compressor speed range is 60−600 rpm. The maximum speed (rated speed) for nitrogen operation at low temperature is 360 rpm, and the maximum speed for normal temperature air operation is 600 rpm. During the operation of the wind tunnel, the strategy of flow adjustment is adjusting the speed for coarse, fan front guide plate for fine adjustment. The Ma number adjustment accuracy of the test section is 1/1000. In order to refine the Ma number control, the required compression ratio can be obtained by changing the inlet deflector to maintain the Mach number required in the test section. In order to shorten the length of the long shaft of the driving compressor hub, the rectifier hood of the compressor makes a turn at the second corner and extends to the space between the first and second corner. At the same time, in order to reduce the aeroacoustic noise of the compressor, the sound liner is installed at the compressor fairing.

(2) European transonic wind tunnel (ETW)

The European Transonic Wind Tunnel (ETW) is also a single-loop, continuous, low-temperature pressurized wind tunnel driven by an axial flow compressor, built in 1993 (as shown in Figs. 6.52, 6.53, 6.54, 6.55, 6.56, 6.57 and 6.58). The wind tunnel test section parameters such as total temperature, total pressure, and fan speed can be independently controlled, so as to study the Ma number (compression), the Re number (viscosity), and aeroelastic (dynamic pressure)

Fig. 6.52 Appearance of the European continuous transonic low-temperature pressurized wind tunnel (ETW)

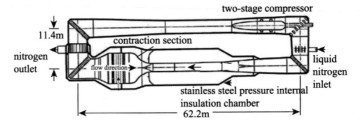

Fig. 6.53 European continuous transonic low-temperature pressurized wind tunnel (ETW)

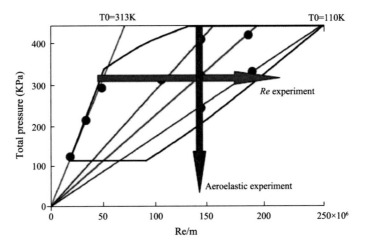

Fig. 6.54 European ETW wind tunnel test section Re number and total pressure curve

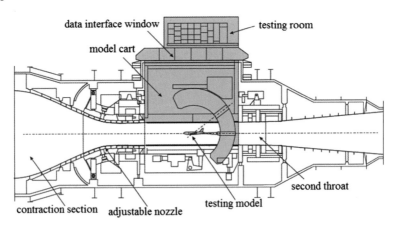

Fig. 6.55 European ETW test section

Fig. 6.56 European ETW wind tunnel test passage

Fig. 6.57 European ETW axial compressor rotor

effect, respectively. By the combination of total temperature, total pressure, and fan speed, the Reynolds number per meter in the test section can be ranged from 6.6×10^6 to 230×10^6/m, Mach number can be ranged from 0.15 to 1.35. The test section is 2.4 m in width, 2.0 m in height, and 9.0 m in length. In order to prevent the transonic flow from blocking, six grooves are opened at the top and bottom of the test section. The wind tunnel liquid nitrogen injection device can make the wind tunnel test section run under a total temperature

Fig. 6.58 ETW axial compressor tail housing

of 110 K, the total temperature of the test section is 110–313 K, and the total pressure of the test section is 115–450 kPa. The power system, located downstream of the second corner section of the wind tunnel, consists of a 50 MW variable frequency ac motor and an axial flow compressor outside the tunnel. The compressor diameter is 4.5 m. Compressor speed ranges from 60 to 830 rpm. The maximum speed is 830 rpm at room temperature. During the operation of the wind tunnel, the strategy of flow adjustment is adjusting the speed for coarse, fan front guide plate for fine adjustment. The Ma number adjustment accuracy of the test section is 1/1000.

(3) 2.4 m × 2.4 m transonic wind tunnel

This wind tunnel is a large sub- and transonic wind tunnel in Mianyang Aerodynamics Research and Development Center, which was put into operation in 1999. The section size of the test section is 2.4 m (width) × 2.4 m (height) × 7 m (length), the Mach number ranges from 0.3 to 1.15, and the maximum Reynolds number is 17×10^6. It can not only carry out the conventional force and pressure tests for full model, half model, and components but also special tests such as buffeting, flutter, dynamic derivative, and TPS can be completed in this wind tunnel, as shown in Figs. 6.59 and 6.60. Currently, it is mainly responsible for high-speed wind tunnel testing of the C919 jumbo jet, the ARJ21-700 regional jet, and the MA-700 regional jet.

Fig. 6.59   2.4 m × 2.4 m transonic wind tunnel

Fig. 6.60   2.4 m × 2.4 m transonic wind tunnel test section

6.7   Hypersonic Wind Tunnel

When the shuttle returned to the ground, it reentered the atmosphere at flight $Ma = 30$ from a flight altitude of about 150 km. The hypersonic flight of the space shuttle has to go through different stages from the free molecular flow zone to the hypersonic continuous flow zone. In the hypersonic continuous flow area, although the flight Mach number is not very high (about 5−12.5), the flight Reynolds number is high, the total temperature and total entropy of the flow are high, and there are real gas effects. The so-called real gas effect mainly refers to the effect of high-temperature heating. When the space shuttle flies in the hypersonic continuous flow zone, both the compression of the strong bow shock wave in front of the body and the airflow caused by the viscous blockage of hypersonic flow in the boundary layer will generate high temperature up to 6000 K or even higher. Since the temperature is higher than 600 K, the temperature rises further, gas molecules can appear and so also the phenomena such as vibration, dissociation, and ionization, which changes the nature of the air. The specific heat ratio no longer remains constant but changes as the temperature changes. The perfect gas equation of state is no longer applicable. Isentropic relationship fails. Enthalpy no longer linearly increases with the temperature and is under the influence of the pressure at the same time. This real gas effect obviously has a considerable impact on the aerodynamic characteristics of the shuttle. It is very difficult to fully simulate the shuttle's return to the various stages of flight in an experimental facility. In the hypersonic continuous flow region, the main parameters for simulation in the space shuttle aerodynamic test are Mach number, Reynolds number, and enthalpy. Hypersonic wind tunnel can simulate high Mach number (5−14) but cannot simulate high enthalpy. Hypersonic wind tunnel refers to a wind tunnel where the Mach number range of the test section is about 5−14, and the flow temperature is high enough to prevent the lique-faction of the test section when the isentropic expansion of the flow reaches the above Mach number, but not high enough to produce real gas effect. Comparing hypersonic wind tunnel with supersonic wind tunnel, the simi-larity is that hypersonic wind tunnel also has two types of continuous and temporary impact. The difference is that an air heater needs to be installed in the hypersonic wind tunnel. If the heater adopts the resistance type, the total air temperature can be heated to about 1500 K. This temperature can prevent the flow from liquefaction in the range of Mach number less than 14, but it can only simulate the enthalpy value when the Mach number is less than 6 and cannot simulate the enthalpy value when the Mach number is higher in the hypersonic range.

Wind tunnels with high enthalpy and hypersonic (*Ma* is about 5–20) are simulated. At present, there is no experimental equipment that can fully simulate the Mach number, Reynolds number, and enthalpy of the space shuttle flight, as shown in Fig. 6.61. The Mach number of the hypersonic wind tunnel is about 5−20 and has a high enthalpy value. Shock wind tunnel is a kind of common hypersonic wind tunnel and is developed on the basis of the shock tube. The shock tube is a device that generates shock waves and compresses the experimental gas. It is a closed tube with a diaphragm (shock diaphragm) separating the tube into two sections called a high-pressure chamber and a low-pressure chamber. Before the experiment, the driven gas with high pressure and the driven gas with low pressure are, respec-tively, filled in the high- and low-pressure chambers to meet the experimental requirements, and the pressure ratio in the two chambers reaches a certain

Fig. 6.61 U.S. x-43 hypersonic vehicle (cruising Mach 7)

value. In the experiment, the shock wave diaphragm is broken first. From the rupture of the diaphragm, a one-dimensional unsteady flow process occurs in the shock wave tube. When the diaphragm is broken, the high-pressure driving gas rushes into the low-pressure chamber, and a contact surface is formed between the two gases. The driven gas on the right side of the contact surface is strongly compressed, forming a positive shock wave moving to the right at supersonic speed. The right side of the right-traveling shock wave is the undisturbed region, and the left side of the right-traveling shock wave is the original low-pressure chamber gas swept by the positive shock wave and moving to the right with a certain velocity. After shock compression, the pressure and temperature of the gas in the original low-pressure chamber are increased obviously. At the same time, the expansion wave traveling to the left propagates through the gas in the original high-pressure chamber. The stability of the high enthalpy flow in the shock tube is very short, usually less than 1 ms. The high-performance shock tube can be used to study the phenomena of gas dissociation and ionization. If the right end of the shock tube is not closed, but the diaphragm (nozzle diaphragm), nozzle, test section, and true empty box are installed at the right end of the shock tube, the shock wind tunnel is formed. Shock wave wind tunnel is a kind of wind tunnel that uses shock wave to compress gas and then produces superhigh velocity airflow by steady expansion method. After the shock diaphragm is broken, the shock wave moves to the right and the driven gas is compressed and heated. When the positive shock wave reaches the right end and meets the nozzle diaphragm, the reflected shock wave in the left row is generated, and at the same time, there is a rupture of the nozzle diaphragm. The test section is connected with the true empty box. Under the action of a fairly high-pressure ratio in the wind tunnel, the driven gas heated by shock wave compression again enters the nozzle, and a high Mach number can be achieved in the test section. If the relevant parameters are properly controlled, the reflected shock waves on the left side will not reflect again after they meet the contact surface, and the speed on the right side of the contact surface will slow down, thus extending the working time of the wind tunnel. The operating time of shock wave wind tunnel is longer than that of shock wave tube and can reach a few milliseconds. The superhigh velocity flow in the shock wind tunnel has a total temperature of 8000 K, a total pressure of 200 MPa, and $Ma = 25$ or more.

Common ground simulation devices include shock tubes (Fig. 6.62), arc-heated wind tunnels (Figs. 6.63 and 6.64), and free-trajectory targets (Fig. 6.65).

The JF12 hypersonic shock wave wind tunnel, developed by the State Key Laboratory of High Temperature Aerodynamics (LHD), Institute of Mechanics, Chinese Academy of Sciences, in 2012, is the world's first shock wave wind tunnel with a speed of nine times the speed of sound in the test section of the wind tunnel. It is called "hyper-dragon" by the media, as shown in Fig. 6.66. The facility is the first of its kind in the world to be able to test the airflow characteristics of an aircraft at an altitude of 25 to 40 km at speeds of five to nine times the speed of sound. By comparison, the wind tunnel used by the U.S. to test the x-51 hypersonic vehicle blows at 7.5 times the speed of sound. The total length of this super ultrasonic shock wave wind tunnel is 265 m. The nozzle exit diameter of the test section is $\Phi2.5/\Phi1.5$ m. The testing gas is clean air, tested for more than 100 ms. The JF12 wind tunnel achieves four key technical indicators at the same time, i.e., "emerging air total temperature and total pressure", "producing pure test gas", "meeting the

Fig. 6.62 Shock tube installation in San Antonio, Texas, US

Fig. 6.63 Nozzle diameter $\Phi = 0.42$ m arc wind tunnel (FD-04, $Ma = 0.6{-}12$, China Academy of Aerospace Aerodynamics)

Fig. 6.64 Nozzle diameter $\Phi = 1.0$ m arc wind tunnel (FD-15, $Ma = 0.6{-}10$. It can undertake all kinds of material ablation properties, particle erosion, cone body heat layer induced ablation heat through waves' roll moments, pneumatic optical transmission, and re-entry physical phenomenon tests. China Academy of Aerospace Aerodynamics.)

demand of basic test time", and "testing full-size or close to full-scale model". The reproducibility of the ground test of hypersonic aircraft has been realized, and it provides an irreplaceable test method for the key technologies of China's major engineering projects and the basic research of high-temperature aerodynamics.

Fig. 6.65 Free ballistic target equipment developed by China Pneumatic and Development Center

Fig. 6.66 JF 12, an ultra-large hypersonic wind tunnel put into operation in 2012, Chinese Academy of Sciences

6.8   Variable Density Wind Tunnel

1. **Basic concepts**

It is well known that in wind tunnel tests, by the similarity criteria, it is generally required to meet the Mach number criterion considering the compression effect and the Reynolds number criterion considering the viscous effect. Because of the correlation between Mach number and Reynolds number, the influence of these parameters cannot be studied independently in conventional wind tunnels. However, if a wind tunnel can be developed to independently control the total temperature, total pressure, fan speed, and other parameters in the test section, then the influence of changes in Ma (compressibility) and Re (viscosity) can be studied independently. For example, different combinations of total temperature, total pressure, and fan speed can be used to study the influence law of different Re at a given Ma. It is also possible to study the influence law of different Ma under a given Re. Such wind tunnels are often called variable density wind tunnels. According to the definition, the Reynolds number per unit length of the test section is defined as

$$Re = \frac{\rho V}{\mu} = \frac{\sqrt{\gamma}}{\mu \sqrt{R}} \frac{p_0}{\sqrt{T_0}} \frac{Ma}{\left(1 + \frac{\gamma-1}{2} Ma^2\right)^{\frac{\gamma+1}{2(\gamma-1)}}}$$

Here, ρ is the gas density (kg^3/m), V is the velocity of the test section (m/s), and μ is the aerodynamic viscosity coefficient of the test section ($kg/m/s$), p_0 is the total pressure of the test section (Pa), T_0 is the total temperature of the test section (K), Ma is the Mach number of the test section, and γ is the gas characteristic coefficient, for air and nitrogen, $\gamma = 1.4$. Nitrogen is a kind of colorless, odorless gas, and generally less dense than air. Nitrogen is the main component of air, accounting for 78.12% (volume fraction) of the total volume of the atmosphere. When cooled to -195.8 °C at normal atmospheric pressure, it becomes a colorless liquid. When cooled to -209.8 °C, liquid nitrogen becomes a snow solid. Nitrogen is not reactive in chemical properties, it is difficult to react with other substances at room temperature, but in high temperature, high-energy conditions can produce chemical changes with some substances, used to make new substances useful to humans. At the standard atmospheric pressure of 101,325 Pa, the temperature Ta $= 288.15$ K, the nitrogen density is 1.1846 kg/m^3 and the air density is 1.225 kg/m^3. The specific heat capacity of nitrogen at constant pressure is cp $= 1038$ J/(kgk), and that of air is cp $= 1004.7$ J/(kgk). The constant specific heat capacity of nitrogen is Cv $= 741$ J/(kgk), and that of air is Cv $= 717.6$ J/(kgk). According to the Sutherland formula, the viscosity coefficient of nitrogen is

$$\mu(T) = \mu_0 \left(\frac{T}{273.16}\right)^{1.5} \frac{273.16 + 104}{T + 104}$$
$$\mu_0 = 1.6606 \times 10^{-5} kg/m/s$$

The curve between Ma and Re is obtained, as shown in Fig. 6.67.

2. **Variable density wind tunnel**

   To independently simulate the effects of Re or Ma, NACA(now NASA) constructed a variable density wind tunnel in 1922, located at the NACA Langley Research Center. Inside the mysterious large tank is a subsonic wind tunnel with a diameter of 5 feet. The pressure inside the tank can reach 20 atmospheres, as shown in Figs. 6.68 and 6.69. Between the 1920 and 1930s, this wind tunnel contributed to the development of the NACA airfoil family. But the device was disabled as a wind tunnel in the 1940s and was used only as a compressor until the 1980s. In 1983, it was the end of use due to the end of life. The wind tunnel is now in the National Museum of American History. Most current wind tunnel tests do not simulate the Re and Ma at the same time, but first simulate the Ma in one wind tunnel, and simulate the Re in the other wind tunnel, and then give the correction according to the test results of these two wind tunnels.

3. **High-altitude airship propeller propulsion system**

   From bottom to top, the atmosphere can be divided into the troposphere, ozone layer, stratosphere, mesosphere, warm layer, and escape layer. A stratospheric airship (also known as a high-altitude airship, as shown in Fig. 6.70) is an airship capable of flying in the stratosphere (at an altitude of $20-50$ km) and capable of manned or autonomous flight. Since its flight altitude is between the highest flight altitude of existing aircraft and the lowest orbital altitude of satellites, it does not belong to the category of aerospace nor to the category of aviation. In recent years, countries

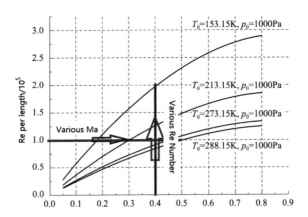

Fig. 6.67 Variation curve between Ma and Re

Fig. 6.68 Outdoor scene of a variable density wind tunnel at the Langley Research Center in NACA

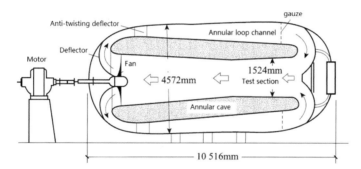

Fig. 6.69 Aerodynamic profile of a variable density wind tunnel at the Langley Research Center in NACA

have been accustomed to classify the airspace at this height as adjacent space. For the high-altitude airship, if the airship can use the low-density air to overcome the gravity, and with the help of the propeller propulsion system to overcome the atmospheric circulation resistance to achieve a fixed point in the air, then the stratospheric airship has a wide range of military and civilian value. In recent years, the United States, the United Kingdom, Germany, France, Russia, Japan, South Korea, and China have put forward corresponding research plans, setting off a worldwide research upsurge of the stratospheric airship. According to the change curve of wind speed with height (as shown in Fig. 6.71) the propeller propulsion system needs to be at the wind-resistant speed of 10−20 m/s, so it is necessary to study the lightweight and efficient propeller system with low forward ratio, low Re, and low Ma. At first glance, such problems need to be studied in low (variable) density wind tunnels, but since the Ma is low in incompressible flow and only Re is similar, it is worth discussing whether such wind tunnels are still needed.

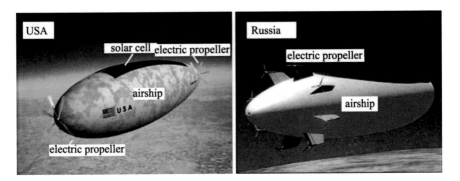

Fig. 6.70 Stratospheric airship

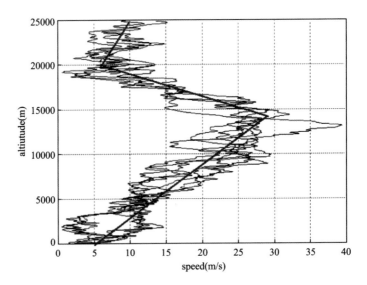

Fig. 6.71 Curve of wind speed and height

6.9   Water Tunnel (or Channel) Equipment

Water tunnel is a kind of pressure tube flow equipment that uses water as the flow medium to study physical phenomena such as boundary layer, wake, turbulence, cavitation, and hydroelasticity. A water tunnel is a water circulation piping system in which velocity and pressure can be controlled separately. The sections of the tunnel are round, square, or rectangular. There are observation windows above, below, before, and after the water tunnel. In contrast to a towing pool, a controlled flow of water moves through the tunnel instead of a test object. The performance of a water tunnel is similar to that of a low-speed wind tunnel, except that the test medium is different. As early as 1896, British scientist C. A. Parsons (1854–1931) established the world's first small cavitation tunnel, which was made of copper, with a total length of about 1 m and a fault area of 15 square centimeters in the working section. Later, a large number of water tunnels were developed to study physical phenomena such as boundary layer, wake, turbulence, cavitation, and hydroelasticity, as well as the force between the water flow and the test object. A water tunnel is a water circulation system in which velocity and pressure can be controlled separately. According to incomplete statistics, nearly 30 countries in the world have built about 220 water caves of various types. China built its first water tunnel in Shanghai in 1957 to study shipping, and there are now about 13 in the country. Figures 6.72, 6.73, 6.74, and 6.75, respectively, show the teaching water tunnel or channel equipment.

Fig. 6.72 Water tunnel equipment (Changzhou University)

Fig. 6.73 Water tunnel equipment (University of Science and Technology of China)

Fig. 6.74 Beihang University towing tank (built in 2015)

Fig. 6.75 Water tunnel at Beihang University (built in 2015)

1. Layout and structure of water tunnels

The layout of the water tunnel is very similar to that of the wind tunnel (as shown in Fig. 6.76). In order to obtain a high-quality flow field in the test section, the stability section, rectifier elements (honeycomb and gauze), and contraction section shall be arranged upstream of the test section. In a piping system, a pump is fitted to circulate water, where the pressure is strong enough to prevent cavitation; the drive motor of the pump can adjust the speed of the water in the hole.

The water tunnel has a pressure control system. The closed chamber upstream of the water tunnel has a free water surface and air on the water surface, which is connected to the vacuum pump. When the air is extracted, the pressure in the test section can be reduced or increased. The tunnel's filtration system keeps the water clean. The tunnel control system regulates the flow velocity and pressure and regulates the test system and data processing system.

The water tunnel can be non-circular, that is, a reservoir or water tank with a certain height of water level is used to discharge water into the test section of the pipeline for testing. This type of water tunnel is called free fall, the disadvantage is that the water speed range is limited, and the advantages are that the background noise of the water tunnel is very small and the turbulence is low, so it is suitable for noise test and flow pattern display test. The tunnel can also be made with a free liquid level, and the section of the test section is rectangular. This kind of water tunnel can be used to simulate the location of objects near the free surface. Some countries have made them large enough to test the combination of hull and propeller. The largest water tunnel in the world is at the Institute

Fig. 6.76 Appearance of the water tunnel at Beihang University (test section size: 1000 mm × 1200 mm)

of Hydrology and Shipbuilding in Berlin, Germany. It has a free surface; the section size of the test section is 5 m by 3 m, and the length is 1 m. The center of the cave is 10.5 m high; the maximum speed of water is 12 m/s. High-velocity water flow (such as cavitation mechanism studies require that the water velocity is greater than 40 m/s) requires a lot of power, so the test section of the high-velocity water tunnel is small, with a diameter of 30−40 mm. The pressure reducing box of the test hydraulic dam, gate, and other buildings is also a kind of circulating pipe, and the test section is box-shaped. The test section of the water tunnel used in hydraulic machineries such as water pumps and water turbines is mostly vertical, which is suitable for installing the wing wheel model. This type of water tunnel is often called a cavitation test bed.

2. **Water tunnel test**

   During the cavitation test, groups of bubbles will be generated in the test section. In order to prevent these bubbles from circulating into the test section, the water tunnel must have a certain height so that the bubbles can flow through a longer loop and a higher pressure zone to eliminate. Special dissolvers can also be installed to eliminate bubbles in the water. Some objects are in the wake of other objects, for example, the propeller is in the wake after the ship. A special grid can be installed in the front of the contraction section to simulate the wake. In recent years, stern models have been installed in the test section to generate wake. In the propeller test, a shaft connected with the external motor of the cave body should be installed upstream or downstream of the test section, and the propeller model should be installed on the shaft end of the test section. Torsimeter, thrust meter, and tachometer can be installed on the shaft outside the cave body, and pressure sensor, balance, and cavitation acoustic observation instrument can be installed in the test section to measure steady pressure, pulsation pressure, various forces, and torques, and determine cavitation initial conditions and various development forms.

7

Flight Mystery and Aerodynamic Principles

7.1 Flying Fantasy

Flying, as the name implies, is flying in the air and flying in the sky. What is the sky like? Qu Yuan (300 BC) wrote in "Tian Wen": "At the beginning of the ancient times, who passed the Tao to this universe? The universe has not yet formed, how to investigate who is the person teaches the Tao?" This indicates that human beings have had an infinite yearning and worship for the mysterious skyline since ancient times. Figure 7.1 shows a picture of Fuxi and Nvwa. In the picture, Fuxi and Nvwa's head are on the sun and their feet are on the moon. This figure not only expresses the desire of human ancestors to travel in space but also reflects the Chinese people's imagination of space. Ancient people have realized that the sky is not just about the space of the atmosphere but also the space between the earth and the moon and the sun. Humans want to fly into the sky by mimicking birds, but it took a long time to recognize the flight principle.

The human dream of flying has a long history that has existed in ancient times. A lot of myths and legends about flying are beautiful and mysterious, and they are very attractive to people. Ancient Greece and ancient Rome have chariot flight, feather flight Cupid flying archery, etc.; China has flying vehicle, Chang E flying to the moon and so on. All kinds of legends show the human desire and imagination of soaring into the sky. Humans are eager to fly like birds!

The ancient Chinese mythology "Chang E flying to the moon" (as shown in Fig. 7.2) is well known to everyone. In the process of flying to the moon, Chang E was very light, and she flew into the air lightly, without the parts

© Science Press 2021
P. Liu, *A General Theory of Fluid Mechanics*,
https://doi.org/10.1007/978-981-33-6660-2_7

Fig. 7.1 Fuxi and Nvwa (Sun and moon)

that produced lift. This shows that people at that time did not know enough about the gravity of the object when it left the ground. They thought that the bird, like human imagination, had little or no gravity as long as it left the ground, so it floated. However, the look of Chang E has a feeling of wind. When the wind blows through the clothes, the belt floats up due to the air resistance. This indicates that people have realized the existence of air resistance when the air bypasses the human body. Therefore, when people's understanding of gravity was still very vague, they did not know that objects will still be affected by gravity when they rose into the air. The flying legends at that time were full of mystery and romance. Chang E did not rely on the instruments that generate lift, just rising from the wind to the sky like weightlessness. Such a scenario can only occur in space.

There is also the legend in ancient Greek mythology, Cupid Archery (as shown in Fig. 7.3). Through the observation of bird flight, the ancient Greeks realized the importance of the wing flapping on the flight, so they added a pair of small wings to Cupid. However, such wings still seem to be incapable of such weight, and these wings have become a symbolic symbol. Therefore, the ancient Greeks' perception of weight is far from sufficient.

Fig. 7.2 Ancient Chinese legend (the end of the Warring States Period)—Chang E flying to the moon

This is the story of the ancient Greek mythology called Daedalus and his son plugging in wings and fleeing (as shown in Fig. 7.4). Daedalus said to his son before he left, "If you fly too low, the wings will touch the sea, and if it gets wet, it will become heavy, and you will be taken to the sea; if you fly too high, the feathers will catch fire because they are close to the sun". It can be seen that people have already had a certain understanding of the impact of gravity in flight, but it is a pity that lcarus still died in the sea because of the scattered wings. The understanding and application of wind power has been in China for 4,000 years, and the idea of wind flying has been recorded as early as BC. Liezi is a legendary figure in the Warring States Period. It is said that he can ride the wind subtly. Usually flying for fifteen days before going home, his free state is enviable. There is no lift device on the body of Liezi (as shown in Fig. 7.5), which indicates that people are not aware of the weight at that time.

In the 100–200 AD, Northern Europe has an ancient mythology called Wayland's feathers (as shown in Fig. 7.6). According to legend, Wayland was

Fig. 7.3 Ancient Greek mythology legend (800 BC)—Cupid (Eros)

a blacksmith who made his first flying wing and began experimenting with his brother Egil, emphasizing: "It is easy to rise against the wind, when you want to descend, it is necessary to fly along the wind". When his brother did according to his method, he found that the feathers were very stable when taking off, but they were not stable when falling down following the wind. This shows that the aircraft has poor performance when flying following the wind and is prone to instability. Later, Wayland told his younger brother, "I forgot to tell you that it is good to go against the wind when you come back". As can be seen from this story, it has been recognized that airplanes, kites, and other aircraft are best, most stable and safest taking off and landing against the wind. Therefore, when you go to the airport to see off, you generally do not say "flying all the way along the wind", but it is better to say "have a pleasant journey".

In 725 AD, Li Bai wrote in the poem "Letter to Li Yong": "Roc took off from the wind and skyrocketed up to 90,000 li. If the wind stops, roc

Fig. 7.4 Ancient Greek mythology legend (800 BC)—Daedalus and his son plugging in wings and fleeing

will stop at the surface of the water, and its power will be able to dry the water". From the flight principle of roc, the first sentence says that when the wind speed reaches a certain value, roc can use the wind power to skyrocket. Because only when the relative speed of the wind and the wing reaches a certain value, it can generate enough lifting force. The second sentence says that if the wind stops, roc stops on the water, and the airflow generated by the wings can still raise the water on the river. That is, although the wind has stopped, the airflow generated by the relative movement between the wings and the air can also stimulate the water. It can be seen that Li Bai had noticed that the relative movement between the wings and the air could generate a lifting force, indicating that Li Bai's observation of the bird flight was very detailed. Here, roc refers to the big bird in "Zhuangzi A Happy Excursion" (as shown in Fig. 7.7). It is recorded in the article: "In the northern ocean there is a fish, called the kun, I do not know how many thousand li in size.

Fig. 7.5 Ancient Chinese's understanding of the wind—Liezi rides the wind

This kun changes into a bird, called the peng. Its back is I do not know how many thousand li in breadth".

In fact, roc which was recorded in ancient times looks like an eagle in appearance, and the flying postures of various hawks are very similar to those of ancient roc. All the stories or things mentioned above are the fantasy of ancient literati about flying. In order to turn flight fantasy into reality, people have experienced long-term flight practices.

7.2   Exploratory Cognition of Flight

Through the deepening of the understanding of the attributes of force, humans have moved from fantasy to practice. In order to fly, first of all, there must be a lifting force acting on the aircraft (also called lift in aerodynamics) to overcome gravity. As mentioned earlier, balloon rising to the sky is the simplest example of the balance between lift and gravity. Knowing this principle, early aircraft such as Kongming lanterns, kites, bamboo dragonflies, etc., solved the problem of lift generation in the simplest way. Kongming

Fig. 7.6 Ancient Northern Europe mythology (100–200 AD)—Wayland's feathers

Fig. 7.7 Roc in "Zhuangzi A Happy Excursion"

lantern was used by Zhuge Liang in the "Romance of the Three Kingdoms". Zhuge Liang put the information in the Kongming lantern. He judged the direction of wind according to meteorological knowledge, and then ignited the wick at the bottom of the lantern. Through the burning of the wick, the gas inside the lantern was heated, the density of the heated gas was lighter, and the density of the cold air outside was larger. This density difference produces upward buoyancy. Under the blow of the wind, Zhuge Liang used this kind of lantern to send the information along the wind direction, so people call this kind of lantern as Kongming lantern (as shown in Fig. 7.8). The lifting principle of the Kongming lantern is realized by the static buoyancy force overcoming the gravity. The density of hot air inside the Kongming lantern is smaller than the density of the cold air outside, this density difference produces buoyancy, just as the object produces buoyancy in the water. So when the buoyancy is greater than the weight of the balloon, the balloon slowly floats up. The density of the air is large on the ground. The higher the altitude, the smaller the density of the air. Therefore, the balloon does not float up infinitely. Instead, the buoyancy is large at the beginning and then it gradually becomes smaller. When the buoyancy and gravity are balanced, the balloon no longer rises, instead it floats at a certain height in the air. Actually, it is very dangerous to use Kongming lantern to send the information out, because if the wind direction changes, the Kongming lantern will run to the enemy.

It is the French Montgolfier brothers who really let the hot air balloon rise to the sky in science. In the eighteenth century (June 1783), the Montgolfier brothers made a hot air balloon with linen (as shown in Fig. 7.9), and completed the launch of the hot air balloon. So the hot air balloon that

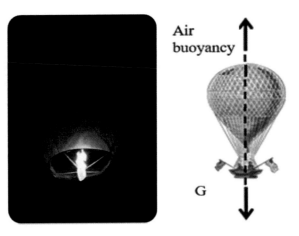

Fig. 7.8 Kongming lantern (the Three Kingdoms Period) and the principle

Fig. 7.9 Hot air balloon made of linen by the French Montgolfier brothers

achieved a successful flight was invented by the Montgolfier brothers. The Montgolfier brothers burned a kind of grass or other substances under the balloon, and the heated flue gas enters the balloon. The density of the hot flue gas inside the balloon is less than the cold air outside, which creates buoyancy and causes the balloon to lift-off. The principle of hot air balloon and Kongming lantern is the same. As long as the buoyancy is greater than gravity, the balloon can fly off.

Another simple device that is driven by the upward movement of hot gas is the Chinese trotting horse lamp (as shown in Fig. 7.10), the principle of which is similar to that of a balloon. People light candles in the lamp, when the candles burn for a certain period of time, the temperature of the gas inside the lamp will rise. At this time, the lamp will rotate. The lampshade is painted with different people riding the horse. People see the painting on the rotating lampshade as if the horse keeps going backwards, so it is called trotting horse lamp. How does the upward movement of the hot air flow rotate the lampshade? From the schematic diagram of the lamp, when the candle is ignited, there will be an upward movement of hot air inside the

Fig. 7.10 Trotting horse lamp

lamp. The airflow enters from under the lamp and then discharges upward from the blade under the hook on the top of the lampshade, this process continually circulates. It has been found that as the airflow passes through the helical blade, it drives the blade to rotate, thereby rotates the lampshade integrated with the axle. Looking out from the outside of the lampshade, it becomes the scene of "Horses gallop and never stop".

7.3   Rapid Development of Aircraft

Humans get inspiration from flying birds and yearn for flight, but they have long been limited by knowledge. The desire for flying can only be pinned on myths and delusions. Thanks to the history of more than 100 years in modern times, human beings have truly realized the dream of soaring into the sky. On December 17, 1903, the American Wright brothers (as shown in Fig. 7.11) invented the first powered aircraft (as shown in Fig. 7.12) and successfully completed the flight. Although the plane flew less than a minute, the descendants gave it a high rating—creating a new era of aviation for humanity. After more than 100 years of hard work, human beings not only mastered the mystery of flight but also overcomed various technical difficulties and developed various aircraft with different uses. The aircraft became the most important scientific and technological achievement of human beings in the twenty-firstcentury and played a huge role in promoting the development of

Fig. 7.11 Wilbur Wright (1867–1912, left) and Orville Wright (1871–1948, right), two American inventors, aircraft manufacturers

social civilization. Aviation science and technology has become an important product of scientific and technological achievement and the development of social civilization.

Although humans have no wings, they have a brain with infinite wisdom and illusions. Through the understanding and practice of the bird's flight mystery, human beings create various aircraft, which in terms of scale, flight altitude, flight speed, flight range, complexity, etc., far exceed those of birds; also far more than the famous verse of Li Shangyin in Tang Dynasty, "We have no wings to fly side by side. Yetone sharp arrow wounded yours and my heart". As shown in Fig. 7.13a, the swan is said to be the most feathered bird, but the sum of feathers and bones is 25,000; the common aircraft Boeing 747-400 has more than 6,000,000 parts (As shown in Fig. 7.13b) is 10^6 times that of train and auto parts. The complexity of the aircraft far exceeds that of birds.

The number of aircraft flying in the atmosphere is huge and there are many types. Aircraft can be divided into two categories, one is the aircraft that generates lift by air buoyancy. The density of such aircraft is lighter than

Fig. 7.12 Aircraft invented by the Wright brothers

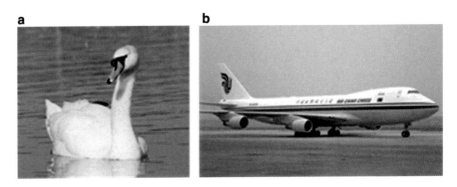

Fig. 7.13 Swan and Boeing 747–400 (**a, b**)

air, such as balloons and airships. Another category is aircraft that generates lift through wings, which are heavier than air. In the second category, fixed-wing aircraft (such as gliders and civil aircraft) and non-fixed-wing aircraft are further divisions. Non-fixed-wing aircraft further includes three types, one is gyroplane, such as helicopter; the other is tilt-rotor aircraft currently in service in Okinawa, Japan and the U. S; the other is flapping-wing aircraft that people have now begun to manufacture, as shown in Fig. 7.14.

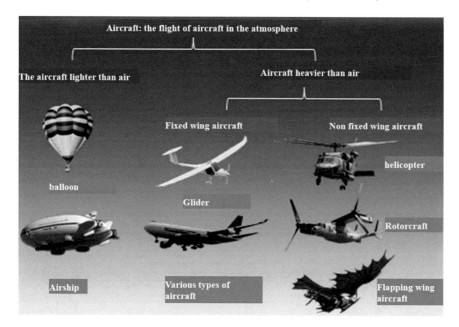

Fig. 7.14 Classification of aircraft

The various aircraft made by humans (as shown in Fig. 7.15) have very different flying heights. The size of the aircraft and birds is represented by the horizontal coordinate, and the height of the flight is represented by the vertical coordinate. Regardless of the scale or height, the birds are concentrated in a small place in the lower left corner (as shown in Fig. 7.16). For example, the world's largest bird (albatross) has a wingspan of up to 4 meters. While the wingspan length of the man-made airplane can reach tens of meters, such as the A380 produced by Airbus, with a wingspan length of 80 meters. From the perspective of flying height, the flying height of ordinary birds is one or two kilometers, even the bird with the highest flying can only fly at an altitude of 9,000 m. The man-made aircraft often cruises at a height of about 10,000 meters. For example, the flight altitude of civil airliners is about 11,000 km, far exceeding the flight height of birds. In addition, high-altitude airships generally fly around 20,000 meters of height, satellites are further away from the atmosphere.

In terms of flight speed, the speed of birds and the speed of man-made aircraft are also not on the same order of magnitude (as shown in Fig. 7.17). Birds usually fly at speeds of a few tens of kilometers per hour, even the fastest flying swift can only reach 170 km/h. The speed of man-made aircraft far exceeds that of birds. The flight speed of civil airliners is 800–900 km/h, the speed of modern supersonic fighters is more than 2000 km/h, and the

Fig. 7.15 Different types of aircraft

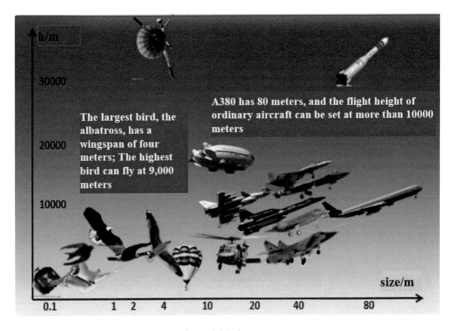

Fig. 7.16 Flight altitude of aircraft and birds

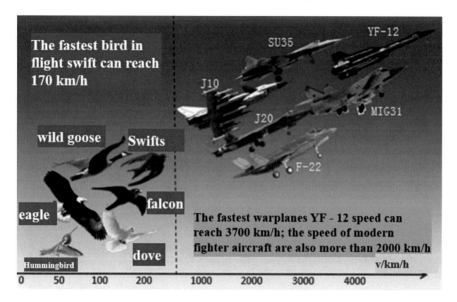

Fig. 7.17 Flight speed of aircraft and birds

fastest military aircraft YF-12 can reach 3700 km/h. This speed is far beyond the reach of birds, and these are the crystallization of human wisdom.

Different types of aircraft play their respective roles in different fields. The level of aviation technology has become an important indicator of a country's economic and technological level, national defense strength, and overall national strength. Skip the difficult technical level, what is the principle of the aircraft? It seems that it is not difficult now, but it took humanity nearly a thousand years to understand this principle. Take the balloon as an example. The weight of the balloon is like the Chinese scale. When the scale is picked up, the upward force and the weight of the scale are equal. At this time, the balloon can fly. This is the principle of balloon flight (as shown in Fig. 7.18). The flight principle of the aircraft is like the Chinese steelyard. The middle part of the aircraft acts on a lifting force. One side is a heavy object and the other side is a scale. When the forces and moments of the three are balanced (as shown in Fig. 7.19), the aircraft can fly in the air. However, even this is a simple principle of force balance human beings have experienced and practiced for thousands of years.

If humans want to be able to fly, they need to answer how the aircraft generates lift during flight, how to overcome air drag, how to maintain balance or stability, and how to control and navigate when voyaging (do not deviate from the route or be unable to return), etc. These questions are answered by people in the modern century. Historically, the history of

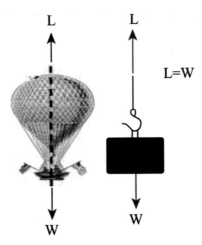

Fig. 7.18 Principle of force balance—spring scale

human flight development can be divided into four stages. The first stage is the flying dream, at which humans rely on illusions to achieve flight. The second stage is the attempt and practice of flight. At this stage, people make a lot of tentative practices in flight in order to turn their dreams into reality. The third stage is the imitation practice of birds, which is called the natural enlightenment flight. The last stage is the crystallization of wisdom. In the course of the modern century, humans used wisdom to creatively invent a large number of complex and different purpose aircraft. At these different stages, aerodynamics plays an important role.

7.4   Flight Principle

In the aerodynamic cognition and application, the windmills in the Netherlands are representative. The principle of windmill rotation is also very simple. When the wind blows over the windmill blades, the windmill will rotate itself (as shown in Fig. 7.20). In the beginning, the windmill was only used for milling. In the sixteenth and seventeenth centuries, windmills played a very important role in the Dutch economy.

In contrast, the flight principle of Chinese bamboo dragonfly (shown in Fig. 7.21) is different from the Dutch windmill. The rotation of the Dutch windmill impeller is driven by the movement of the wind. However, the Chinese bamboo dragonfly is in the still air, and impeller is rotated by twisting the impeller shaft with both hands, which causes the bamboo dragonfly to move upward. Initially there is an inclination angle between the

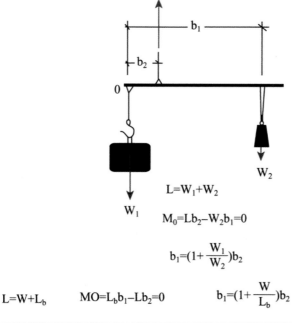

$$L = W_1 + W_2$$

$$M_0 = Lb_2 - W_2 b_1 = 0$$

$$b_1 = \left(1 + \frac{W_1}{W_2}\right) b_2$$

$$L = W + L_b \qquad MO = L_b b_1 - L b_2 = 0 \qquad b_1 = \left(1 + \frac{W}{L_b}\right) b_2$$

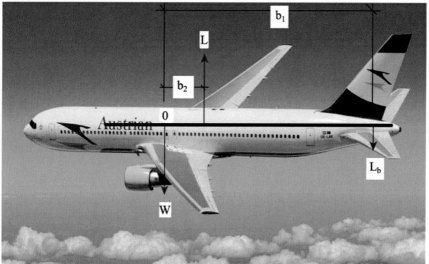

Fig. 7.19 Principle of force balance—Chinese steelyard balance

blade of the bamboo dragonfly and the horizontal rotating surface. When the impeller rotates, the rotating blade discharges the airflow from the upper side to the lower side, that is, gives the air a downward thrust, thereby causing the air to also act on the rotating blade with an upward reaction force, i.e.,

Fig. 7.20 Dutch windmill (1200)—torque principle

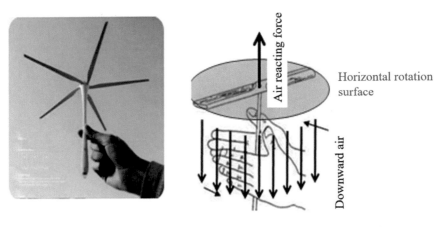

Fig. 7.21 Bamboo dragonfly and its flight principle

an upward lift. When the lift is greater than the weight of the bamboo dragonfly, the bamboo dragonfly can fly upwards. On the basis of this principle, humans invented the propeller later (as shown in Fig. 7.22). The bamboo dragonfly is also the predecessor of the helicopter rotor.

Fig. 7.22 Aircraft piston propeller

The early discussion of the flight mechanism of the aircraft involved almost no control of flight attitude. How can humans fly freely like birds? Careful observers have found that when birds need to pitch or turn, they either lean the body or twist the tail. This leads to the concept of torque. The above flight principle of the balloon with upward buoyancy equal to gravity is very simple. In contrast, the flight principle of the airplane is much more complicated. In addition to the necessary force balance conditions, there is also a torque balance requirement for the center of gravity of the aircraft. Otherwise, the airplane will bow or fly sideways, the flight process cannot be stable; so, the flight principle of the airplane should be the leverage principle of the Chinese scale. For example, when the scale hook lifts the weight, the weight of the device to be tested can be measured by moving the sliding weight on the weigh beam. If the airplane is to remain stable during flight, it needs to maintain a moment balance (as shown in Fig. 7.23). This can be achieved by adjusting the aerodynamic force of the tail so that the aircraft can fly for a long time without falling. A lot of observations have been made on the bird's flight, paying particular attention to how the bird's feather shape is changed to maintain the balance of the body during pitching and steering. In order to maintain the balance of the body during flight, all external forces acting on the bird or aircraft should have a torque equal to zero (torque balance principle) to achieve balance, just like the operating principle of the scale.

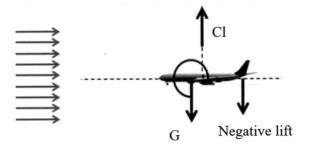

Fig. 7.23 Principle of torque balance during flight

In order to solve the problem of flight stability, many great scientists in history have conducted research. According to the data, it is clearly stated that the concept of torque in flight principle should be derived from the Italian Renaissance. In the fifteenth century, the most studied person in flight principle is the well-known Italian all-round scientist Da Vinci, who was a famous painter, a great natural scientist, and a biologist. People call him an all-round scientist because he was involved in almost all major fields at the time, especially the scientific research on the flight principle of birds for the first time. In history, Italian all-round scientist Da Vinci was the first person to study the principle of bird flight with scientific ideas. Da Vinci has done a lot of research on the bird's feathers, bones, skeletons, and the balance during flight. Nowadays, Da Vinci's manuscript on bird research is still in the world, and today this manuscript is valueless (as shown in Fig. 7.24). Da Vinci studied the principle of flapping wing for nearly 20 years (as shown in Fig. 7.25). The final conclusion is that, limited by the body weight, it is impossible to lift-off by human arms driving the wings. Because the human body is too heavy, the lift generated by the two wings driven by human back cannot overcome the gravity. Da Vinci's self-designed aircraft, which mimicked the principle of flapping wings, was powered by the pilot's arm, called the "Flapping Aircraft" (as shown in Fig. 7.26)—its two wings can flap up and down.

In addition, those who contributed greatly to the study of flight principles include well-known scientists such as Langley in the U.S., Kelly in the U.K., and Otto Lilienthal in Germany. American astronomer and flight pioneer Langley (Samuel Pierpont, 1834–1906, as shown in Fig. 2.5) carefully studied the aerodynamics principle (as shown in Fig. 7.27) and illustrated how birds glide through the wings, the lift calculation formula proposed by him is still used today.

British scientist George Kelly (1773–1857), as shown in Figs. 2.4 and 7.28) is the father of classical aerodynamics. He was 300 years behind Da

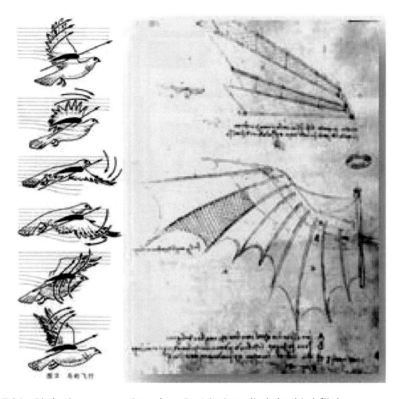

Fig. 7.24   Bird wing manuscript when Da Vinci studied the bird flight

Fig. 7.25   The imagination of human arms drive wing to fly

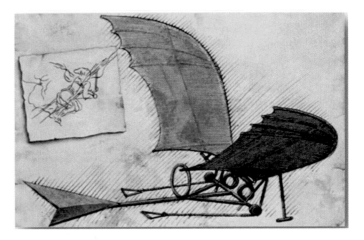

Fig. 7.26 The flapping aircraft designed by Da Vinci

Fig. 7.27 American astronomer and flight pioneer Langley's (Samuel Pierpont, 1834–1906) glider

Vinci. George Kelly also conducted extensive research on bird flight, estimating the relationship between speed, wing area and lift by observing the bird's wing area, weight and speed. In his famous treatise "On Aerial Navigation" published in 1809, he put forward a very important conclusion, "The man-made aircraft should consider the propulsion power and the lift surface separately, and do not consider the trust and lift together". This judgment

Fig. 7.28 Glider from George Kelly (1773–1857)

made people abandon the flapping flight that mimics the bird and gradually accept and practice the correct principle of generating lift with fixed wing. It has also become an important concept for humans to manufacture fixed-wing aircraft.

Kelly's judgment may have been inspired by Chinese kites. For a real kite (as shown in Fig. 7.29), the flow around the kite and its aerodynamic principle indicate that the total aerodynamic force generated perpendicular to the surface of the kite when the airflow bypasses the kite can be decomposed into vertical upward and horizontal drag. The tension of the rope on the kite in the horizontal direction overcomes the aerodynamic drag, and the vertical

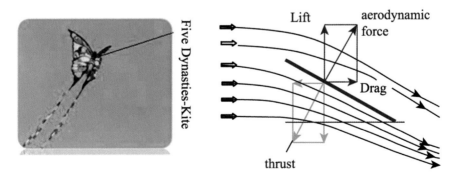

Fig. 7.29 Flow around the kite and its aerodynamic forces

downward component and gravity are balanced with the aerodynamic lift. If the horizontal tension of the rope is replaced by the thrust of the engine, which is balanced with the aerodynamic drag of the kite, then the pull-down force in the vertical direction of the rope is replaced by the gravity of the aircraft, and it is balanced with the aerodynamic lift of the kite. This achieves the concept of balancing lift and drag separately. That is to say, when building an aircraft, the parts that generate lift and the parts that generate thrust can be considered separately, unlike the flapping-wing aircraft, which is considered together (the wing tip portion is both the lift surface and the thrust surface). The decomposition principle proposed by George Kelly played an important role in guiding the later Wright brothers to invent the powered fixed-wing aircraft. Although it was not possible to manufacture a suitable thrust component (the engine) under the technical conditions at that time, he did a lot of practice on the parts that generate lift. He made many gliders, so he is also called the inventor of the glider.

The gilder made by George Kelly has two layers of wings (as shown in Fig. 7.30). Because the lift generated by one layer of the wing is not enough, the two layers of wings can meet the lift requirements. There is a tail behind the glider, and the tail that stands up is called the vertical tail, which plays the role of yaw or side slip control. The horizontal tail has the function of controlling the pitch. What is the principle of glider flight? How can the fixed wing glide? George Kelly analyzed the bird's gliding and got the flight principle of the glider.

Fig. 7.30 Glider made by George Kelly

The eagle glides without power, the open mouth can never function as an engine (as shown in Fig. 7.31). From the appearance point of view, the bird should be the most labor-saving when gliding. If there is no attempt, it can close the eyes and raise the spirit. How can it fly so well? What balances the gravity of the bird and what overcomes the aerodynamic drag? These questions relate to how the aerodynamic forces of fixed-wing aircraft are generated. It has been found that when the bird is gliding, the wings are as straight as possible, and the airflow bypasses at a relatively high speed relative to the wings. At this time, the airflow around the wing is discharged backwards and downwards, so that the airflow acts on the wing with an aerodynamic force. The component of the aerodynamic force perpendicular to the flight direction is called the lift of the wing, and the component of the aerodynamic force parallel to the flight direction is called the aerodynamic drag. Obviously, the bird's gravity is balanced with the lift of the wing, but what is used to overcome the drag. It has been explained that the bird has no power when gliding. In this case, where does the bird's power come from? A careful observation reveals that in this case the bird is not flying horizontally, but flying at a certain downtilt because the component of bird's gravity in parallel with the flight direction is consistent with the flight direction. Its magnitude is just balanced with the drag generated by airflow. When using

Fig. 7.31 Mechanism of eagle gliding

the glider, the driver and the glider stand at a high position, and then the glider floats all the way down. At this time, the force component of gravity overcomes the aerodynamic drag. But sometimes we can see that the glider can also float upwards. What is the reason? It turns out that if the glider flies over the airflow with the upward velocity component, the glider is lifted up. But if there is no upward airflow, the glider can only slide down from the high platform.

After nearly 40 years, German engineer and glider aviator Otto Lilienthal (1848–1896), as shown in Fig. 7.32 began manufacturing gliders. He was one of the pioneers in the manufacture and practice of fixed-wing gliders. In 1896, he died due to a plane crash in a flight test. Lilienthal built a number

Fig. 7.32 Otto Lilienthal (1848–1896, German engineer) and his glider

of single-wing or double-wing fixed-wing gliders and tested more than 2000 times near Berlin. He accumulated a wealth of information. Although he eventually failed to achieve power flight, his valuable flight experience and a large amount of data accumulated have provided a lot of valuable experience for the successful power flight of the American Wright brothers.

Under the guidance of the aviation experience and knowledge of the predecessors George Kelly and Otto Lilienthal, the American Wright brothers built their own powered aircraft based on their experimental data. On December 17, 1903, the Wright brothers successfully tested their first powered, steerable aircraft in the world (as shown in Fig. 7.33). The takeoff weight of the Flyer No. 1 was only 360 kilograms. It could barely carry one person off the ground. The speed was slower than the car, only 48 km/h, the most successful flight lasted only 59 s, and the flight distance was 260 meters. But it was such an inconspicuous small plane that opened an important page in the history of human aviation. From then on, humans have realized the powered flight of fixed-wing aircraft, allowing humans to enter the era of aviation civilization. The Wright brothers did not go to college, they were engaged in bicycle repair work; their mechanical principles were well mastered. So, the Wright brothers were able to turn the scientific principles into reality through hands-on practice. Such strong hands-on practical ability is necessary for people to

Fig. 7.33 The historic moment by John Daniels

carry out innovative work. So, it is necessary for the students to strengthen hands-on practical ability training.

The Wright brothers' aircraft is a biplane (as shown in Fig. 7.34). There is a horizontal tail in front of the plane, and a vertical tail in the back, the pilot lies in the middle, and the propeller pushes the plane behind. In this

Fig. 7.34 Aircraft design and physical map of the Wright brothers

plane, a person lies on the wing, and the horizontal tail is placed in front of the wing; this arrangement is called the canard layout. The layout of the Wright brothers' aircraft is different from the aircraft that is often seen today. If the horizontal tail is placed behind the wing, it is called the conventional layout; if the horizontal tail is placed in front of the wing, it is the layout of the Wright brothers, called the canard layout. The flight control principle of these two layouts is different during flight. The flight stability of the conventional layout is better, so the civil aircraft uses this layout. The canard layout, although poor in stability, has good maneuverability, so fighters often use this layout. For example, the Chinese J-10 fighter placed the horizontal tail in front of the wing. The tail that is placed in front is called the canard, and the tail placed behind is called the horizontal tail.

The speed of the early aircraft was relatively small, so the first problem encountered was the insufficient lift generated. How to generate a large enough lift to fly the aircraft is the key to the early design and manufacture of the aircraft. Since the aerodynamic force of the aircraft is proportional to the square of the flight speed, when the flight speed is small, the aerodynamic force generated under the same wing area is also small, but the weight of the aircraft cannot be reduced, so measures to increase the wing area are taken. But considering the factors in the structure of the wing, the layout of two-layer wing and even the three-layer wing appeared in the early aircraft. The three-layer wing layout was created because the two-layer wing was not enough, so the three-layer was created (as shown in Fig. 7.35). Because the flight speed is relatively small, the drag is not highlighted compared to the lift.

According to the definition of Langley et al., the lift expression of the aircraft is

$$L = \frac{1}{2}\rho V_\infty^2 C_L S$$

V_∞ is the inflow velocity (the flying speed of the aircraft), S is the characteristic area of the wing, C_L is the lift coefficient of the wing. The drag expression is

$$D = \frac{1}{2}\rho V_\infty^2 C_D S$$

where D is drag of the wing (parallel to the direction of flow), C_D is the drag coefficient of the wing.

Fig. 7.35 Two-layer and three-layer wing aircraft (German Fokker Dr.I three-layer wing fighter)

The wing lift-to-drag ratio defined by Lilienthal is

$$K = \frac{L}{D} = \frac{C_L}{C_D}$$

This is a measure of wing performance and efficiency and plays an important role in wing design. The definition of characteristic area is different in aerodynamics and general fluid mechanics. In aerodynamics, the characteristic area S of the wing generally refers to the horizontal projected area of the wing (or the gross area). For general fluid mechanics textbooks, the area S used to calculate the drag coefficient is generally the windward area (refers to the projected area of the object perpendicular to the flow direction).

A large number of wind tunnel tests have shown that for a given wing profile, the wing lift is primarily a function of the following variables without side slip.

$$L = f(V_\infty, \rho, S, \alpha, \mu, a)$$

In the formula, α is the angle of attack, a is the sound speed in the air, and the rest of the symbols have the same physical meaning as before. According to the dimensional analysis, the lift and lift coefficient expressions of the wing can be obtained as

$$L = \frac{1}{2}\rho V_\infty^2 S C_L$$
$$C_L = f(Ma, \text{Re}, \alpha)$$

where Ma $(=V_\infty/a)$ is the Mach number, $\text{Re}(=V_\infty b/\nu$, b is the characteristic chord length of the wing, ν is the kinematic viscosity coefficient of the air) is inflow Reynolds number. For low-speed wing flow, the compressibility of the air is negligible, but the viscosity of the air must be considered. Therefore, the aerodynamic coefficient is actually a function of the inflow angle of attack and the Re number. The specific form of the function can be given by experimental or theoretical analysis. For high-speed flow, the effects of compressibility must be considered. Ma number becomes an important variable for theoretical analysis and experimental simulation. When studying the wing drag, the above lift expression can be written as a drag expression (as shown in Fig. 7.36).

$$D = \frac{1}{2}\rho V_\infty^2 S C_D$$
$$C_D = g(Ma, \text{Re}, \alpha)$$

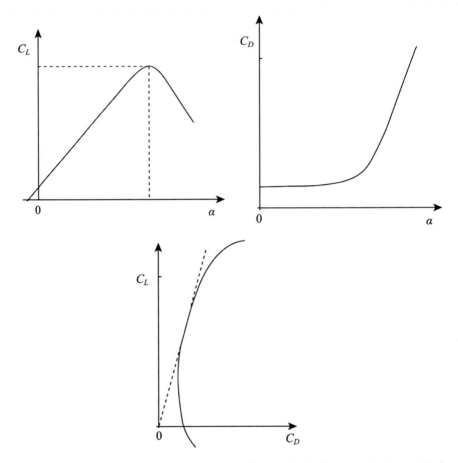

Fig. 7.36 The curve of the lift and drag coefficient of the low-speed wing with the angle of attack and the polar curve

7.5   Wing Shape and Aerodynamic Coefficient

1. *Wing shape*

The shape of the aircraft wing is varied, ranging from straight to triangular, swept to forward (as shown in Fig. 7.37). However, no matter what shape is used, the designer must make the aircraft have a good aerodynamic shape and make the structural weight as light as possible. The so-called good aerodynamic shape means that the wing has large lift, low drag, and good maneuverability. For low-speed wing, in order to reduce the induced drag, a straight wing with a large aspect ratio is often adopted. For transport aircraft, most of them use upper wing (easy to load). For high subsonic airliner, in

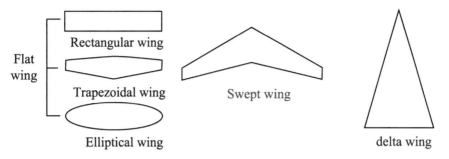

Fig. 7.37 Plane shape of the wing

order to suppress shock waves, generally adopting a conventional swept-back lower wing (large lift-to-drag ratio, good economy, low cabin noise, and wide field of view), placing goods in the lower half of the fuselage; for fighters, most of them use medium or lower single-wing, delta wing, conventional or canard layout with large sweep angle (fast speed, low drag, flexible maneuver and large stall angle of attack).

For example, the American tactical transport aircraft C-130 (as shown in Fig. 7.38) has a cruising speed of 540 km/h and a maximum takeoff weight of 70.3 tons. It adopts the layout of the single upper rectangular straight

Fig. 7.38 American C-130 tactical transport aircraft (cruising speed 540 km/h) (upper wing, straight wing, four-engine wing hanging arrangement, conventional layout)

wing, four engines, and large cargo door at the tail. This layout laid the design "standard" for the medium-sized transport aircraft after the Second World War. The most striking feature of the C-130 is its design to thoroughly meet the actual requirements of tactical air transport, so it is very suitable for a variety of air transport tasks, for example, a large tail cargo door of an aluminum alloy semi-monocoque structure fuselage. The airfoil of the wing root is NACA 64A318, the airfoil of the wing tip is NACA 64A412, the dihedral angle is 2°30′, the wing root installation angle is 3°, the wing tip installation angle is 0°, and the ¼ chord sweep angle is 0°. The all-metal double-beam forced skin structure and the machined integral reinforced variable thickness skin panels with a length of 14.63 m are adopted. The ailerons are made of ordinary aluminum alloy. A tandem hydraulic booster is used to supply pressure from two separate hydraulic systems. There are tabs on the aileron. The Fowler aluminum alloy trailing edge flaps are used, and the leading edge of the wing uses engine bleed air to prevent ice.

In Chinese transport Y-20 (as shown in Fig. 7.39), the cruising speed is 800 km/h and the maximum takeoff weight is 220 tons. The conventional layout is adopted. The wing is a cantilevered upper wing. The main wing is a large aspect ratio, medium-swept supercritical wing, the leading edge sweep angle of the wing is constant (the 1/4 chord sweep angle is about 24 to 26°), the trailing edge of the wing adopts two types of sweeping (the trailing edge of the middle and outer wing segments has a larger sweep angle, while the inner wing segment has a significantly reduced sweep angle). The

Fig. 7.39 Chinese transport aircraft Y-20 (cruising speed 800 km/h) (upper wing, four-engine wing hanging arrangement, swept supercritical wing, conventional high horizontal tail layout)

aircraft has no winglets, and it is equipped with a lifting device including a leading edge slat and a push-back Fowler flap system. The cantilever T-tail is used, and the small dorsal fin is extended forward at the joint between the vertical stabilizer and the fuselage. The embedded rudder is divided into upper and lower sections, and the elevator is divided into two sections. The hydraulic retractable front three-point landing gear can be freely lowered by gravity during emergency. The front landing gear has two wheels and the main landing gear has six wheels. The front landing gear retracts forward into the fuselage, and the main landing gear rotates 90° and retracts into the fairing on both sides of the fuselage. The Y-20 is based on the design of the IL-76 transport aircraft, but it is larger in size, more capable of carrying and the electronic equipment is also very advanced.

The large airliner C919 being developed in China (as shown in Fig. 7.40) is a single-aisle narrow-body 150 seat high subsonic trunk airliner with a cruising speed of 850 km/h and a maximum takeoff weight of 72.5 tons. The aircraft has a total length of 38 meters, a wingspan of 33 meters, and a height of 12 meters. Its basic layout has 168 seats. The standard voyage is 4,075 km, the increased range is 5,555 km, and the economic life is 90,000 flight hours. The aircraft is a conventional layout of the lower wing with wing

Fig. 7.40 large airliner C919 being developed in China (cruising speed 850 km/h) (lower single wing, two-engine wing hanging arrangement, swept-back supercritical wing, conventional low horizontal tail layout)

crane engine. The main wing is a large aspect ratio, medium-swept supercritical wing. The lifting device is a leading edge slat and a push-back Fowler flap, and the wing tip is equipped with a winglet. It achieves better cruise aerodynamic efficiency than the same type of aircraft in service, and has comparable cruise aerodynamic efficiency with competitive aircraft in the market ten years later. Advanced engines are used to reduce fuel consumption, noise, and emissions. A large number of advanced composite materials, advanced aluminum-lithium alloys, etc., are used. The composite material usage will reach 20%, which will reduce the structural weight of the aircraft. Advanced fly-by-wire and active control technologies are used to enhance the overall performance of the aircraft and improve human factor and comfort. The advanced integrated avionics technology is adopted to reduce pilot burden, improve navigation performance and perfect man-machine interface.

2. *Aerodynamic coefficient of the wing*

If the incoming flow V_∞ is parallel to the symmetry plane of the wing, the flow along the direction of the incoming flow is referred to as the longitudinal flow of the wing. The angle between V_∞ and chord of the wing profile (wing root profile) at the symmetry plane is defined as the angle of attack α. The aerodynamic forces acting on the wing in longitudinal direction are lift L (perpendicular to V_∞), drag D (parallel to V_∞), longitudinal moment M_z (the pitching moment of a reference point). The dimensionless aerodynamic coefficients defined in longitudinal flow are

Lift coefficient $C_L = \dfrac{L}{\frac{1}{2}\rho_\infty V_\infty^2 S}$

Drag coefficient $C_d = \dfrac{D}{\frac{1}{2}\rho_\infty V_\infty^2 S}$

Longitudinal moment coefficient $m_z = \dfrac{M_z}{\frac{1}{2}\rho_\infty V_\infty^2 S b_A}$

where S is the area of the wing, b_A is the average aerodynamic chord of the wing. The average aerodynamic chord length is the chord length of an imaginary rectangular wing. The area S of this imaginary wing is equal to the area of the actual wing, and its torque characteristics are the same as the actual wing (as shown in Fig. 7.41).

$$b_A = \frac{2}{S} \int_0^{l/2} b^2(z)dz$$

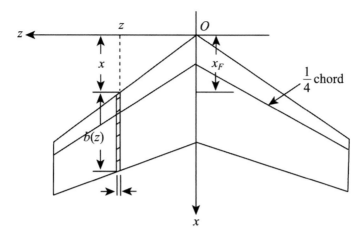

Fig. 7.41 Plane shape of the wing

3. *Low-speed aerodynamic characteristics of large aspect ratio straight wing*

From 1911 to 1918, Prandtl discovered through wind tunnel tests that the flow around the large aspect ratio straight wing (the leading edge sweep angle of the wing is less than 20 degrees and the aspect ratio is greater than 5) is affected by the spanwise flow, and the flow around the wing can be substituted by a model of straight flow superimposed attached vortex (filament) and free vortex sheet (as shown in Fig. 7.42). The attached vortex and the free vortex sheet are connected by an infinite number of Π-shaped horseshoe vortices, called the lift line model. This is because, for a large aspect ratio straight wing, the rolling and bending of the free vortex sheet occurs mainly away from the wing (approximately one wingspan length from the trailing edge of the wing). In the trailing edge region of the wing, it can be assumed that the free vortex sheet is neither rolled up nor dissipated, and extends along the incoming flow direction to infinity. The reason why the aerodynamic model conforms to the actual flow is as follows:

(1) The model conforms to ideal fluid vortex invariance theorem that the vortex intensity along a vortex line does not change and cannot be interrupted in the fluid.
(2) The part of the Π-shaped horseshoe vortex that vertical to the incoming flow is the attached vortex, which can replace the lift of the wing. The number of vortex lines passing through each section of the span is different. The middle section passes the most vortex lines and the largest amount of circulation; the end section of the wing has no vortex line

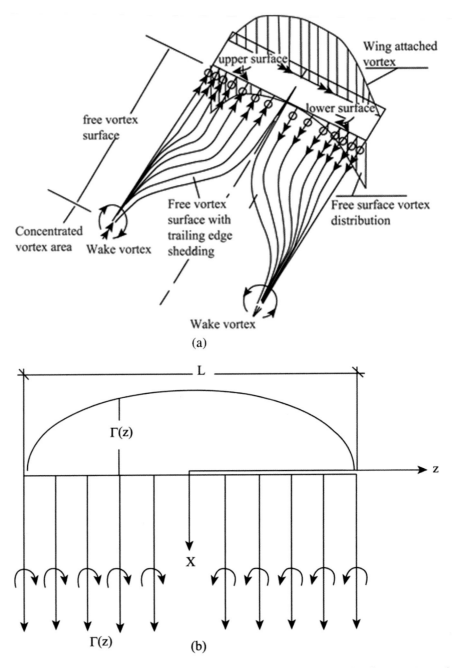

Fig. 7.42 **a** The relationship between the attached vortices and the free vortices of the large aspect ratio straight wing. **b** Lift line model for large aspect ratio straight wing

passing through, and the circulation is zero, which simulates the span-wise distribution of the circulation and lift (the elliptical distribution is the best). It can be seen that the strength of the attached vortex varies along the span, and is the same as the profile lift distribution which is zero at the wing tip and largest at the wing root.

(3) The Π-shaped horseshoe vortices are parallel to the inflow and drag infinity downstream, simulating the free vortex sheet. Since the free vortex intensity drawn between the spanwise adjacent two sections is equal to the circulation difference of the attached vortices on the two sections, the relationship between the spanwise free vortex line intensity and the circulation of attached vortex on the wing is established.

(4) For the large aspect ratio straight wing, since the chord length is much smaller than the span length, the attached vortices on the wing can be approximately combined into a vortex line with variable strength along spanwise. When the lift of each section acts on the line, it is called the lift line hypothesis. Because the lift increasement of the low-speed airfoil acts on the focus, about the quarter chord point, the attached vortex line can be placed on the line connecting the quarter chord point of each section, which is the lift line.

The lift coefficient of the three-dimensional wing obtained using this model is

$$
C_L = C_L^\alpha (\alpha - \alpha_{0\infty}) = \frac{C_{L\infty}^\alpha}{1 + \frac{C_{L\infty}^\alpha}{\pi \lambda}(1 + \tau)}(\alpha - \alpha_{0\infty})
$$

The induced drag coefficient is

$$
C_{Di} = \frac{C_L^2}{\pi \lambda}(1 + \delta)
$$

where τ and δ represent the aerodynamic correction coefficients of the non-elliptical wing relative to the elliptical wing, indicating the extent of other planar-shaped wings that deviate from the optimal planar shape. $\alpha_{0\infty}$ is the zero-lift angle of attack of the two-dimensional airfoil. $C_{L\infty}^\alpha$ is the slope of the lift line for a two-dimensional airfoil. $\lambda = L^2/S$ is the aspect ratio of the wing. The above equations show that at the same angle of attack, the slope of the lift line of the three-dimensional wing is smaller than that of the infinite span wing (two-dimensional airfoil), and the slope of the lift line decreases as the aspect ratio decreases.

The lift line theory is an approximate potential flow theory for solving large aspect ratio straight wing flow. After the known wing plane shape and airfoil aerodynamic data, the circulation distribution, the lift coefficient distribution of the profile, the lift coefficient, slope of the lift line, and induced drag coefficient of the entire wing can be obtained. The outstanding advantage is that the influence of the wing plane shape parameters on the aerodynamic characteristics of the wing can be clearly given. The application conditions of this theory are as follows:

1. The angle of attack cannot be too large ($\alpha < 10°$). The lift line theory does not consider the viscosity of the air, and there is a significant separation of the flow at high angle of attack.
2. The aspect ratio cannot be too small ($\lambda \geq 5$).
3. The sweep angle cannot be too large ($\beta \leq 20°$).

7.6   Supercritical Wing

The concept of supercritical airfoil (shown in Fig. 7.43) was proposed by Richard T. Whitcomb (1921–2009, who is called "a person who lives by talking to the airflow", as shown in Fig. 7.44), director of the NASA Langley Research Center in the United States, in 1967 to increase the drag divergence Mach number of the subsonic transporter (as shown in Fig. 7.45). It was first applied in the large passenger aircraft A320 in the 1980s, and is currently the core technology for the design of the large passenger aircraft wing (supercritical wing). His other two famous creative research results are the area rule proposed in 1955 (refers to the distribution relationship between the zero-lift shock resistance and the cross-sectional area distribution along the longitudinal axis during transonic or supersonic flight. According to the area rule,

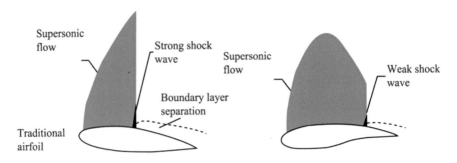

Fig. 7.43   Transonic airfoil flow (normal airfoil and supercritical airfoil)

Fig. 7.44   Richard T. Whitcomb (1921~2009, American aerodynamicist)

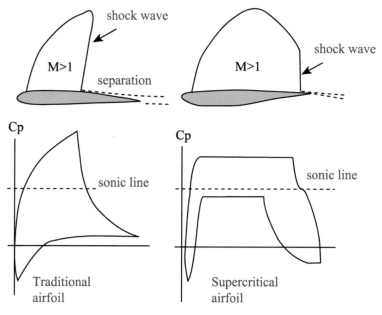

Fig. 7.45   Comparison of transonic flow between normal airfoil and supercritical airfoil

it is possible to reduce the shock resistance during transonic or supersonic flight when designing the aircraft, and improve the flight performance. The area rule also provides a simplified method of estimating the shock resistance of an aircraft, replacing the calculation of the shock resistance of a complex aircraft by a simple equivalent rotating body. Therefore, the area rule is widely used in the design of transonic and supersonic aircraft, as shown in Fig. 7.46) and the winglet proposed in the 1970s (as shown in Fig. 7.47); these play a significant role in modern aircraft design.

Supercritical wings are widely used in the new generation civil aircraft and military transport aircraft due to their high aerodynamic efficiency, high cruise Mach number (as shown in Fig. 7.48), and large relative thickness of the wing (as shown in Fig. 7.49). This type of airfoil was also used in the design of supercritical fighters which was adopted by almost all transonic aircraft after the 1980s.

Along with the vigorous development of the world civil aviation market and the progress of transonic aerodynamics, the concept of supercritical airfoil has experienced the whole stage of concept development, theoretical maturity, design and experiment to model application since its introduction, and has perfected and accumulated a lot of successful model usage experience. NASA has publicly released three generations of supercritical airfoil (as shown

Fig. 7.46 Indented fuselage of supersonic aircraft (area rule)

Fig. 7.47 Winglets of various large passenger aircraft

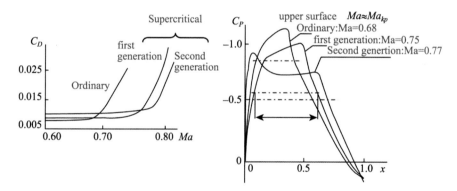

Fig. 7.48 Drag divergence characteristics and pressure distribution of supercritical and normal airfoil

in Fig. 7.50). And the airfoil families developed and used by major aircraft design companies including Boeing and Airbus are more advanced. Since the first successful use of supercritical wing on the A320 by Airbus, the subsequent A330, A340, A380, and A350 adopted this type of wing (as shown in Fig. 7.51, 7.52, 7.53 and 7.54). Boeing's B737-800, B747-8, B787, etc., also use supercritical wings (as shown in Figs. 7.55, 7.56, 7.57). The large

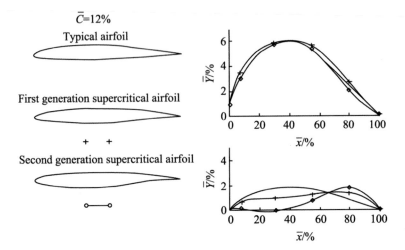

Fig. 7.49 Comparison of shape, thickness, and camber of normal and supercritical airfoil

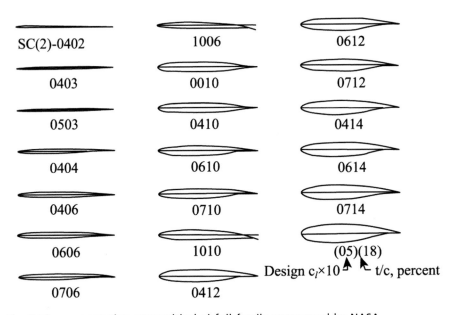

Fig. 7.50 s generation supercritical airfoil family announced by NASA

passenger aircraft family is shown in Figs. 7.58, 7.59 and 7.60. Figure 7.61 shows the magical flock of birds flying.

The geometric characteristics and flow mechanisms of the supercritical airfoil (as shown in Figs. 7.43 and 7.62) are as follows: (1) The curvature of the upper airfoil is relatively flat. When the incoming Mach number exceeds

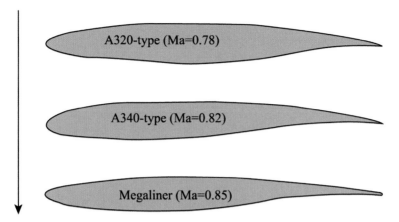

Fig. 7.51 Typical wing profiles of Airbus aircraft

Fig. 7.52 A320 (supercritical wing)

the critical Mach number, starting from 5% chord length from the leading edge, there is a uniform supersonic flow with almost no acceleration along the upper surface. Thus, the supersonic Mach number before the shock wave at the trailing edge is low, the shock intensity is weak, and the stretching range is not large, the back pressure gradient after the shock wave is small, the boundary is not easy to separate, thereby alleviating the drag divergence phenomenon. (2) In order to compensate for the lack of lift in front section of the upper airfoil, the lower airfoil near the trailing edge is made concave to

Fig. 7.53 A350 (supercritical wing)

Fig. 7.54 A380 (supercritical wing)

increase the curvature of the rear section of the airfoil so that the rear section can generate a large lift (post-loading effect).

Compared to the conventional peaky airfoil, the supercritical airfoil can increase the drag divergence Mach number by about 0.05–0.12, or increase

Fig. 7.55 B737-800 (supercritical wing)

Fig. 7.56 B787 (supercritical wing)

Fig. 7.57   B747-8 (supercritical wing)

Fig. 7.58   Large passenger aircraft family (takeoff configuration)

the maximum relative thickness of the airfoil by 2–5%. The use of a thicker airfoil can increase the wing aspect ratio by 2.5–3.0, or reduce the wing sweep angle by about 5–10 degrees while maintaining the drag divergence Mach number.

Fig. 7.59 Large passenger aircraft family (landing configuration)

Fig. 7.60 Large passenger aircraft family (climb)

Fig. 7.61 Magical flock of birds flying

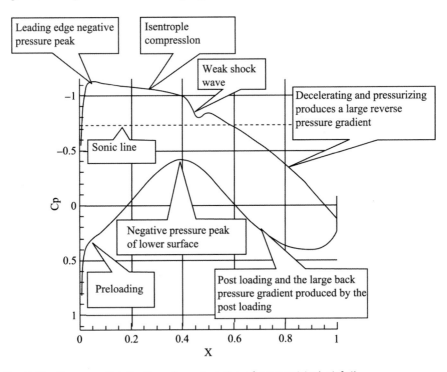

Fig. 7.62 Pressure distribution characteristics of supercritical airfoil

7.7 Winglet

A closer look at the shape of the bird's wing as it landed and soared (as shown in Fig. 7.63) reveals that when the birds are soaring, the wings extend as far as possible to the ends to obtain low drag. When landing (as shown in Fig. 7.63), the wings will expand as much as possible for greater lift.

When the swan is soaring, the feathers at the wing tip also appear in a raised shape (as shown in Fig. 7.64). Modeling this shape, in the 1970s, Whitcomb, director of the NASA Langley Wind Tunnel Laboratory in the United States, invented a wing tip raised device called the winglet. The winglets are mainly used to reduce the induced drag during cruising. At present, the large aircraft produced by Boeing and European Airbus is basically equipped with winglets. The Chinese ARJ21 and C919 passenger aircraft also have installed winglets.

Fig. 7.63 Comparison of the wing shape when flying and taking off

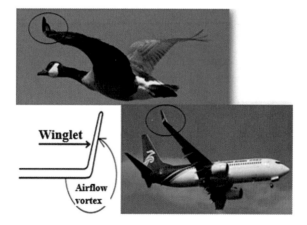

Fig. 7.64 Winglets of plane flying in horizontal and swan soaring

Due to the pressure difference between the upper and lower surfaces of the wing tip, the air tends to flow along the lower surface to the upper surface. After the winglet is added, it will have an end plate effect on the spanwise flow, and the winglet vortex will diffuse the wing tip vortex, which will weaken the downwash effect of the wing tail vortex, lower the downwash angle, and reduce the induced drag. At present, the winglets used in the civil aircraft include ordinary winglets, wing tip vortex diffusers, shark fin winglets, wing tip sail, etc.

The main features of the winglet include (1) end plate effect, blocking the flow from the lower surface to the upper surface, weakening the wing tip vortex strength, thereby increasing the effective aspect ratio of the wing; (2) Dissipating the wing tip vortex of the main wing, because the winglet itself is also a small wing, it can also generate a wing tip vortex, its direction is the same as the wing tip vortex of the main wing, but because the distance is very close, at the intersection of the two vortices the shearing effect is very strong, casing a large viscous dissipation, preventing the winding of the main vortex. The wing tip vortex plays the role of diffusing the main vortex and also reducing the induced drag (as shown in Fig. 7.65); (3) Increasing the wing lift and forward thrust. The upper winglet can generate lift and thrust components using the ternary distortion flow field (as shown in Fig. 7.66); (4) Delay the premature separation of the wing tip flow and increase the stall angle of attack. Generally, the wing tip of the swept wing is a ternary effect zone, the flow tube contracts, the airflow firstly accelerates rapidly, the pressure decreases, and then experiences an intense pressure gradient recovery, enters a steep reverse pressure gradient zone, causing the wing tip boundary layer separate prematurely, resulting in stall. However, the winglet mounted on the wing tip can use its forward pressure field to counteract the partial wing tip reverse pressure field, so that the pressure distribution is gentle and the reverse pressure gradient is reduced. If properly designed, the airflow separation at the wing tip can be delayed, and the aircraft stall angle of attack and buffeting lift coefficient can be improved.

The effect of the winglet: the winglet has single upper winglet, upper and lower winglet and many other forms. The single upper winglet is used more because of its simple structure. The induced drag of the aircraft accounts for about 30% of the cruise drag. Reducing the induced drag is important for improving cruise economy. The larger the aspect ratio of the wing, the smaller the induced drag. However, an excessively large aspect ratio will make the wing too heavy, thus there is a limit to increasing the wing aspect ratio. In addition to being a wing tip end plate to increase the effective aspect ratio of the wing, the winglet also utilizes the "pull effect" produced by the deflection

Fig. 7.65 Shear dissipation of wing tip vortex and winglet vortex

of the wing tip airflow to reduce the induced drag. Wind tunnel experiments and flight test results show that the winglet can reduce the induced drag of the whole aircraft by 20%–30%, which is equivalent to a 5% increase in lift-to-drag ratio. As an advanced aerodynamic design to improve flight economy and save fuel, the winglet has been adopted in many aircraft. The types of winglets are wing tip vortex diffusers (as shown in Fig. 7.67), shark fin winglets (as shown in Fig. 7.68), and wing tip sail (Chinese Y-5, as shown in Fig. 7.69).

Fig. 7.66 Winglet

7.8   Slender Fuselage

The aircraft fuselage is a component used to load people, cargo, weapons and airborne equipment, as well as connector for components such as wings, tails, and landing gears. On light aircraft and fighters, the engine is often installed in the fuselage. The drag of the fuselage in flight accounts for about 30%~40% of the total drag. Therefore, the slender streamlined fuselage plays an important role in reducing drag and improving flight performance. Since the driver, passengers, cargo, and airborne equipment are concentrated on the fuselage, most of the requirements related to the use of the aircraft (such as the driver's vision, the environmental requirements of the cockpit, the loading and unloading of goods and weapons, the inspection and maintenance of system equipment, etc.) have a direct impact on the shape and structure of

Fig. 7.67 Wing tip vortex diffusers

the fuselage. Similarly, when birds are flying, they also pursue a low-resistance shape. At this time, the bird puts its legs under the stomach, and after the retraction, they are covered with feathers to form a slender cone (as shown in Fig. 7.70) to reduce air resistance.

When the seagull soars, the shape of the body is a cone with a large slenderness ratio. The slenderness ratio refers to the ratio of the length of the body to the maximum diameter. This ratio is generally between 6 and 10 when the air resistance is small. Imitating the bird's body shape, the fuselage of the human-made aircraft is also a large slenderness ratio cone, so that the smallest drag can be obtained. The slender body of a large passenger aircraft can accommodate 150–200 people, but its drag coefficient is much smaller than that of a car, which is only 1/8–1/7 of the best car. This magnitude

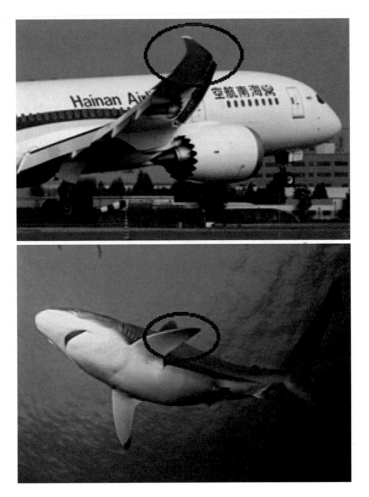

Fig. 7.68 Shark fin winglets

is very magical. The car can accommodate five people, but its drag coefficient is 7–8 times higher than that of a huge fuselage that can accommodate 150 people, this is an aerodynamic achievement. For the fuselage of different aspect ratios, in the 1950s, German aerodynamicist D.Küchemann (Prandtl's student, 1911–1976), Theodore von Kármán (1881–1963) et al. carried out systematic research and obtained many results of great value.

In daily life, people also use the knowledge of aerodynamic drag reduction. For example, when riding a bicycle, if there is a headwind, in order to reduce the drag, people will consciously bend the waist down to achieve the purpose of reducing the windward area. Although the big helmet worn by the racing athlete looks a bit clumsy, wearing such a helmet can reduce the drag. Compared with not wearing the helmet, the drag is reduced. At the

Fig. 7.69 Wing tip sail (Chinese Y-5)

extreme moments, as long as the drag is reduced a little bit, you can win the game. Even if the bird is flying with a fish, it will carry the fish in the direction of the airflow (as shown in Fig. 7.71). The geometrical characteristics of different aircraft fuselages are shown in Figs. 7.72, 7.73, 7.74, 7.75, 7.76, 7.77, 7.78 and 7.79.

7.9 Moment in Stable Flight and Tail

In order to fly stably, in addition to the balance between lift and gravity, engine thrust and drag, there is also a balance of torque (as shown in Fig. 7.80). The lift acts on the wing, and the center of gravity of the airplane

Fig. 7.70 Flying seagull

Fig. 7.71 Practice of drag reduction

is in front of the lift. The lift is not at the same point as the gravity. This will result in a nose down moment around the center of gravity. If there is no horizontal tail behind the fuselage, the airplane cannot fly stably. When making a kite, in order to avoid the nose down flight, the knot of the rope must be

Fig. 7.72 Auxiliary tank

Fig. 7.73 Large passenger aircraft in cruising flight

Fig. 7.74 Fuselage section of the B787

Fig. 7.75 Head and front fuselage section of the Airbus 380

Fig. 7.76 Fuselage of Cessna 510

Fig. 7.77 Fuselage of Airbus 340

tied to the action point of the aerodynamic force in order to maintain the balance of the moment. In order to make the head of a plane be not down when flying, there must be a moment that raises the plane's head up, so that a horizontal tail is placed at the rear of the fuselage to produce a downward force, like the scale of the Chinese steelyard, generating the head up torque.

Fig. 7.78 New concept blended wing body layout

Fig. 7.79 Fuselage of the SH-5

When the plane is flying, the horizontal tail produces a small negative lift, the upward lift and the downward gravity are balanced with the negative lift, and the total moment of the center of gravity is zero, so the plane can fly smoothly without bowing.

When designing an aircraft, it is reasonable to match the relative position and area of the horizontal tail. If the horizontal tail area is large, although the arm is short (the fuselage is relatively short), the negative lift is large. On the one hand, the large negative lift will reduce the wing lift too much so

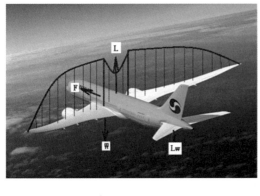

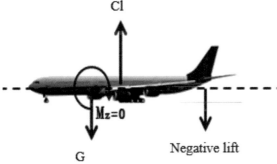

Fig. 7.80 Schematic diagram of the forces when the plane is cruising

that the total lift of the aircraft is insufficient; on the other hand, the large area will generate a large horizontal tail drag. On the contrary, if the area of the horizontal tail is too small, the arm is too large, resulting in the fuselage being too long and not easy to take off. Therefore, the aircraft designer must March the relative position and size of the main wing and the tail so that the aircraft can maintain a good torque balance in various flight attitudes.

7.10 Demand of Aircraft Power (Engine)

First, let us see how the bird generates power. When the bird's wings are open, it is divided into inner and outer sections. No matter how the bird waves its wings, the inner part only generates lift and with almost no thrust, so it is called the arm wing. The outer section of the wing is called the bow wing. When the bow wing moves, it goes up in the direction of the inflow and comes back against the direction of the inflow; at this time, the airflow will generate a thrust on the wing. This force is the power used by the bird to fly. Therefore, through the connection of the feathers, muscles, and bones, the

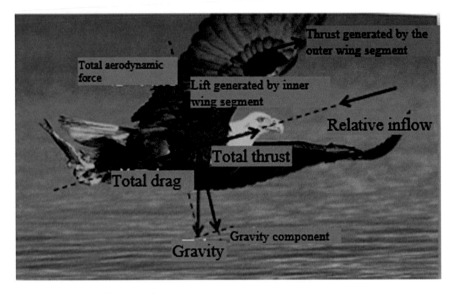

Fig. 7.81 Thrust of the bird wing

bow wing can generate enough thrust, which allows the bird to climb into the air and also to carry out various maneuvers (as shown in Fig. 7.81).

So how does a man-made aircraft generate power? Because humans have difficulty in the manufacture of musculoskeletal joints, they are unable to generate enough aerodynamic forces, so Kelly proposed to consider power and lift separately. Separating the power from the lift (as shown in Fig. 7.82) is equivalent to cutting off the bird's bow wing, leaving the arm wing to generate lift. In this way, how can we generate power? Thus, humans invented the engine to replace the bow wing. The wing only provides lift to balance the aircraft's gravity, and the engine only provides thrust to overcome the aircraft's drag.

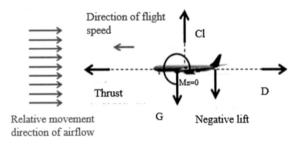

Fig. 7.82 The principle of the aircraft flying forward

The earliest device used to generate thrust can be regarded as the bamboo dragonfly invented by the Chinese. The principle that the bamboo dragonfly can fly is to make the blades mounted on the hub rotate rapidly. At this time, a relative motion occurs between the airflow and the blade. This relative motion generates aerodynamic force. There is lift and drag on the rotating blade relative to the direction of the airflow. If the aerodynamic forces on all blades are projected in the direction of flight and summed, the thrust of the propeller is obtained. If the hub shaft is connected to the engine behind it to drive the propeller to rotate, the propeller can generate thrust. Obviously, the propeller here is to change the rotational kinetic energy of the engine into the kinetic energy required for the level fight of the plane (as shown in Fig. 7.83). In general, the maximum speed of a propeller aircraft can reach 700 km/h, usually 500–600 km/h. For example, Chinese Y-8 and Y-7 are such a flying speed.

In terms of propeller theory, in 1878, W. Froude first proposed the concept of blade element theory, which divided the blade into a finite number of tiny segments (called blade elements), and then calculated aerodynamic forces on each element according to the airfoil theory. It is considered that the airflow passing each blade element is two-dimensional, so the blade elements do not affect each other, which is equivalent to assuming that there is no radical

Fig. 7.83 Turboprop engine

flow on the propeller and no interference between the blades. Finally, the total aerodynamic force on the blades is obtained in the radial direction. Compared with the momentum theory, the blade element theory deals with the aerodynamic forces acting on the blades. The blade element theory does not account for the downwash effect produced by the propeller blades, nor does it account for interference between the blades. In order to improve the blade element theory, Joukowski and his postgraduate students created the propeller vortex theory in four papers published between 1912 and 1915, but in order to simplify the problem, Joukowski assumed that the circulation of attached vortex is constant along spanwise, the number of blades is infinite, forming a vortex basket composed of an intermediate vortex belt, a cylindrical vortex side surface, and an attached vortex chassis. Considering the actual circulation that varies along the spanwise of the blade, the British fluid dynamicist H.Glauert proposed a theoretical model of the propeller vortex. This method applies Prandtl's finite span theory to the aerodynamic design of the propeller. According to the finite span theory, when the airflow passes a finite span wing that generates lift, it will change direction, causing the downwash of the airflow, The downwash angle depends on the lift and span of the wing. The finite wingspan flow can be seen as a two-dimensional airfoil flow (infinite wingspan flow) superimposed downwash flow. The effect of the downwash reduces the angle of attack of the incoming flow. The application of airfoil theory for propeller blade element analysis uses the lift and drag characteristic data of the airfoil, and avoids the problem of the aspect ratio of the finite wing. The interference flow is calculated by the momentum theorem, depending on the number of blades, the spacing, and the aerodynamic forces acting on each blade. Therefore, the problem of blade interference is also considered. This theory is also called the standard strip analysis.

The later invented turbofan engine (as shown in Fig. 7.84) became the main engine of the modern large transport aircraft. Formally, the front of the engine is a large fan, followed by a high-speed rotating turbine, and the cold airflow generated by such engine fans is the main contributor to the thrust. In general, the economical cruising speed of a turbofan engine aircraft can reach 800–900 km/h, which is much faster than that of a propeller aircraft.

In order to increase the speed and thrust, compared to the turbofan engine, the front of the turbine is not a low-speed large diameter fan, but a small diameter medium pressure compressor that forms a turbojet engine (as shown in Fig. 7.85). The main thrust of this type of engine comes from the jet behind the turbine. In general, the cruising speed of the aircraft equipped with this type of engine can be supersonic, such as the Concorde mid-range supersonic passenger aircraft developed by France and U. K. (This aircraft

Fig. 7.84 Turbofan engine

Fig. 7.85 Turbojet engine

and the Tu-144 developed by former Soviet Union Tupolev design bureau are the few commercial supersonic passenger aircraft in the world that have been put into commercial use, first flight in 1969), the cruise speed is 2150 km/h, and the maximum cruise Mach number is 2.04. Although the speed of the propeller engine is the lowest, it has the highest propulsive efficiency in the three types of engines, which can reach 88–89%. Therefore, humans first took off the aircraft and realized the dream of flying by propeller engine. But with the depletion of oil, if there is no alternative energy source, it may be possible to return to the propeller engine era because it is the most fuel efficient. If other energy sources can replace it, such as replacing the fuel engine with a nuclear battery, the nuclear battery supplies energy to the electromotor and the electromotor drives the propeller. Due to the difficulty of aero-engine technology, there are only a few companies in the world that can produce it, such as Pratt & Whitney Group in the U. S. and Rolls-Royce in the U. K. These are the most famous aero-engine manufacturing companies in the world.

China's large airliner C919 uses the LEAP-X1C engine (This engine is similar to CFM56 series which is assembled on the production line in China jointly established by Aviation Industry Corporation of China and CFM International Company. Its thrust is 10 tones.). The CFM56 series engines are supplied by the CFM International Company which is a joint venture company between General Electric of the U. S. and SNECMA of France. The GE90-115B installed on the Boeing 777-300ER is the most powerful civil engine with a single engine thrust of more than 56 tons. If the PW4090 engine of P&W (Pratt & Whitney Group) Company is selected, then the thrust is about 40 tons. If the Boeing 777-200 series aircraft uses the Trent 800 series engine produced by the Rolls-Royce Company of the U. K., then the thrust is 41 tons.

The aviation industry has experienced a leap from the "propeller area" to the "jet era". The birth of jet engines has brought a revolution to the world aviation industry. The life of the jet engine founder Sir Frank Whittle (1907–1996, as shown in Fig. 7.86) is full of hardships and legends.

Various types of engines and installations are given in Figs. 7.87, 7.88, 7.89, 7.90, 7.91, 7.92, 7.93, 7.94, 7.95, 7.96, 7.97.

Fig. 7.86 Jet engine founder Frank Whittle (1907–1996, British aeronautical engineer)

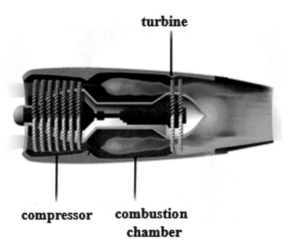

Fig. 7.87 Turbojet engine

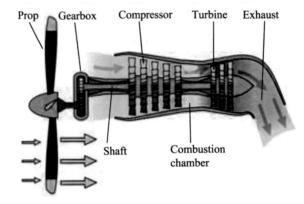

Fig. 7.88 Turboprop engine

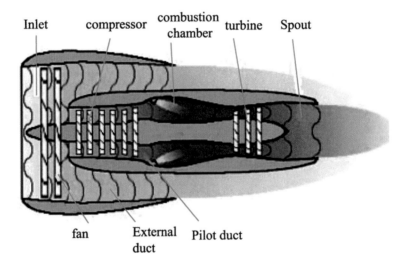

Fig. 7.89 Turbofan engine

7.11 High-Lift Device of an Aircraft

In addition to the albatross taking off from the ground, the swan can take off on the water surface (as shown in Fig. 7.98). The takeoff attitude and action, etc., of these birds are worthy of human-designed aircraft imitation. It is conceivable that if the bird can speak, it may have to compete with humans for the invention of the aircraft. When the aircraft is taking off or landing, the pitch angle (also called the angle of attack) is relatively large, but the angle of attack is relatively small when cruising. Why? This is because the speed of the aircraft is fast when cruising. Because the lift is proportional to the square of the speed, there is no need to increase the lift by increasing the angle of

Fig. 7.90 Counter-rotating propeller engine

Fig. 7.91 European A400M propeller transporter

attack, at which point it can fly smoothly at a small angle of attack. However, when taking off or landing, because the speed of the aircraft is small, in order to achieve the required lift, on the one hand, the angle of attack of the aircraft needs to be enlarged, on the other hand, the airflow bypassing the aircraft is in the acceleration and deceleration zone, and the aircraft is near the ground. At this time, the control is more difficult than the cruise, and the aircraft accident rate is much higher than the cruise state. According to statistics, the accident rate of the aircraft in the takeoff and landing state can reach about

Fig. 7.92 American C130 propeller transporter

Fig. 7.93 Airbus A320 airliner (large bypass ratio turbofan engine)

70%. It can be seen that, for the sake of safety, the airworthiness regulations are that passengers are not allowed to walk around when taking off and landing, and the seat belt must be fastened.

In addition to the takeoff state, the change in the shape of the wing when the bird is landing is also the object of imitation. When the eagle is landing (as shown in Fig. 7.99), the feathers of its wings expand as much as possible. Compared with the shape of the wing in the soaring state, not only the area is enlarged, but also the degree of feather bending is increased. Why? Because

Fig. 7.94 Boeing 787 airliner (large bypass ratio turbofan engine)

Fig. 7.95 Airbus A350 airliner (large bypass ratio turbofan engine)

when the eagle is soaring, the speed is fast, the lift is high, the wing area is small and the curvature is small. However, when taking off and landing, the speed is low. Under a certain weight, maintaining the original wing area will result in a serious lack of lift. For this reason, the eagle actively expands the feather as much as possible and increases the downward camber to increase the area of the wing (lift is proportional to area), at the same time increases the lift by increasing the angle of attack and camber. For the aircraft, it also

Fig. 7.96   Airbus A380 airliner (large bypass ratio turbofan engine)

Fig. 7.97   Concorde (Turbojet engine)

has a similar behavior when landing. When taking a plane, it can be found that when the aircraft is landing (the same is true for takeoff), the active surfaces of the front and rear edges of the wing will all open, the area of the wing increases and the downward chamber increases to enhance wing lift, while also increasing the angle of attack to further increase the lift (as shown in Fig. 7.99). This means that when the speed of the aircraft changes, the lift can be altering by changing the wing area and attitude angle. The speed of the modern large airliner during takeoff and landing is quite different from the

Fig. 7.98 Swan and fighter takeoff

cruising speed. Under normal circumstances, the takeoff and landing speed of large airliner is about 220–240 km/h, and the cruising speed can reach 800–900 km/h. The speed ratio of the two is nearly 3 to 4. In order to achieve the above behavior, the active surfaces of the front and rear edges of the wing are called high-lift devices. The design techniques of these movable surfaces are called low-speed configuration design technology in aircraft wing design, which is one of the most important core technologies in aircraft wing design and one of the important core technologies is also one of the hotspots of aerodynamic research, involving almost all the complex problems of modern viscous fluid mechanics.

Careful observation of the bird's takeoff method reveals that in addition to the takeoff roll, there are also high platform gliding takeoff, ground bounce takeoff, etc. For example, pigeons often take off by bouncing. When the pigeon wants to take off, the two legs are struggling to bounce off the body,

Fig. 7.99 Comparison of the landing status of the eagle and B747

and at the same time, the wings are deployed with a rapid downward air discharge, so that they can take off. Obviously, no matter whether the ground run-up takeoff or bouncing takeoff, the bird has to exert great strength during the takeoff process, which consumes a lot of energy. In comparison, when the bird is gilding, the speed of the airflow around the wing is high, and the bird is flying in the upward airflow as much as possible, so the effort is small. Some people may think that if the high-lift device is not installed on the wing, that is, the wing area of the aircraft in cruising is the same as that of taking off and landing, which is equivalent to the speed of takeoff and landing is similar to the speed of cruising. What will happen? At this time, if the aircraft takes off and lands at the cruising speed, it not only greatly increases the length of the runway, but also has extremely poor safety. When the aircraft is taking off normally, due to the excessive ground speed, the aircraft taxies a large distance on the ground runway. If the aircraft taxies to the lift-off speed, the engine stalls and the aircraft cannot take off normally, the aircraft needs to slow down. Obviously the greater the lift-off speed, the longer the distance required for the deceleration, the safety requires the actual runway length to be about twice the takeoff length, so the takeoff speed has a decisive relationship to the length of the runway. Actually, for the sake of safety and to reduce the cost of airport runways, it is reasonable to reduce the takeoff and landing speed of large airliner as much as possible. If the wing area is not reduced, then it will cruise at a high speed in the area of takeoff and landing. At this time, the frictional drag of the wing is too large, and the aircraft cannot achieve economic cruising.

The design of high-efficiency lifting devices for large airliner remains a challenging issue worldwide. Large airliner requires complex multi-element high-lift systems to meet aircraft takeoff and landing performance requirements. In today's competitive civil aircraft market, the design trend of large airliner requires more efficient lifting devices to maximize the lift coefficient and lift-to-drag ratio at a given angle of attack and flap deflection angle. The design of the lifting device belongs to a multi-objective, multi-technology integrated design problem. The overall technical requirements of the aircraft must be met, including aircraft performance, safety, reliability, maintainability, noise, etc. In terms of aerodynamics, the aircraft needs to meet the requirements for takeoff, short landing distance, and climb gradient; In terms of structure, it requires less components, lightweight, simple connection and sufficient strength and rigidity; in terms of maneuverability, it requires easy to maintain, reliable, low cost, and meet the requirements of damage tolerance.

Research on the lifting device has been carried out long ago in foreign countries. First of all, in the high-lift design, there are quite a lot of theoretical and experimental research, which provides a direct theoretical basis and data support for the design of the lifting device. The aerodynamicist A.M.O. Smith of Douglas Corporation in the U.S. published the paper "High Lift Aerodynamics" to reveal the high-lift mechanism of multi-element wing (as shown in Fig. 7.100 and Fig. 7.101) in 1975.

From the aerodynamic point of view, A.M.O. Smith et al. carried out a lot of basic research on the mechanism of high lift and the flow of multi-element wing flow, and carried out in-depth excavation of the nature of the flow to maximize the lift. For two-dimensional multi-element airfoil, various flow phenomena that may occur include boundary layer transition, shock/boundary layer interference, wake/boundary layer blending, boundary layer separation, laminar separation bubble, separated concave angular flow, the streamline is greatly curved, etc.

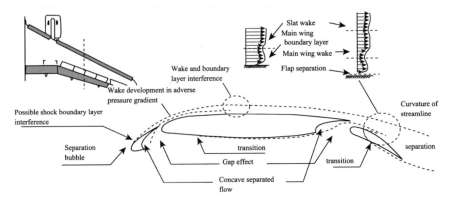

Fig. 7.100 Multi-element airfoil flow

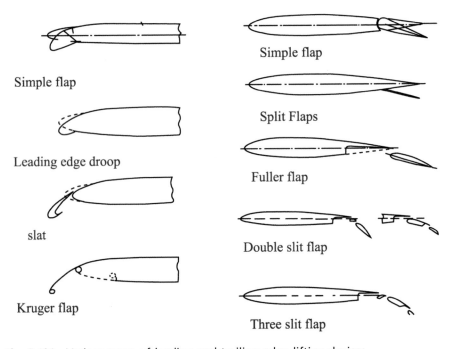

Fig. 7.101 Various types of leading and trailing edge lifting devices

In the 1950s, NASA in the United States began to study the above problems. Until 1975, there were major breakthroughs in multi-element wing flow problems. A.M.O. Smith proposed that the beneficial effects of slats (flap) on multi-element wing flow are the gap effect of the front wing; circulation effect; dumping effect; pressure recovery after leaving the object surface; and each wing element begins a new boundary layer. The main effects on improving lift are (as shown in Fig. 7.102) as follows:

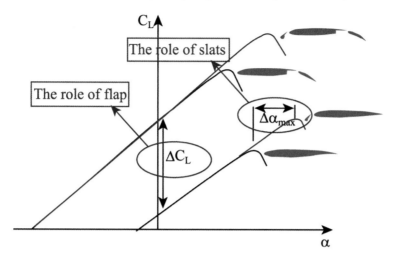

Fig. 7.102 Influence of leading and trailing edge lifting device on lift coefficient

Increase the camber effect of the wing

Increasing the camber of the wing, that is, increasing the circulation, which will result in a large nose down moment. Especially when landing on the ground, the horizontal stabilizer or the trailing edge of the elevator needs to deflect upward to meet the trim requirement.

Increase the Effective Area of the Wing

Most lifting devices are moved in such a way as to increase the basic chord length of the wing. With the same nominal area where the shape of the wing profile is not changed, the effective wing area is increased and the lift is increased. In this case, the nominal area is constant, which is equivalent to increasing the zero angle of attack lift coefficient, thus increasing the maximum lift coefficient.

Improve the Flow Quality of the Seam

By improving the flow quality of the seam between the wing elements and the boundary layer state on the airfoil surface to enhance the boundary layer's ability to withstand the back pressure gradient and delay separation, increase the stall angle of attack, and increase the maximum lift coefficient, as shown in Fig. 7.103. Figure 7.104 shows.

In the design of the actual aircraft lifting device, it is also necessary to carry out the design of the retracting mechanism and structure. After careful aerodynamic design, the leading edge slat and trailing edge flap require a

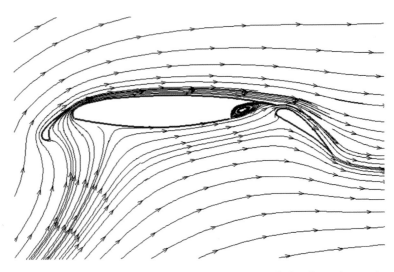

Fig. 7.103 Flow field around three-segment wing with leading slat and trailing Fowler flap

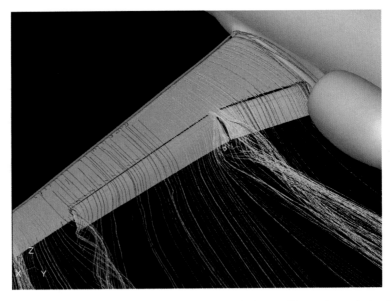

Fig. 7.104 Detached wake from discontinuity of the trailing edge of the lifting device

mechanical retraction mechanism to guide them to the appropriate design position. Airbus's Peter K. C. Rudolph et al. wrote "Transonic Airliner Lifting Device Design" in 1996 to provide a detailed description of the transonic airliner's lifting device before 1996. It can be seen that different aircraft adopt different retracting mechanisms. Peter K. C. Rudolph is more inclined to the description of the mechanism of the lifting device, but the specific aerodynamic performance of the aircraft lifting device is not described in detail. There are also no specific data to explain the specific effects of various lifting mechanisms on aerodynamic performance. Both A.M.O. Smith and Peter K. C. Rudolph have conducted in-depth research on the lifting devices from different fields.

Airbus' Daniel Reckzeh has been designing high-lift devices for more than a decade and has been involved in the high-lift devices design of A380, A400M, and A350XWB aircraft. He used the practical experience to explain the advanced methods and means of Airbus in designing the lifting device. In his several papers, the lifting devices design of the A380, A400M, and A350 were described, and the concepts and methods of joint design of aerodynamics and mechanisms were described. In foreign countries, due to the success of Boeing and Airbus in the development of large airliners, they have gained rich research and design experience in the design and manufacturing technology of the lifting device, as well as long-term technical accumulation.

In order to avoid following the complicated lifting mechanism of Boeing, Airbus carried out a bold and innovative design from the simplification, using a multi-objective and comprehensive design concept of aerodynamics, mechanism, structure, strength, maintenance, economy, etc. The simplified mechanism form was successful on the A320, especially under the premise of meeting the aerodynamic requirements, the support and drive systems were boldly reformed (as shown in Fig. 7.105). Airbus broke the gap between aerodynamic and mechanical design in the design of A380 lifting device and proposed the integrated design concept of aerodynamic, mechanism and drive system, and completed the integrated design platform in CATIA environment, becoming the most advanced lifting device design platform at present. The lifting devices for different types of aircraft are given in Figs. 7.106, 7.107, 7.108, 7.109, 7.122.

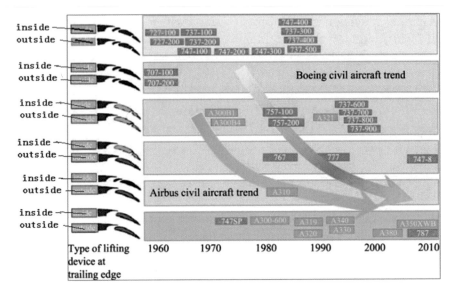

Fig. 7.105 Development trend of large aircraft lifting device

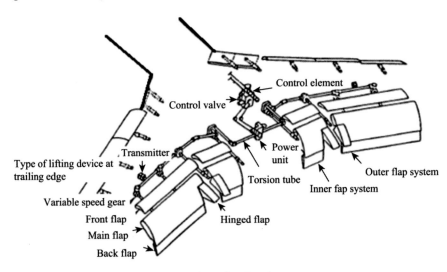

Fig. 7.106 -300 lifting device control mechanism layout

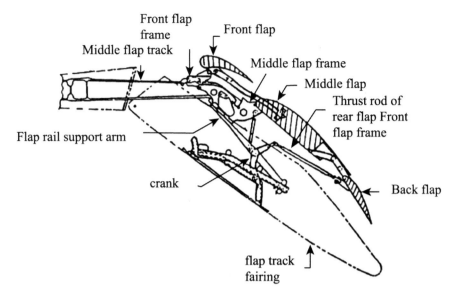

Fig. 7.107 B737 slide rail and pulley mechanism

Fig. 7.108 B737 slide rail and pulley type flap

7.12 Aircraft Landing Gear

Another important component produced by mimicking bird is the landing gear, which is used when the aircraft is taking off and landing. As mentioned above, the bird will retract its legs when soaring, indicating that the bird has realized that the aerodynamic drag generated by the two legs if they are not retracted will be very large, resulting in the bird not being able to fly

Fig. 7.109 B737 large airliner lifting device (trailing edge)

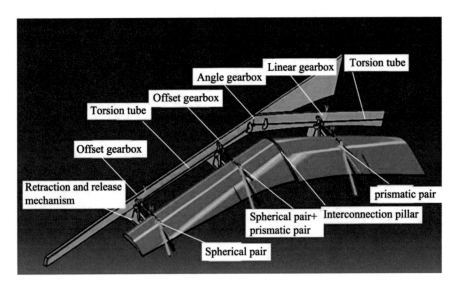

Fig. 7.110 Trailing edge lifting device mechanisms and constraints

quickly. This reminds people that the landing gear of a man-made aircraft should also be retracted after the aircraft is lifted off. The speed of the early aircraft was relatively small. For the sake of simplicity, the landing gear was often not retracted, so the landing gear of the early aircraft would not be retracted. Modern aircraft landing gear is retractable. For the A380 airliner, it

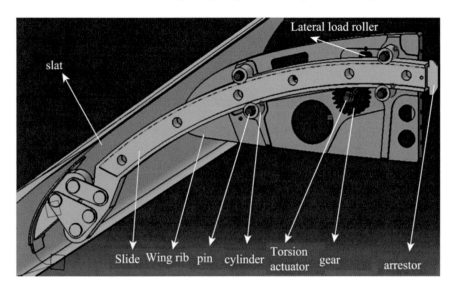

Fig. 7.111 Leading edge slat gear and rack mechanism

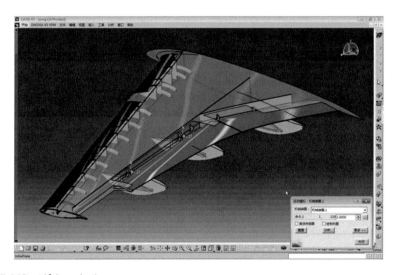

Fig. 7.112 Lifting device

has double-deck cabin, three-class cockpit (first class–business class–economy class) configuration which can carry 555 passengers, maximum takeoff weight is 560 tons, wingspan is 79.8 meters, fuselage length is 73 meters, wing area is 845 square meters, the cruising speed is 902 km per hour. Although the bird cannot be compared with this plane, the careful people will find that when the plane takes off, lands or cruises, its various postures imitate the big birds

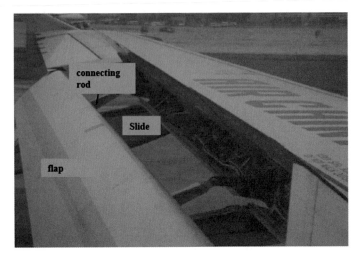

Fig. 7.113 A320 Large airliner lifting device (trailing edge connecting rod slide mechanism)

Fig. 7.114 A320 large aircraft lifting device (leading edge)

like albatross, including the wing area and shape variations and retraction of the landing gear, etc., in different states (as shown in Figs. 7.123, 7.124, 7.125).

It has been observed that when the albatross is taking off, the wings are spread as far as possible, and they have to run a long distance on the spacious ground. When the running speed reaches a certain value, they can take off from the ground. Nowadays fixed-wing aircraft use this type of run-up takeoff. For security, the aircraft needs to taxi on a fixed runway before taking

Fig. 7.115   B747 Lifting device

Fig. 7.116   7.116 B747 Lifting device (maintenance)

Fig. 7.117 A380 lifting device in takeoff state

Fig. 7.118 A380 lifting device in landing state

off. When the speed reaches a certain value, the pilot starts to pull the rod and the plane can take off. After the albatross takes off, in order to speed up, first it retracts the two legs, the two legs come close to the body, and are covered with feathers to keep the body smooth, and then it flies high. You can see it flies freely in the air and does not spend too much power (as shown in Figs. 7.126, 7.127 7.128).

Fig. 7.119 B787 lifting device

Fig. 7.120 B787 lifting device

Similarly, after the plane leaves the ground, the landing gear is first retracted. If the landing gear is not retracted, the drag of the plane will be great, the plane cannot fly fast. In addition, the airflow is also very noisy. Sitting in the cabin of the plane, it is not quiet at all.

Since the landing gear of the aircraft mainly bears the impact load and gravity when the aircraft is landing (as shown in Fig. 7.129), it is required to

Fig. 7.121 A350 lifting device

Fig. 7.122 A350 lifting device

have good impact resistance. The general prop is made of steel, the current widely used landing gear material is low alloy ultra-high-strength steels, such as 300 M in the United States, 35NCD16 in France, and 30XΓCH2A in Russia, their distinctive feature is extraordinary high strength. The high strength of the material can make the landing gear light and the weight loss has always been an important indicator of the landing gear design. At the same time, the material should have excellent overall performance to

Fig. 7.123 Airbus A380 landing gear puts down

Fig. 7.124 A380 landing gear puts down

ensure the reliability of the landing gear. With the development of material technology and manufacturing technology, 300 M steel with strength class of 1900–2100 MPa and its anti-fatigue manufacturing technology have become the main application technology of American aircraft landing gear. The landing gear and wheel type of different types of airliners are given in Figs. 7.130, 7.131, 7.132, 7.133, 7.134, 7.135, 7.136, 7.137.

Fig. 7.125 A380 landing gear (fuselage single pillar 6 wheels and wing single pillar 4 wheels)

Fig. 7.126 Albatross cruising shape (landing gear retracted)

Fig. 7.127 Albatross takes off on water and the albatross landing state (landing gear puts down)

7.13 Aircraft Aerodynamic Noise

With the development of society and the continuous advancement of industrial technology, people have put forward more and more strict environmental protection requirements for the civil aviation industry. How to design a more environmentally friendly aircraft is the focus of the aviation industry. The International Civil Aviation Organization (ICAO) has developed recommendations for aircraft aerodynamic noise certification. The United States and Europe have developed a series of aircraft aerodynamic noise airworthiness regulations based on this, which limits the aerodynamic noise level of civil aircraft. The fourth phase requires after 2006, the noise level of the new civil aircraft proposed for airworthiness application should be 10EPNdB lower than the third stage (where EPN is the abbreviation of Effective Perceived Noise). The effective perceived noise level is the perceived noise level in terms of duration and pure tone correction, and its unit is EPNdB. When

Fig. 7.128 The albatross landing state (landing gear puts down)

the airplane flies over the observer's head, the pitch of the noise changes. In addition, the pure tone component or narrowband component of the aircraft noise and the duration of flying over the observer's head affect the degree of annoyance that people feel about the aircraft noise. Therefore, the effective perceived noise is proposed for the evaluation of aircraft noise. The calculation and measurement of the effective perceived noise level is very complicated, and can generally be estimated by adding 15 dB to the sound level. NASA proposed in 1997 to reduce noise by 10EPNdB in 10 years and 20EPNdB in 20 years. In addition, the noise level of civilian passenger aircraft has gradually become an important indicator for airlines to consider when purchasing aircraft. These are undoubtedly huge challenges for the large civilian passenger aircraft that China is developing. The noise level is one of the key factors for its ability to obtain airworthiness certificates and to occupy a place in the world aviation field in the future.

Fig. 7.129   Modern large aircraft nose landing gear and main landing gear

Fig. 7.130   Front three-point landing gear for modern large aircraft

Fig. 7.131 Four wheels main landing gear and six wheels main landing gear

Fig. 7.132 An-225 landing state

Fig. 7.133 An-225 multi-wheel multi-pillar main landing gear

The external noise of the aircraft mainly includes propulsion system noise, airframe noise, and interference noise between the propulsion system and the airframe, as shown in Fig. 7.138. Propulsion system noise, i.e., engine noise, including fan noise, compressor/turbine noise, combustion noise, and jet noise, is a dynamic noise, all of which are dynamic noise. The airframe noise includes the lifting device and the landing gear noise, it and the interference noise between the propulsion system and the airframe are all non-dynamic noise.

With the use of a large bypass ratio turbofan engines and the effective application of noise reduction technologies such as muffler nacelles and V-shaped petal nozzles, the proportion of engine noise in overall noise is decreasing. Especially in the landing stage of the aircraft, when the engine is in a low power state, the lifting device and the landing gear are all open,

Fig. 7.134 B747-8 landing gear puts down landing state

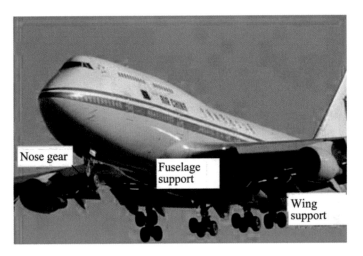

Fig. 7.135 B747-8 landing gear (fuselage single pillar 4 wheels and wing single pillar 4 wheels)

the airframe noise is equivalent to the engine noise, even exceeding the engine noise, as shown in Figs. 7.139 and 7.140.

Since most of the modern large airliners use multi-element airfoils with leading edge slats and trailing edge flaps as the lifting device, during the landing of the aircraft, the leading edge slat, trailing edge flap, and landing

Fig. 7.136 A350 landing gear puts down

Fig. 7.137 B787 landing gear puts down

gear are important sources for the aerodynamic noise. For the leading edge slats, the shed trailing edge vortex and the unstable impulse in the groove area are the main noise sources. The unstable impulse includes the interaction between the vortex in the shear layer, and the interaction between the vortex and the solid wall in the recirculation zone and the reattachment zone (as shown in Figs. 7.141 and 7.142). For the trailing edge flap, when the flaps are open, due to the sudden change in the spanwise lift, a strong vortex is generated at the side edges of the flaps, including high-frequency small-scale unstable vortices and low-frequency large-scale vortices. These two different

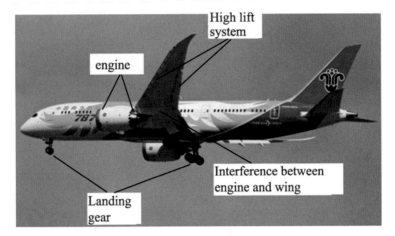

Fig. 7.138 Main components of aircraft aerodynamic noise sources

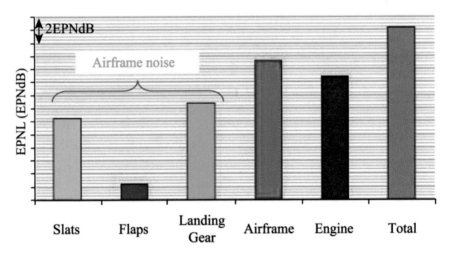

Fig. 7.139 Proportion of aerodynamic noise of the airframe and engine components during the approach phase of the aircraft

scale vortices form the main source of noise (as shown in Fig. 7.143). For the landing gear, the landing gear bay is a typical cavity structure. The cavity flow-induced oscillation not only generates additional aerodynamic noise but also causes unsteady load, so bluff body separation is the main reason for its aerodynamic noise (as shown in Fig. 7.144).

The relative noise magnitude of these noise sources depends on the specific configuration. Wind tunnel experiments (as shown in Fig. 7.145 and 7.146) and flight tests consistently show that the noise of the lifting device accounts for a large proportion of the airframe noise and cannot be ignored. Specific

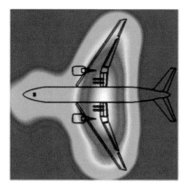

Fig. 7.140 Aircraft aerodynamic noise n distribution (In September 2001, Boeing conducted a flight test on the full-size B777-200 in Montana, USA, using a free-range microphone and microphone array to test aerodynamic noise sources in different states, such as trailing edge flap opening and engine idling.)

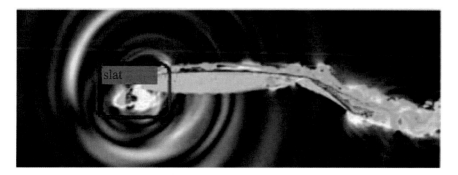

Fig. 7.141 Leading edge slat aerodynamic noise

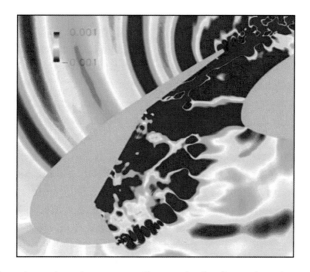

Fig. 7.142 Aerodynamic noise propagation at the leading edge slat

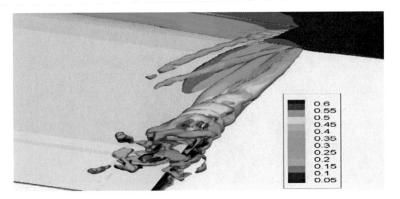

Fig. 7.143 Vortex and aerodynamic noise on the side of the trailing edge flap

to the noise of the lifting device, NASA, Boeing, Airbus, and some university research institutes have conducted in-depth research on the noise principle through experiments and numerical simulations.

In terms of noise reduction technology, some effective measures have been proposed for the lifting device and the landing gear. For example, the landing gear with a fairing or a surface-opening fairing can reduce the aerodynamic noise by 2–3 dB.

For leading edge slats, noise reduction technology can be divided into two categories: one is to reduce high frequency and narrow frequency noise, such as slat trailing edge serration, slat surface active flow control, etc. Mainly to reduce the vortex shedding of the trailing edge of the slat; the other is to reduce the low and wide frequency noise, such as the leading edge slat groove shielding, groove filling, liner on the lower surface of the leading edge slat and the main wing, the drooping leading edge structure and the porous permeation structure on the lower surface of the leading edge slat, etc., which mainly reduce the unstable impulse inside the groove.

For trailing edge flaps, the main starting point for side noise control is to weaken the vortex structure and the mutual interference between the flow field and the wall. According to whether to inject energy into the flow field, the noise reduction technology can be divided into passive control technology and active control technology. Passive control technology refers to a means to reduce noise by changing or correcting the side shape without injecting energy into the flow field. There are mainly the following types: the addition of porous material to the sides of the flaps, the use of fences, and the continuous line method on the sides of the flaps. Active control technology is directed to the injection of energy into the flow field to blow the vortex away from the wall, reducing the mutual interference between the vortex and

Fig. 7.144   Landing gear aerodynamic noise

the wall to reduce the side noise (blowing control). According to the effective noise reduction measures of existing components, NASA predicts the future low-noise aircraft as shown in Fig. 7.147.

Fig. 7.145 B777 scaled model test (26% scaled, NASA Ames wind tunnel aerodynamic noise test)

7.14 Supersonic Aircraft

The flow around the supersonic aircraft is a supersonic external flow. Usually supersonic external flow refers to the situation where the entire flow field or most of the flow field is supersonic. After the flying Mach number is greater than 1.4, a series of flow phenomena controlled by shock waves and expansion waves will appear. Generally, the Mach number of supersonic flow is between 1.5 and 5.0.

An important feature of steady supersonic flow is that the range of influence of any disturbance in the flow field is bounded, and any disturbance is in the form of waves. When the supersonic flow expands or is subjected to a series of weak compressions, the beginning and end of the disturbance are all Mach lines (see Prandtl–Meyer flow).

The supersonic flow around the double-arc airfoil is illustrated in Fig. 7.148. When the incoming angle of attack is smaller than the half apex angle of the airfoil, the upper and lower surfaces of the leading edge are compressed to form a oblique shock wave with different strength; when the incoming angle of attack is greater than the half apex angle of the airfoil, an expansion wave is formed on the upper surface of the leading edge and a

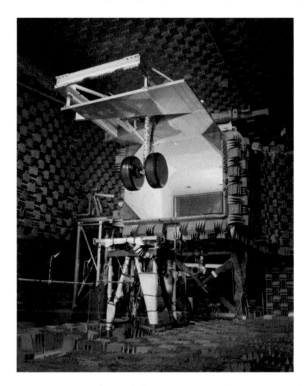

Fig. 7.146 Aeroacoustic test of the full-size B737 landing gear model (NASA low-speed aeroacoustic wind tunnel)

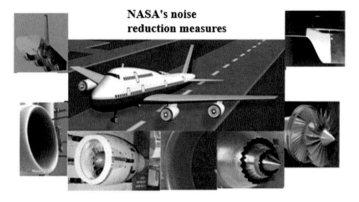

Fig. 7.147 Future low-noise aircraft given by NASA

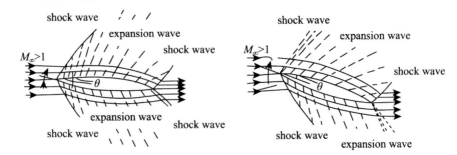

Fig. 7.148 Double-arc airfoil supersonic flow (small angle of attack, high angle of attack)

oblique shock wave is formed on the lower surface. After a series of expansion waves, due to the inconsistent flow direction and pressure at the trailing edge, two oblique shock waves, or one oblique shock wave and a series of expansion waves are formed, in order to make the airflow after the trailing edge has the same direction and equal pressure. (Approximately considered to be the same as the incoming flow).

In supersonic flow, the magnitude of the shock wave drag around the wing is closely related to the bluntness of the wing head. Since the flow around the blunt object will generate a detached shock wave, the shock wave drag is large; and the flow around the sharp tip body will generate an attached shock, the shock wave drag is small. Therefore, for the supersonic airfoil, the leading edge is preferably sharpened, such as diamond, quadrilateral, double curved, etc. (as shown in Figs. 7.149 and 7.150). However, for supersonic aircraft, it is always necessary to experience the low-speed phase of take-off and landing. When the sharp tip airfoil is flowing at a low speed, the airflow is separated at a small angle of attack, which deteriorates the aerodynamic performance of the airfoil. To this end, in order to balance the low-speed characteristics of the supersonic aircraft, the airfoil of the current low-sonic aircraft adopts a symmetrical thin wing with a small round head.

For the steady, non-viscous and adiabatic two-dimensional non-rotating supersonic flow, the lift coefficient, the wave drag coefficient, and the pitching moment coefficient of the leading edge obtained by the small disturbance linearization theory are consistent with the wind tunnel test results. For the supersonic flow around a three-dimensional wing, the flow conditions can be divided into the followings based on the relationship between the Mach number and the leading edge sweep angle or the trailing edge sweep angle:①supersonic leading and trailing edge;② subsonic leading and trailing edge; ③ supersonic leading edge and subsonic trailing edge; and ④ subsonic

Fig. 7.149 Flow around supersonic objects (detached and attached shock wave)

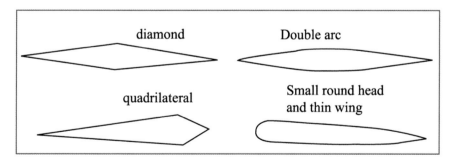

Fig. 7.150 Supersonic airfoil

leading edge and supersonic trailing edge. The pressure distribution and aero-dynamic characteristics corresponding to these different flow conditions are different. The establishment of the theory provides an important basis for understanding the supersonic flow. At the same time, a wealth of information has been accumulated based on a large number of wind tunnel tests, which provides an important support for the development of supersonic aircraft.

The shock wave drag of a supersonic aircraft accounts for the majority of the total drag. Therefore, the surface of the wing is made into a smooth, thin and short swept or triangle shape, and the fuselage is made into a sharp tip slender shape. The characteristics of supersonic flight are the aerodynamic center moves backward, and the longitudinal static stability increases; the drag of the aircraft decreases as the Mach number increases. Both of them cause the disturbance attenuation to slow down, the maneuverability becomes worse, and the yaw stability becomes worse. Therefore, it is necessary to increase the vertical tail area or use an automatic device. When the Mach number is less than 2.5, the strength of the aluminum alloy can maintain the aerodynamic heating caused by high-speed flight. After the Mach number reaches 3.0, the aerodynamic heating is intensified and heat-resistant materials are required. To prevent sonic booms and noise hazards, many countries prohibit supersonic flight over residential areas.

The flight speed of a supersonic aircraft exceeds the speed of sound. On October 14, 1947, the U.S. Air Force captain Charles Yeager drove the X-1 at an altitude of 12,800 meters to a speed of 1,278 km/h, $M = 1.1015$, which is the first time humans broke the sound barrier. The representative of the civilian supersonic aircraft is the medium-range supersonic passenger aircraft Concorde (as shown in Figs. 7.151 and 7.152) jointly developed by French Aerospace and British Aircraft Corporation. The aircraft adopts a no horizontal tail layout. In order to adapt to supersonic flight, the Concorde's wing adopts a delta wing, and the leading edge of the wing is S-shaped. The Concorde has a total of four turbojet engines. The engines were developed by Rolls-Royce and the SNECMA. The engine has an afterburner that is typically used on supersonic fighters. The Concorde can fly at speed up to twice the speed of sound, with a maximum flight speed of Mach 2.04, a cruising altitude of 18,000 meters, and a cruising speed of 2,150 km per hour. Serviced in 1976, it is mainly used to carry out transatlantic regular routes from London Heathrow Airport (British Airways) and Paris Charles de Gaulle International Airport (Air France) to and from the John F. Kennedy International Airport in New York. The aircraft can cruise at a speed of 2.02 sound speed at an altitude of 15,000 meters. It takes only about 3 h and 20 min to fly from Paris to New York. It saves more than half of the time

Fig. 7.151 Concorde takes off (Started on January 21, 1976, retired on October 24, 2003)

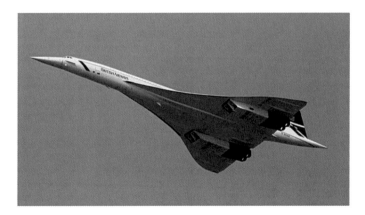

Fig. 7.152 Concorde climbs

compared to ordinary civil airliners, so it is still popular among business travelers although it is expensive. On February 7, 1996, the Concorde flew from London to New York in just 2 h, 52 min and 59 s, which set the fastest flight record.

The Concorde only produced a total of 20. British Airways and Air France used Concorde to operate transatlantic routes. By 2003, there were still 12 Concorde for commercial flights. On October 24, 2003, the Concorde carried out its last flight and completely retired.

Fig. 7.153 J-10 (canard layout)

The J-10 developed in China is a medium-sized, multi-functional, super-sonic, all-weather air superiority fighter (fourth-generation fighter) with a maximum flying Mach number of 2.2.

J-10 uses a canard layout and blended wing body design (as shown in Fig. 7.153). By carefully designing the curved surface of the joint between the main wing and the middle of the fuselage, not only the volume inside the aircraft (for oil loading, equipment, and space for future development) increases, but also effectively utilizes the aerodynamic increase brought by it. There are no other structures arranged on the sides of the fuselage at the rear of the main wing, which once again reflects the design concept of the blended wing body, except that two outer oblique ventral fins are installed under the abdomen in front of the tail nozzle. These two ventral fins are used to maintain the stability of the aircraft with a tall vertical tail when flying at a high angle of attack.

The F22 "Raptor" is a single-seat, twin-engine, high stealth, fifth-generation fighter developed by Lockheed Martin and Boeing in the U.S. (see Fig. 7.154). The F22 is the first fifth-generation fighter in service in the world, with a maximum flight Mach number of 2.25. At the beginning of this century, the F22 successively entered the U.S. Air Force to replace the previous generation of the main model F15 "Eagle" fighter. Lockheed Martin is the prime contractor responsible for most of the fuselage, weapon systems, and the final assembly of the F22. The partner Boeing provides wings, rear fuselage, avionics integrated systems, and training systems. Lockheed Martin

Fig. 7.154 F22 (trapezoidal wing layout)

claims that Raptor combines stealth performance, agility, precision, and situational awareness to create air-to-air and air-to-ground combat capabilities, making it the world's best performing fighter.

The F22 adopts a twin-vertical tail, twin-engine, single-seat layout, with a vertical tail that slopes out 27 degrees, just at the edge of a general stealth design. The air inlets on both sides are mounted below the extended surface of the leading edge of the wing (strake wing). Like the nozzles, a stealth design that suppresses infrared radiation is adopted. The main wing and the horizontal stabilizer have the same sweepback angle and the trailing edge sweep forward angle, a trapezoidal plane with a small aspect ratio. The blister-type cockpit cover protrudes from the upper part of the front fuselage, and all the weapons are concealed in four internal bomb bays. The future supersonic aircraft given by NASA in the United States is shown in Figs. 7.155, 7.156 and 7.157.

7.15 Drag Reduction Technology for Large Transport Aircraft

1. *Drag of the aircraft*

When the airflow passes around the aircraft, the drag of the aircraft is defined as the component of the resultant force of the positive pressure and friction

Fig. 7.155 The future supersonic aircraft of NASA in the U.S

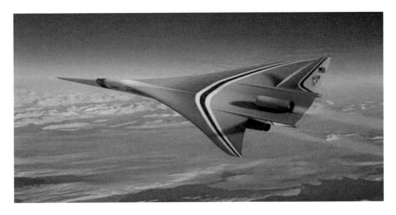

Fig. 7.156 Next-generation supersonic airliner

shear stress on the surface in the direction of the incoming flow (as shown in Fig. 7.158). The component force perpendicular to incoming flow direction is called lift. When the aircraft is cruising, the gravity and lift of the aircraft are balanced, and the thrust of the aircraft engine and the resistance are balanced.

Fig. 7.157 Next-generation supersonic airliner

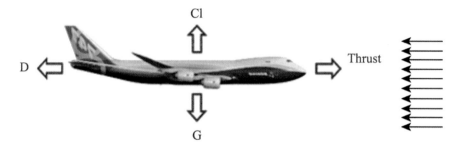

Fig. 7.158 Schematic diagram of the force of the aircraft during cruise flight

Because the pressure and frictional shear stress on the aircraft surface are related to the aircraft's flight speed, attitude angle, size, surface shape, and roughness, etc., the drag of the aircraft must be affected by these factors. Aircraft drag is divided into two categories from the large aspect. One is the drag caused by the integral of the pressure projected in the direction of the incoming flow; the other is the drag generated by the integral of the surface friction shear stress, which is called friction drag.

Specifically, according to the main cause for the drag, the drag obtained from the integration of the surface pressure can be divided into the induced drag due to the downwash induced by the free wake from the trailing edge of the wing, the pressure drag caused by the different aircraft shapes (including wing body interference drag, bottom drag, drag around exposed components, etc.). For high subsonic aircraft, there is additional shock wave drag due to the existence of a supersonic zone on the upper wing surface.

In the cruising state of the aircraft, along the direction of the incoming flow, the thrust generated by the engine is balanced with the drag of the aircraft; while perpendicular to the direction of the incoming flow, the lift generated by the aircraft wing is balanced with its own gravity. For drag, if it is used to express with incoming flow velocity and characteristic area of the wing, it can be written as

$$D = \frac{1}{2}\rho_\infty V_\infty^2 S C_D$$

where D is the total drag of the aircraft, ρ_∞ is the density of air, V_∞ is the velocity of incoming flow, S is the characteristic area of the wing (the exposed area of the wing), and C_D is the total drag coefficient of the aircraft. According to the causes for the drag, the total drag coefficient can be divided into

$$C_D = C_f + C_{dp} + C_i + C_{sw}$$

Among them, C_f is the frictional drag coefficient, C_{dp} is the pressure drag coefficient caused by different viscous boundary layers, C_i is the induced drag coefficient, and C_{sw} is the shock wave drag coefficient. In aircraft design, the sum of frictional drag and viscous pressure drag is also called parasitic drag, or waste drag, or additional drag.

The range of the aircraft can be estimated by the famous Breguet relation, namely

$$R = \frac{C_L}{C_D} \frac{V_\infty}{SFC} \ln\left(\frac{W_L + W_F}{W_L}\right)$$

where R is range, C_L/C_D is lift-to-drag ratio, V_∞ is flight speed, SFC is fuel consumption ratio (fuel required to generate unit thrust per unit time), W_F is total fuel weight, and W_L is basic weight. It can be seen from the above expression that the larger the lift-drag ratio, the longer the aircraft's range, so reducing the aircraft drag will directly help improve flight performance. Generally, the total drag coefficient of a large aircraft cruise configuration is about 0.03–0.04, and the "unit" of the aircraft drag coefficient is usually defined as count, that is, 1 count = 0.0001. Studies have shown that even a small change in one count will cause changes in aircraft flight performance. Therefore, this drag unit has been accepted by the aviation industry as a standard for aircraft drag accuracy design, which is enough to see the importance of reducing aircraft drag.

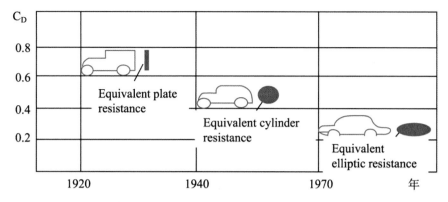

Fig. 7.159 Drag coefficients for different shape cars

Actual flight statistics for large aircraft indicate that drag is closely related to fuel economy. For different aircraft, under a typical usage rate, increasing the drag by 1% is equivalent to consuming more aviation fuel each year as follows (approximately): B737 is 15,000 gal, B757 is 25,000 gal, B767 is 30,000 gal, and B777 is 70,000 gal, B747 is 100,000 gal. It can be seen that the reduction in aircraft drag can increase the range, reduce the takeoff weight, increase the cruise lift-to-drag ratio, save fuel, increase effective payload, and reduce the direct operating costs of the aircraft. In addition, by reducing the drag to reduce the emissions of fuel exhaust, thereby reducing air environmental pollution, this is particularly important in the current era of rapid development of science and technology.

During cruising of high subsonic large aircraft, the surface frictional drag accounts for 50% of the total drag, the induced drag accounts for 30%, shock wave drag accounts for 5%, and the pressure drag accounts for 15%. Figures 7.159 and 7.160 show the drag coefficients of typical cars and large airplanes, respectively. However, it should be noted that the characteristic area for the drag coefficient of the car is the maximum windward area, and the characteristic area of the aircraft is the exposed area of the wings. The following will introduce the relevant technologies and flow mechanisms for reducing frictional drag, induced drag, and shock wave drag, respectively.

2. *Reduce frictional drag*

First consider the aircraft surface frictional drag which accounts for the largest proportion of total drag. The frictional drag is directly related to the boundary layer on the aircraft surface. The boundary layer is caused by the viscosity of the air and the relative motion between the aircraft and the

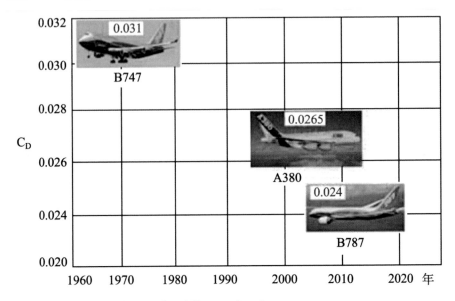

Fig. 7.160 Drag coefficients for different aircraft

air. Depending on the Reynolds number, a transition from laminar to turbulent flow will occur in the boundary layer. According to the boundary layer theory, the inertia and viscous forces of air particles in the laminar boundary layer are of the same magnitude. If the Reynolds number Rex is calculated using the surface length x in the flow direction as the characteristic length and the boundary layer outward velocity V∞ as the characteristic velocity, the boundary layer thickness is inversely proportional to the square of the Reynolds number Rex. For a flat plateflow, it was found that the Rex of the laminar boundary layer transition was located at 3.5×10^5–3.5×10^6. The research shows that the frictional shear stress on the surface of the aircraft is related to the flow state in the boundary layer. Generally, the frictional shear stress in the laminar boundary layer is 1/7 to 1/8 of the frictional shear stress in the turbulent boundary layer, as shown in Figs. 7.161 and 7.162. Therefore, the best way to reduce the drag is to delay the transition of the boundary layer and to maintain laminar flow on the surface of the wing and fuselage as much as possible. Therefore, a technique for reducing drag through laminar flow control is proposed.

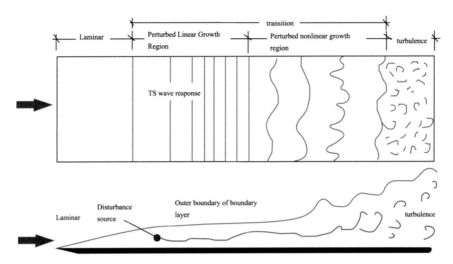

Fig. 7.161 Flat plate transition

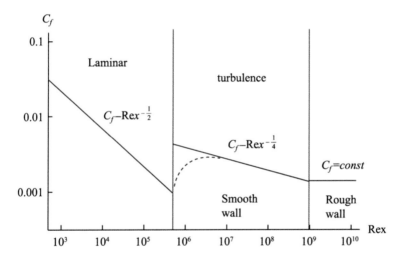

Fig. 7.162 Frictional drag coefficient of laminar and turbulent boundary layers

Laminar flow control and drag reduction

The most effective way to reduce aircraft drag is to reduce frictional drag on the surface of the aircraft. Since the frictional drag of the turbulent boundary layer is much larger than that of the laminar boundary layer, the basic idea of reducing the frictional drag includes two aspects: one is to delay the occurrence of transitions as much as possible and expand the laminar flow area on the surface; the second is to reduce the frictional drag in the flow region of the turbulent boundary layer. In the past few decades, domestic and foreign scholars have proposed many control technologies to reduce frictional drag, and have done a lot of research on it. However, these technologies are still in the research stage, and almost no control technology is used in actual aircraft. Among the many control technologies, the laminar flow control is one of the effective methods to reduce frictional drag. This technology adopts control measures to delay the transition of the boundary layer and enlarge the laminar flow area on the object surface, thereby achieving the purpose of drag reduction. For an aircraft, the surface of wing, engine pod, nose, horizontal tail, and vertical tail are the main areas to obtain laminar flow, as shown in Fig. 7.163. According to estimates by Arcara et al., if 50% chordal laminar flow coverage can be achieved on the main wing's upper airfoil, vertical tail and horizontal tail, and 40% laminar flow coverage on the engine case, then the total takeoff weight (TOGW) can be reduced by 9.9%, the operating empty weight (OEW) can be reduced by 5.7%, and the lift-to-drag ratio can be increased by 14.7%.

The research on laminar flow control technology has more than 70 years of history. Existing research shows that in the design of laminar airfoil and wing, the purpose of disturbance control in the boundary layer is to delay

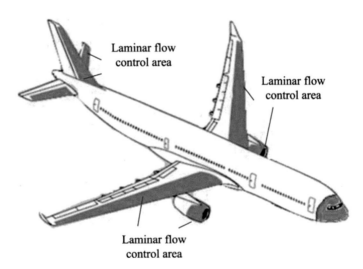

Fig. 7.163 The main laminar flow control area on the aircraft surface (including the outer surface of the aircraft main wing, vertical tail, horizontal tail, and engine nacelle)

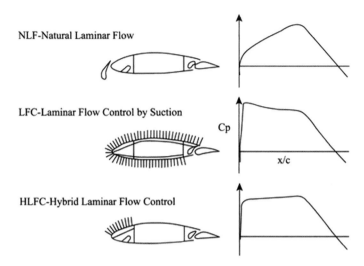

Fig. 7.164 Conceptual design of NLF, LFC, and HLFC

the transition position and minimize the influence on aerodynamic performance and structure. According to the different control methods, there are three control technologies (as shown in Fig. 7.164): one is passive control or natural laminar flow control (NLF), which adjusts the shape to increase the range of the positive pressure gradient on the surface, thereby delaying the transition. This method has poor aerodynamic performance in the non-design state; the second is active control or laminar flow control (LFC), that is, control at a specific location in the boundary layer (such as suction technology) to delay the transition; the third is the hybrid laminar flow control technology (HLFC), which combines the advantages of natural laminar flow control (passive control) and laminar flow control (active control, such as suction technology), can effectively reduce the volume of suction and the complexity of the control system. The characteristics of hybrid laminar flow control are as follows: (1) only suction air is required at the leading edge; (2) only surface geometry modification is required near the leading edge to achieve a favorable pressure gradient; (3) the hybrid laminar flow control wing has good turbulence performance. As shown in Fig. 7.165, the active control technologies mainly include suction air, wall cooling, and active compliant wall technology. The passive control technologies mainly include wall surface modification, surface roughness distribution, passive compliant wall porous wall surface technology, etc. The current development trend is the hybrid laminar flow control technology (HLFC). The most widely used is the combination of wall surface modification (to maintain a good positive pressure gradient) and suction technology.

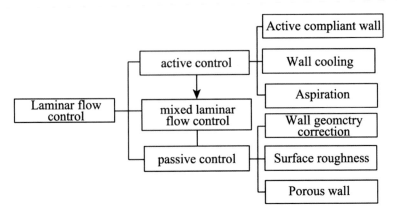

Fig. 7.165 Classification of laminar flow control technologies

In terms of mechanism research, for the aircraft wing, there are four main mechanisms that can cause the transition of the boundary layer, namely TS instability (Tollmien–Schlichting instability), cross-flow instability, attachment-line instability, and TG instability (Taylor-Gortler instability). TS instability is one of the most common boundary layer transition mechanisms, which refers to the TS wave (Tollmien–Schlichting wave) that appears in the boundary layer and continues to grow through the receptive mechanism, and finally lead to the boundary layer transition process through the nonlinear interaction of different modes. Generally, the TS wave is stable in the negative pressure gradient region on the wing (negative increase in disturbance intensity), and unstable in the positive pressure gradient region (disturbance increases rapidly). Therefore, the instability of TS wave can be effectively controlled by the reasonable design of wing pressure distribution or the reasonable control of boundary layer velocity profile.

Cross-flow instability occurs in the area of negative pressure gradient on the swept wing. When the aircraft is cruising, the turbulence of the incoming flow is usually low, and the cross-flow instability is mainly characterized by the amplification of stationary cross-flow wave and their evolution into cross-flow vortices. Then, the secondary instability on the cross-flow vortex is excited, which leads to the breaking and transition of the cross-flow vortex. When the wing sweep angle exceeds 25°–30°, the cross-flow instability will replace TS instability as the dominant factor for boundary layer transition.

TG instability occurs at locations where the wall surface has recesses toward the interior of the airfoil, such as at the rear of the lower surface of some wings. In this case, the disturbance in the boundary layer develops into Görtler vortices under the action of centrifugal force. Unlike the cross-flow vortex, the adjacent Görtler vortices rotate in the opposite direction while the cross-flow vortices rotate in the same direction. Similar to the cross-flow instability, the breaking of Görtler vortex is also caused by the rapid amplification of the secondary instability on the vortex, which causes the transition of boundary layer.

Attachment-line instability usually means that in the three-dimensional flow of the wing leading edge, the instability wave develops, amplifies along the spanwise, and leads to the transition of the boundary layer. Generally, the connecting part of the wing and the fuselage does not isolate the turbulent boundary layer on the fuselage, so the turbulent fluctuation on the fuselage surface will propagate to the wing surface flow, which is easy to cause the flow on the wing surface to be polluted and then transition.

Natural laminar flow control

For example, NACA-6 series airfoils designed in the early stage are representative of natural laminar flow control. The early results of laminar airfoil design are usually unsatisfactory. For example, although the NACA 632-215 airfoil can achieve low drag, the lift range that can be used is much smaller than that of the turbulent NACA 23015 airfoil. Of course, with the development of airfoil design technology, the performance of natural laminar flow airfoil is getting better and better. For example, NLF (2)–0415 airfoil is often used for boundary layer transition test because of its very effective negative pressure gradient region design on the upper wing surface. As shown in Fig. 7.166, the engine nacelle of B787 is designed with special laminar flow control,

Fig. 7.166 Laminar flow control of engine nacelle of B787

which is the first kind of laminar flow nacelle control that has been put into commercial operation.

However, the design of laminar airfoil alone cannot solve the laminar flow control problem of large transport aircraft. For cruising at a higher speed, such as Mach number M = 0.8, the swept wing is usually used to control the shock wave so as to increase the drag divergence Mach number, so the cross-flow instability becomes the dominant factor of the boundary layer transition. However, the cross-flow instability develops rapidly in the negative pressure gradient region, so other ways are needed to control it.

Active Laminar Flow Control

The active laminar flow control methods include suction control, temperature control, active flexible wall control, plasma control, etc. Among them, the development of suction control is more mature, and through a large number of flight tests, it is proved that the drag reduction effect is obvious. The principle of suction control can be simply understood as changing the average velocity profile of the local boundary layer, and then restraining the growth of the related instability disturbance. There are usually two ways of suction control, one is channel suction, the other is hole suction. In order to test the actual effect of suction control, NASA carried out the Leading-Edge Flight Test (L.E.F.T.), in which two suction control devices were installed on the leading edges of both wings of C-140 Jetstar aircraft (as shown in Fig. 7.167), and a large number of flight tests were carried out.

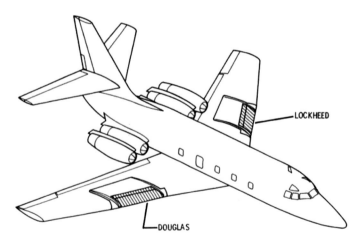

Fig. 7.167 NASA C-140 Jetstar aircraft and suction control devices installed on the left and right sides

The two suction controls were designed and manufactured by McDonnell Douglas and Lockheed, respectively. McDonnell Douglas has designed a set of porous suction devices (including functional components for deworming and anti-icing, etc.), as shown in Fig. 7.168, which was installed on the right wing of Jetstar aircraft, accounting for about 20% of the span length. The suction part is from the leading edge to the front beam of the wing (12% chord length position), in addition, fairing and wing transition extending to 65% chord length are added. The flight test shows that the laminar flow coverage of nearly 65–75% chord can be achieved in most flight states.

Lockheed's suction control was mounted on the left wing of the aircraft. The device adopted channel suction to control laminar flow, as shown in Fig. 7.169. The actual flight test showed that there was little difference between the channel suction control effect and the porous suction control effect. It was estimated that if the laminar flow coverage can be achieved on the upper and lower wing surfaces of the main wing by 75%, and the laminar flow coverage can be achieved on the tail wing by 65%, then 60% of the wing drag can be reduced, and 15% of the whole aircraft drag can be reduced.

If the active control is added to the natural laminar flow, this control method is called hybrid laminar flow control (HLFC). The concept of hybrid laminar flow control was tested on the B757 (Fig. 7.170). From Fig. 7.171, it can be clearly seen that the flow has become turbulent at the edge of the laminar flow control area while maintaining laminar flow in the active control area. The flight test was carried out at Mach number M = 0.8. The results

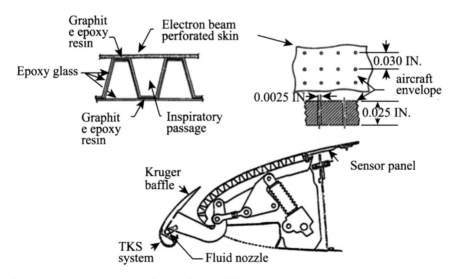

Fig. 7.168 Porous suction device designed by McDonnell Douglas

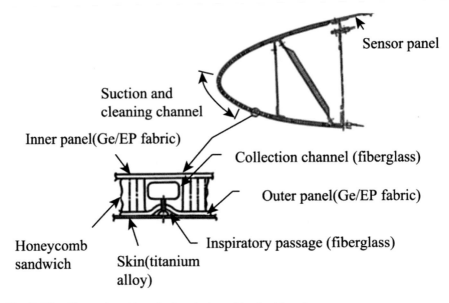

Fig. 7.169 Channel suction device designed by Lockheed

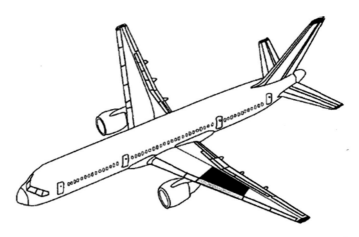

Fig. 7.170 B757 aircraft (the black area on the left wing is the laminar flow control test area)

showed that only 1/3 of the designed suction capacity was needed to achieve 65% chord length laminar flow coverage. The results were more ideal than expected. After calculation, the hybrid laminar flow control can reduce 29% of the wing drag and 6% of the whole aircraft drag.

In addition to the main wing, laminar flow control tests were carried out on the outer surface of the nacelle and on the vertical tail. For example, on the vertical tail of A320 aircraft, applying laminar flow control to the

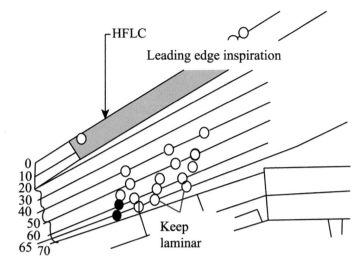

Fig. 7.171 Hybrid laminar flow control results of B757 aircraft wing (hollow dots represent laminar flow state, while solid dots represent turbulent state, and the number on the left represents the percentage of position relative to chord length)

basic vertical tail can obtain 40% chord length laminar flow coverage, and on the HLFC vertical tail it can obtain 50% chord length coverage, which can reduce the drag of the whole aircraft by about 1–1.5%; the outer surface of the GEAE CF6-50C2 engine nacelle of an A300/B2 aircraft has been modified and the laminar flow control test has been carried out, the results show that porous suction can achieve 43% laminar flow coverage.

Wall roughness and flexible wall control

The research on improved wall structure control and delayed boundary layer transition is relatively late, and most of the techniques are still in the experimental stage. In 1998, Saric et al. proposed to use distributed roughness elements (DRE) to control the cross-flow transition of swept wing, and achieved success in wind tunnel test for the first time. Subsequently, Saric's team carried out a lot of theoretical, computational, and wind tunnel experimental research. This laminar flow control is based on the cross-flow transition principle of the swept wing. The development of the cross-flow vortex is directly related to the spanwise wavelength, which is determined by the small rough structure on the wall. If no control is applied, the most unstable wavelength will be dominant, and then the transition will be earlier.

If the distributed roughness elements are added, the development of cross-flow vortex is determined by the roughness elements. Then, we can find roughness elements which can delay the transition by adjusting the parameters. The roughness here is different from that of the wall itself, which is close to a very flat small cylinder. With the deepening of the research, Saric's team carried out laminar flow control test under actual flight conditions on a Cessna O-2A Skymaster aircraft. The test results show that the distributed roughness elements can keep more than 80% chord length covered by laminar flow under certain conditions.

Later, Fransson et al. proposed to use distributed roughness elements to control the development of TS wave. Although distributed roughness is used, the principle used here is not the same as that used to control cross-flow transition. Here, the periodic change along spanwise direction is generated by distributed roughness elements, and the gradient along spanwise is introduced into the two-dimensional boundary layer flow, then the negative production term of Reynolds stress is formed during the evolution of disturbance, which restrains the further growth of disturbance in the flow.

Based on the same principle, the experiment of using micro-vortex generators to produce strips in the laminar boundary layer to restrain the growth of TS wave and thus delay the transition has also been successful, and if multiple groups of vortex generators are used at the same time, the laminar flow coverage can be continuously extended. After careful design, the plasma actuator has been successfully used to delay the transition caused by TS wave. These new control methods are still in the stage of experimental research and test, and have not been tested in the field.

Regarding delaying the transition of the boundary layer, there are also some ideas based on observations of nature. For example, Kramer proposed the use of flexible walls to delay the transition based on the study of dolphins (as shown in Fig. 7.172). The research in this area is relatively tortuous. For a long time, no one could repeat the effect of the delayed transition mentioned by Kramer until Gaster carefully designed the test of unstable disturbance growth on the flexible wall and carried out the corresponding theoretical analysis, which is consistent with the experimental results, thus bringing new vitality to the research on transition control of flexible walls. Although the laminar flow control method of the flexible wall surface is currently under research, it is considered to be very promising for laminar drag reduction control of aircraft.

Fig. 7.172   Laminar drag reduction in dolphin skin flexible wall surface

Turbulence control and drag reduction

A large number of studies have shown that improving the turbulent structure in the near-wall region is an effective method to reduce the wall
surface frictional drag in the turbulent boundary layer region. Turbulence
drag reduction is the result of controlling the turbulent vortex structure in
the near-wall region, specifically the control of the pseudo-sequential large-
scale vortex structure in the turbulent boundary layer. The main features of
pseudo-sequential structures in near-wall turbulence are as follows:

(1) Low-speed strips in the viscous sublayer; (2) the jet behavior of the
low-speed flow in the wall area causes the low-speed strip to rise; (3) the
high-speed fluid from the outer edge of the boundary layer rushes and
sweeps toward the wall surface, making the outer area flow in; (4) various
forms of turbulent vortex structure; (5) the inclination of the near-wall shear
structure, which shows the concentration of spanwise vorticity; (6) the near-
wall "vortex package" structure; (7) the large-scale turbulent vortex structure
covered by the three-dimensional bulge on the contact surface between the
outer boundary layer and the potential flow; (8) the shear layer moves "backward" caused by the large-scale turbulent vortex structure movement outside
the boundary layer, resulting in the discontinuity of inflow velocity.

Because of these complexes near-wall turbulent structures, the physical
parameters in the boundary layer are uncertain in space and time, so the
control of turbulent boundary layer is much more difficult than that of
laminar boundary layer. Generally speaking, turbulence control methods are

also divided into active control and passive control, which are described below.

Passive Turbulence Control

The frictional drag of turbulent boundary layer can be reduced by controlling the spanwise fluctuation effectively. At present, the wall riblet control is a passive control technology. In the study of turbulent boundary layer control, the mechanism of shark skin riblet drag reduction has been widely concerned. As shown in Fig. 7.173, the microstructure of shark skin is actually a complex riblet.

Walsh was the first to study the turbulent structure on the surface of the riblets. After measurement, it was found that under some riblet parameters, the drag reduction effect of about 8% could be achieved, but it was also noted that the riblets did not have a significant impact on the frequency of turbulent burst events. Suzuki and Kasagi found in experiments that the existence of riblets can significantly inhibit the spanwise pulsating energy exchange in the turbulent boundary layer. As shown in Fig. 7.174 is a typical wall riblet, this control technology does not need energy input, so it is a passive control technology of boundary layer. It is found through experiments that the dimensionless riblet spacing with good drag reduction effect is $S^+=10\sim20$.

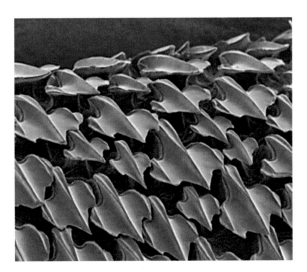

Fig. 7.173 Shark skin riblet structure

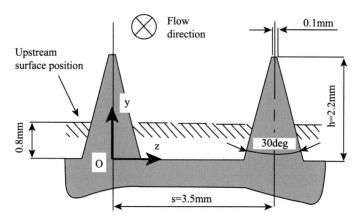

Fig. 7.174 Cross section of riblet surface

Among

$$S^+ = \frac{u^* s}{v}, u^* = \sqrt{\tau_W / \rho}$$

where u^* is the wall friction velocity, v is the air kinematic viscosity coefficient, τ_w is the wall friction drag, and ρ is the air density. Under flight conditions, the actual spacing is generally 25–75 μm (human hair is about 70 μm). From the existing research, it is found that the riblet drag reduction is about 5–15%.

Symmetrical V-shaped riblet (height equals spacing) produced by 3 M company of the United States are widely used in the research and flight test of riblet drag reduction. For example, the flight test results on NACA0012 airfoil show that riblet can reduce drag of straight wing by 5% to 8%. For swept wing, if the angle of riblet can be within a small angle with the local potential flow direction, then similar results can be achieved. Boeing carried out the actual flight test of turbulent drag reduction by riblets on the t-33 trainer. The surface of the wing covered by the riblets was distributed in the range of 7–83% chord length, and the wake of the boundary layer on the riblet surface was measured at 83% chord length. It was found that the maximum drag reduction was 6–7%, and the effects of 0.033 mm height and 0.076 mm height were similar. In the flight test of A340-300, the riblet can reduce the wall friction drag by about 5–8%.

Walsh conducted a riblet test on a Learjet Model 28/29 twin-engine business jet. The results of flight test and wind tunnel test were basically the same, and it was confirmed that the drag reduction effect can be predicted by the height of the riblets, but the maximum drag reduction effect was slightly lower than that of wind tunnel test. Szodruch carried out the flight test of drag reduction by riblets on A320. Under the flight condition of Mach number M = 0.77–0.79, the riblet can reduce drag by about 2%, which is similar to the expected effect.

In addition, there is a passive control device for reducing the turbulent boundary layer drag, which is called the large eddy breakup device. This device can reduce the drag of the boundary layer near it, but it will introduce a lot of additional drag, which may cause the net drag to rise. This device is still under study.

Active turbulence control

The results show that it is a very effective turbulence control method to apply the spanwise vibration in the turbulent boundary layer. The spanwise vibration control method can reduce the wall drag by generating the spanwise vorticity in the Stokes layer and then affecting the time-averaged velocity gradient in the viscous sublayer. The spanwise vibration of the turbulent flow can be directly achieved by the vibration of the wall surface. In the experiment by Choi et al. the control method can reduce drag by 45%. Another way to reduce frictional drag by periodically blowing air in the turbulent boundary layer (as shown in Fig. 7.175) is a potential active control technology that is currently being studied. It has been found that periodic blown air in the turbulent boundary layer will form local re-laminarization and

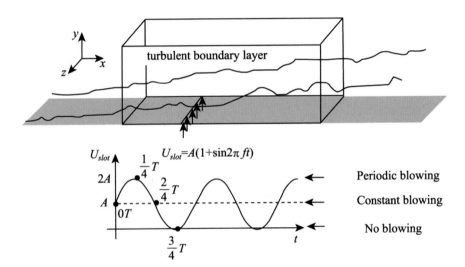

Fig. 7.175 Periodic blowing control of turbulent boundary layer

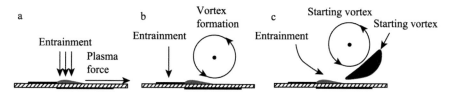

Fig. 7.176 The process of vortices generated by the plasma actuator

affect the pseudo-sequential structure downstream, which will then produce drag reduction effects.

Plasma actuator is a method that can effectively realize the spanwise vibration. The process of generating force on the fluid and forming vortices is shown in Fig. 7.176.

Jukes et al. investigated the effect of spanwise vibration control of the plasma actuator at $Re_\tau = 400$ in turbulent boundary layer. It was found that when the vibration period is $T^+ = T u_\tau^2 / v = 16$, the plasma jet velocity is $W^+ = W / u_\tau \approx 10$, and the electrode spacing is $s^+ = s u_\tau / v = 20$, a 45% drag reduction effect can be achieved. In addition to directly vibrating fluids, plasma actuators are also used to generate spanwise traveling waves to reduce drag. Spanwise traveling waves cause nearby low-speed strips to converge into a wider strip and form a streamwise vortex, accompanied by jets on one side and downwash on the other.

3. *Reduce induced drag*

In the cruise state of large aircraft, the ratio of wing induced drag to total drag is only next to frictional drag, which is caused by the downwash induced by the vortex shed from the trailing edge of the wing. The reduction in induced drag can be achieved by enlarging the wing span, but the expansion of the span will be limited by the wing structure. In the 1970s, Whitcomb, director of Langley wind tunnel laboratory of NASA, designed a device attached to the wing tip of the main wing, which can effectively reduce the induced drag of the aircraft during cruising. This device is called the winglet (as shown in Fig. 7.177). Shortly after the invention of winglets, the U.S. Air Force tested the effects of winglets on the KC-135 tanker. The test results show that after adding the winglets, the cruise drag can be reduced by about 7%, and it is estimated that this improvement can save the KC-135 fleet billions of dollars in the next two decades.

Bourdin et al. designed the winglet as a movable device, and found that the movable winglet has obvious effect on improving the low-speed performance. A large number of studies show that winglet is an effective device to reduce induced drag. Now large aircraft produced by Boeing and Airbus are

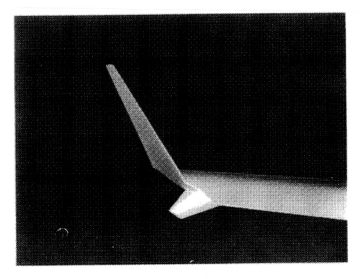

Fig. 7.177 Winglets designed by Whitcomb

basically equipped with such winglets. The flow mechanism is that under the pressure difference between the upper and lower surfaces of the wingtip, the airflow tends to flow outward along the lower surface and inward along the upper surface. After installing a winglet, it will play an end plate effect on the spanwise flow of the wing. Under the shear effect of the winglet vortex and the wingtip vortex, it will play a diffusion effect on the wingtip vortex, so that the downwash of the wing tail vortex will be weakened, the downwash angle will be reduced, and the induced drag will be reduced.

In addition to serving as the wing tip end plate to increase the effective aspect ratio of the wing, the winglet reduces the drag using the "pull effect" caused by the deflection of the wing tip airflow, as shown in Fig. 7.178. The results of wind tunnel test and flight test show that the winglet can reduce the induced drag of the whole aircraft by 20–30%, which is equivalent to 5% increase in lift-to-drag ratio. Winglets, as an advanced aerodynamic design measure to improve flight economy and save fuel, have been used in many aircraft.

4. *Reduce shock drag*

For a long time, it has been the technical difficulty of aircraft design to reduce the flight drag of transonic aircraft and increase the drag divergence Mach number. As early as in the 1950s, Whitcomb et al. of NASA in the United States found through wind tunnel tests that when the flight speed is near the speed of sound, the zero-lift wave drag of the aircraft will be greatly affected

Fig. 7.178 Winglet lift increment and pull effect

Fig. 7.179 Wasp waist body design of H-6

by the longitudinal distribution of its cross-sectional area, and is the same as the revolutionary body with the same cross-sectional area distribution. That is to say, the shape of the cross-sectional area of the aircraft in the longitudinal position has no effect on the wave drag. What has an effect is the change mode of the size of the cross-sectional area in the longitudinal direction. When the traditional straight fuselage passes through the wing, the wave drag will increase obviously. If the wasp waist structure is adopted, the wave drag can be greatly reduced. Therefore, the transonic area law is proposed to reduce the zero-lift wave drag by modifying the fuselage. It is found through experiments that the area law can reduce the zero-lift drag by 25–30%, but with the increase in Mach number, the drag reduction effect of the area law decreases gradually. When Mach number is between 1.8 and 2.0, the effect of area law is almost zero.

As shown in Fig. 1.179, the cruising Mach number of H-6 in China is 0.75 which adopts cantilever middle single wing, double beam box structure. The sweep angle of focus line is 35°, the negative dihedral angle of wing chord plane is 3°, and the installation angle is 1°. The wing trailing edge is equipped with inner and outer flaps and ailerons. The flap is a backward slotted type with a maximum deflection angle of 35°. The aileron is equipped with an internal aerodynamic axial compensation and adjustment plate. All metal semi-rigid shell body structures and wasp waist streamline body are adopted.

In addition to considering the reduction in shock wave drag in the overall design of the aircraft, a new technology has been developed in recent years to control the shock intensity on the wing by increasing the shock bump on the airfoil, thereby reducing the shock wave drag. In the design of supercritical wing, the swept wing with a weak shock wave is given in the cruise design state, but when the flight state deviates from the design state, the shock wave drag will increase sharply. Tai et al. studied the aerodynamic performance of airfoil with bump at transonic speed, and proposed to improve the transonic drag characteristics of airfoil by adding bump (as shown in Fig. 7.180). The following research showed that the range of shock wave bump can be extended from 20% to 40% of the chord length, and the bump shape can be adjusted dynamically as required.

Europe (Euroshock project) and the United States (NASA aircraft deformation project) have carried out systematic research on the control of shock wave intensity by the bump. The results of numerical simulation and experimental test show that the reasonable design of the bump can effectively improve the lift-to-drag ratio of the aircraft when the flight speed is close

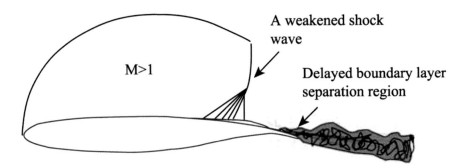

Fig. 7.180 Shock intensity controlled by shock bump (red line is shock bump)

to the sound speed. The design of shock bump has developed from two-dimensional to three-dimensional bump array. Although it is still in the experimental research stage, the shock wave bump is hopeful to be applied to the future aircraft.

8

Introduction to Celebrities in Fluid Mechanics

In order to facilitate the study and understanding of the development of fluid mechanics, this chapter mainly collects 52 scientists who have made important contributions to fluid mechanics in the world (sorted according to the year of birth) from *the Encyclopedia of China* and *baike.baidu.com*, and introduces their main scientific achievements, so as to provide a reference for beginners.

8.1 Archimedes (287–212 B.C.)

Archimedes (287–212 B.C.), a great Greek philosopher, scientist, mathematician, physicist, mechanic, founder of static mechanics and hydrostatics, enjoys the reputation of "the father of mechanics". Archimedes, Gauss, and Newton are the three greatest mathematicians in the world. Archimedes once said, "Give me a fulcrum, and I will lift the whole earth". He established the basic principles of statics and hydrostatics and gave many methods to find the center of gravity of a geometric figure, including that of a figure enclosed by a parabola and its parallel chords. Archimedes proved that the buoyancy of an object in a liquid is equal to the weight of the liquid it displaces, which was later called Archimedes principle. He also gave a criterion for the equilibrium stability of a positive parabolic rotating body floating in a liquid. The machines invented by Archimedes include a water screw for water diversion, a lever pulley machine that can move a full ship, and an Earth-Moon-Sun operation model that can explain the phenomenon of solar and lunar eclipse. But he thought that mechanical invention was inferior to pure mathematics, so

© Science Press 2021
P. Liu, *A General Theory of Fluid Mechanics*,
https://doi.org/10.1007/978-981-33-6660-2_8

Fig. 8.1 Archimedes (287–212 B.C.)

he did not write any work on it. Archimedes also used the method of continuous division to calculate the volume of ellipsoid, rotating parabolic body, etc., which had the rudiments of integral calculation. He has more than 10 kinds of works, most of which are Greek manuscripts. His works focus on the problem of quadrature, mainly the area of curved edge figure and the volume of curved cube and their style is deeply influenced by Euclid's *The Original of Geometry*, which is first assumed and then proved by rigorous logical inference. He constantly sought general principles for special projects. His work always blended mathematics and physics (Fig. 8.1).

8.2 Leonardo Da Vinci (1451–1519)

Leonardo Da Vinci (1452–1519), a European Renaissance genius scientist, inventor and painter, who is called "the most perfect representative of the Renaissance" and "the only all-round talent in human history" by the modern scholars. His greatest achievement was painting and his masterpieces *Mona Lisa, The Last Supper, The Virgin of the Rocks,* and other works reflect his exquisite artistic attainment. He thought that the most beautiful object of study in nature is the human body, which is a wonderful work of nature and painters should take humans as the object of painting. He was a painter, astronomer, inventor, and architectural engineer with profound thoughts,

Fig. 8.2 Leonardo Da Vinci (1452–1519)

knowledge, and versatile skills. He was also good at sculpture, music, invention, architecture, and was familiar with mathematics, physiology, physics, astronomy, geology, and other disciplines. He was versatile, diligent, and prolific. His manuscript has about 6000 pages. All his scientific research results are preserved in his manuscript. Einstein believed that if Leonardo da Vinci's scientific research results were published at that time, science and technology could be advanced 30–50 years (Fig. 8.2).

8.3   Galileo (1564–1642)

Galileo (1564–1642), an Italian mathematician, physicist, astronomer, pioneer of scientific revolution, invented the pendulum and thermometer, and made great contributions to human science. He was one of the founders of modern experimental science. In history, he first integrated mathematics, physics, and astronomy on the basis of scientific experiments, deepening and changing human understanding of matter movement and the universe. Galileo summed up the law of free fall, the law of inertia, and the principle of Galileo relativity from the experiment, which overthrew many assumptions of Aristotle's physics, and laid the foundation of classical mechanics. He refuted

Ptolemy's geocentric system, and strongly supported Copernicus's heliocentric theory. With systematic experiments and observations, he overthrew the traditional natural view of speculation and created a modern science based on experimental facts and a strict logical system. Therefore, he is known as "the father of modern mechanics" and "the father of modern science". His work laid a foundation for the establishment of Newton's theoretical system. Galileo advocated the research method of combining mathematics with experiment, which is the source of his great achievements in science and the most important contribution to modern science. According to Galileo, experience is the only source of knowledge. He advocated the experimental mathematical method to study the laws of nature and opposed the mysterious speculation of scholasticism. He believed that the book of nature is written in mathematical language. Only the shape, size, and speed that can be attributed to quantitative characteristics are the objective properties of objects. He was the first person to make a lot of achievements in observing celestial bodies with telescopes. Galileo played an important role in the development of natural science and world outlook in the seventeenth century. Experimental science, which was initiated by Galileo and Newton, is the beginning of modern natural science (Fig. 8.3).

Fig. 8.3 Galileo (1564–1642)

8.4 Pascal (1623–1662)

Blaise Pascal (1623–1662) was a French mathematician, physicist, philosopher, and essayist. At the age of 16, he discovered the famous hexagon theorem and he completed *The Theory of Conic Curve* at the age of 17. In 1642, he designed and manufactured an automatic carry addition and subtraction calculation device, which is called the first digital calculator in the world and provides the basic principle for later computer design. In 1654, he began to study several aspects of mathematical problems, deeply discussed the principle of indivisibility in infinitesimal analysis, obtained the general method to find the area and center of gravity enclosed by different curves, solved the cycloid problem with the principle of integral theory, and completed *The Theory of Cycloid* in 1658. His manuscript of the paper has a great inspiration for Leibniz to establish calculus. When studying the properties of binomial coefficients, he wrote *Arithmetical Triangle*, which was included in his complete works and published in 1665, and submitted it to the Paris Academy of Sciences. The expansion binomial coefficient is called "Pascal triangle", which was known by Chinese mathematician Jia Xian in 1100. He also made a mercury barometer (1646), wrote papers on liquid balance, air weight and density (1651–1654) and classic works such as *Pensees* (1658) (Fig. 8.4).

Fig. 8.4 Pascal (1623–1662)

8.5   Newton (1643–1727)

Isaac Newton (1643–1727), graduated from Trinity College of Cambridge University, is a famous British physicist and president of the Royal Society. He is encyclopedic "all-rounder" and has written *Mathematical Principles of Natural Philosophy* and *Optics*. In his paper *The Law of Nature*, published in 1687, he described gravity and the three laws of motion. These descriptions laid the scientific view of the physical world for the next three centuries and became the foundation of modern engineering. By demonstrating the consistency between Kepler's laws of motion of planets and his theory of gravitation, he showed that the motion of objects on the ground and celestial bodies follow the same laws of nature, which provides strong theoretical support for the theory of sun center and promotes the scientific revolution. In mechanics, Newton clarified the principle of conservation of momentum and angular momentum, and proposed Newton's law of motion. In optics, he invented the reflection telescope and developed the color theory based on the observation of the white light emitted by the prism into the visible spectrum. He also formulated the cooling law systematically and studied the speed of sound. In mathematics, Newton shared the honor of developing calculus with Leibniz. He also proved the generalized binomial theorem, proposed "Newton method" to approach the zero point of function, and made contribution to the research of power series (Fig. 8.5).

Fig. 8.5   Newton (1643–1727)

8.6   Leibniz (1646–1716)

Gottfried Wilhelm Leibniz (1646–1716), German philosopher and mathematician, a rare generalist in history, was known as Aristotle in the seventeenth century. He was a lawyer, often back and forth in the major cities, and many formulas were completed in the bumpy carriage. He also claimed to have the Baron's noble identity. Leibniz played an important role in the history of mathematics and philosophy. In mathematics, he and Newton have independently invented calculus, and the mathematical symbols of calculus he used are more widely used. The symbols invented by Leibniz are generally considered to be more comprehensive and applicable. Leibniz also contributed to the development of binary. In philosophy, Leibniz's rationalism is the most famous. Leibniz, Descartes, and Baruch de Spinoza are regarded as the three greatest rationalist philosophers in the seventeenth century. Leibniz's work in philosophy foresees the birth of modern logic and analytical philosophy, but it was also deeply influenced by the tradition of scholasticism. He used more the first principle or transcendental definition than experimental evidence to derive conclusions. Leibniz has left many works in politics, law, ethics, theology, philosophy, history linguistics, etc (Fig. 8.6).

Fig. 8.6   Leibniz (1646–1716)

8.7 Bernoulli (1700–1782)

Daniel Bernoulli (1700–1782) was one of the important members of Bernoulli family, a famous scientific family in Switzerland. From 1726 to 1733, he presided over the department of mathematics at the Academy of Sciences in St. Petersburg, Russia. Daniel Bernoulli had a solid mathematical foundation and keen insight, and often showed his originality in solving problems. From 1725 to 1749, he won ten bonuses from the French Academy of Sciences. His research fields included mathematics, mechanics, magnetism, tide, ocean current, planetary orbit, etc. He has collaborated with Swiss mathematician L. Euler and Scottish mathematician C. Makrolin to write papers on tides. In 1738, he published the book *Hydrodynamics*, which laid the foundation of this subject, and thus gained a high reputation. He put forward the law of conservation of energy of ideal fluid, that is, the sum of position potential energy, pressure potential energy, and kinetic energy per unit weight of liquid remains constant, which is called Bernoulli's theorem. On this basis, the relationship between pressure and velocity of water was expounded, and the important conclusion that the pressure decreases when the fluid velocity increases was put forward. Daniel Bernoulli also has many works on solid mechanics. For example, the vibration equation of cantilever beam was proposed in 1735, and the superposition principle of elastic vibration theory was proposed in 1742 (Fig. 8.7).

Fig. 8.7 Bernoulli (1700–1782)

8.8   Euler (1707–1783)

Leonhard Euler (1707–1783) was a Swiss mathematician and mechanic who wrote the most in the eighteenth century. His works involved all fields of mathematics at that time and many mathematical terms were named after Euler, such as Euler integral, Euler number, various Euler formulas, etc. He and his successor Lagrange completed the transition from comprehensive method to analytical method in mathematics, but they are quite different in style. Euler is famous for his concreteness and delicacy, while Lagrange is good at abstraction and generalization. Euler applied mathematical methods to mechanics, and made outstanding contributions in all fields of mechanics. He was the founder of rigid body dynamics and hydrodynamics, and the founder of stability theory of elastic system. In the two volumes of *Analysis and Explanation of Mechanics or Motion Science* published in 1736, he considered the differential equations of motion of free particles and constrained particles. In terms of mechanical principles, in the study of rigid body kinematics and rigid body dynamics, he obtained the most basic results, among which the finite motion of rigid body fixed point is equivalent to the rotation around a certain axis of fixed point; the movement of rigid body fixed point can be described by the change of three angles (called Euler angle); the relationship between the change of angular velocity and the external moment when the rigid body fixed point rotates; the motion law of the fixed-point rigid body without external force moment and the differential equation of free rigid body. Euler believed that the differential equations of particle dynamics could be applied to liquids (1750). He used two methods to describe the motion of fluid, that is, to describe the velocity field of fluid according to the fixed point of space (1755) and the particle of fluid (1759), respectively. These two methods are usually called Eulerian representation and Lagrangian representation. Euler laid the theoretical foundation for the motion of ideal fluid (assuming that the fluid is incompressible and its viscosity can be ignored). The continuity equation reflecting the conservation of mass (1752) and the hydrodynamic equation reflecting the law of momentum change (1755) were given. Euler studied the vibration of elastic systems such as strings and bars. Together with Daniel Bernoulli, he analyzed the vibrations of the heavy chains suspended at the top and the corresponding discrete models (lines with a string of masses). With the help of Daniel Bernoulli, he got the exact solution of the deflection curve of the thin rod under elastic compression. The minimum pressure which can make the thin rod produce this kind of deflection is called the Euler critical load. Euler had also studied applied mechanics

Fig. 8.8 Euler (1707–1783)

such as ballistics, ship theory, moon motion theory, and so on. Euler wrote more than 800 books and papers (Fig. 8.8).

8.9   D'Alembert (1717–1783)

Jean Le Rond d'Alembert (1717–1783) is a famous French physicist, mathematician, and astronomer. He was the main pioneer and founder of mathematical analysis. He gave a better definition of limit, but he did not formulate this expression. Boyle made such a comment: it is impossible for d'Alembert to express the limit in strict form without getting rid of the influence of traditional geometric methods, but he was almost the only mathematician who regarded differential as the limit of function at that time.

D'Alembert is one of the few mathematicians who separated convergent series from divergent series in the eighteenth century, and he also proposed a method to judge the absolute convergence of series—d'Alembert discriminant method—which is the ratio discriminant method still used now. He was also the founder of the theory of trigonometric series, and he also made a great contribution to the emergence of partial differential equations. In 1746, he published the paper *The Study of the Curve Formed by the String Vibration under Tension*, in which he first proposed the wave equation, and proved the functional relationship in 1750. In 1763, he further discussed the vibration of non-uniform strings and proposed a generalized wave equation. In addition,

Fig. 8.9 D'Alembert (1717–1783)

d'Alembert has also studied the properties of complex numbers and probability theory, and he had proved the basic theorem of algebra for a long time. D'Alembert had made great achievements in all aspects of mathematics, but he had not carried out rigorous and systematic in-depth research, and he even believed that the knowledge of mathematics is almost exhausted. In any case, the rapid development of mathematics in the nineteenth century was based on the research of their generation of scientists, and d'Alembert had made an important contribution to the development of mathematics (Fig. 8.9).

8.10 Lagrange (1736–1813)

Joseph-Louis Lagrange (1736–1813) was a French mathematician. He was appointed senator and Earl by Napoleon. Lagrange is the founder of analytical mechanics. In his *Analytical Mechanics* (1788), he absorbed and developed the research results of Euler and d'Alembert, and applied mathematical analysis to solve the mechanical problems of particles and particle systems (including rigid bodies and fluids). After Euler, Lagrange studied the motion equation of ideal fluid, and first put forward the concepts of velocity potential and flow function, which became the basis of the theory of fluid irrotational motion. In *Analytical Mechanics*, the fluid motion equation derived from the general equation of dynamics focuses on the fluid particles and describes the

Fig. 8.10   Lagrange (1736–1813)

motion process of each fluid particle from beginning to end. This method is now called the Lagrange method to distinguish the Euler method focusing on space points, but in reality, this method had also been applied by Euler. From 1764 to 1778, he was awarded the prize of French Academy of Sciences five times for his research on celestial mechanics such as lunar translation. In mathematics, Lagrange is one of the founders of the variational method. His research on algebraic equations has played a leading role in the establishment of Galois group theory (Fig. 8.10).

8.11   Laplace (1749–1827)

Pierre-Simon Laplace (1749–1827) was a French analyst, probability theorist and physicist, and academician of the French Academy of Sciences. In 1812, his important book *Theory of Probability Analysis* was published, in which the whole research of probability theory at that time was summarized. He had long been engaged in the study of the theory of planet motion and the theory of moon motion, especially in the study of the solar system's celestial perturbation, the general stability of the solar system and the dynamics of the solar system's stability. In the five volumes of *Celestial Mechanics*, a representative work of classical celestial mechanics published from 1799 to

Fig. 8.11 Laplace (1749–1827)

1825, the term "celestial mechanics" was first proposed. So he is known as "the father of celestial mechanics" and "Newton in France". In 1814, Laplace put forward the scientific hypothesis that if there is an intelligent creature that can determine the current state of motion from the largest celestial body to the lightest atom, the past state and future state of the whole universe can be calculated according to the mechanical law. Later generations called his supposed intelligent creature the Laplacian demon. He has published more than 270 papers on astronomy, mathematics and physics, with a total of more than 4006 pages. Among them, the most representative monographs are *Celestial Mechanics, Cosmology System Theory, and Theory of Probability Analysis* (Fig. 8.11).

8.12 Kelly (1773–1857)

George Kelly (1773–1857) was a British aerodynamicist and the father of classical aerodynamics. In 1804, Kelly studied the driving force of birds and tested a glider model on a rotating arm. Soon, he launched the winged projectile to sea. Almost at the same time, he designed a composite aircraft with fixed wings on the wheel car and flapping wings on the wing tip. In 1807, Kelly studied the steam engine and another engine using gunpowder. In 1808, Kelly developed "rotor" and "impeller" aircraft, and designed a flapping wing aircraft in the same year. In 1809, Kelly began to study the relationship between fish and what we call streamline today, successfully built the first full-scale glider in the history of aviation and carried out test flight. In 1809, his paper *On Air Navigation* was published in the Journal of Natural Philosophy, in which he put forward very important scientific conclusions: (1) the definition of four kinds of forces acting on the air heavier aircraft: lift, gravity, thrust, and drag; (2) the mechanism of determining lift is separated from that of thrust. At that time, Kelly had realized that bird wings not only have the function of propulsion, but also have the function of generating lift and it is much easier for human aircraft to realize the above functions with different devices than to simply imitate the flight actions of birds. This important discovery laid the basic idea and theoretical foundation of the fixed-wing aircraft. He described fixed wing, tail, fuselage, and elevator, explained the role of wings, and pointed out the appropriate stability; then he mentioned that the aircraft must be windward, and there must be vertical and horizontal rudder surfaces. Kelly's paper also described the relationship between speed and lift, and at the same time he pointed out that the reduction of wing load, tension, and gravity, even the principle of internal combustion engine and the streamline appearance are important for aircraft design. He had pointed out that the whole problem of mechanical flight is to provide power to a flat plate, make it generate lift force in the air flow, and support a certain mass. His paper *On Air Navigation* is regarded as the starting line of aviation theory by later generations (Fig. 8.12).

8.13 Gauss (1777–1855)

Gauss (1777–1855) was a German mathematician, physicist, astronomer, and geodesist. Gauss is the founder of modern mathematics, considered to be one of the most important mathematicians in history, and enjoys the title of "Prince of mathematics". Gauss, Archimedes, and Newton are the three greatest mathematicians in the world. He has made great achievements in his life and there are 110 achievements named after his name Gauss, which is the most among mathematicians. Gauss has a great influence on history, which can be paralleled with that of Archimedes, Newton, and Euler. Gauss's mathematical research is almost all over the field, and he had made pioneering contributions in number theory, algebra, non-Euclidean geometry, complex function, and differential geometry. He also applied mathematics to astronomy, geodesy, and magnetism, and invented the principle of least square method. Gauss had published 155 papers in his whole life. He

Fig. 8.12 Kelly (1773–1857)

was very strict with knowledge and only published works that he thought were very mature. Gauss's important contribution to algebra was to prove the basic theorem of algebra, and his proof of existence opens up a new way of mathematical research. In fact, before Gauss, many mathematicians thought that they had proved this result, but none of them was strict. Gauss pointed out the lack of previous proof one by one, and then put forward his own opinions. In his life, he gave four different proofs. Gauss got the principle of non-Euclidean geometry around 1816. He also made a deep study of complex functions, established some basic concepts and discovered the famous Cauchy integral theorem. He also found the biperiodic nature of elliptic functions, but these works were not published before his death. In physics, Gauss's most remarkable achievement was the invention of cable telegraph with physicist Weber in 1833, which made Gauss's reputation go beyond the academic circle and enter the public society. In addition, Gauss had made outstanding contributions in mechanics, geodesy, hydraulics, electricity, magnetism, and optics (Fig. 8.13).

Fig. 8.13   Gauss (1777–1855)

8.14   Poisson (1781–1840)

Simeon-Denis Poisson (1781–1840), French mathematician, geometer, and physicist, graduated from the comprehensive engineering school of Paris and was appreciated by Laplace and Lagrange. Poisson's scientific career began with the study of differential equations and their application in the theory of pendulum motion and acoustics. His work was characterized by the application of mathematical methods to study all kinds of physical problems, and then he obtained mathematical discoveries. He had made important contributions to integral theory, planetary motion theory, thermophysics, elasticity theory, electromagnetic theory, potential theory, and probability theory. He was also an outstanding figure in the field of probability and statistics in the nineteenth century. He improved the application of probability theory, especially in statistics, and established a probability distribution, Poisson distribution, to describe random phenomena. He generalized "the law of large numbers" and derived the Poisson integral which has important application in probability theory and mathematical equation. His *Mechanical Course* was used as a standard textbook for a long time. In terms of celestial mechanics, he extended Lagrange's and Laplace's studies on the orbital stability of planets, and calculated the gravity between the sphere and the ellipsoid. In his report *On the General Equations of Equilibrium and Motion of Elastic Solids and Fluids* published in 1831, he gave the first complete equation describing the physical properties of viscous fluids, that is, constitutive relation. Poisson

Fig. 8.14 Poisson (1781–1840)

solved many problems of heat conduction employing the expansions of trigonometric series, Legendre polynomials, and Laplace surface harmonic functions. Many achievements of heat conduction are included in his monograph *The Mathematical Theory of Heat*. In his works, the general integral method of elastic theoretical equation was proposed, and Poisson constant was introduced. He also used the variational method to solve the problem of elasticity (Fig. 8.14).

8.15 Navier (1785–1836)

Claude-Louis Navier (1785–1836) was a French engineer, whose main contribution was to establish basic equations for hydrodynamics and elasticity. In 1821, he extended L. Euler's equations of fluid motion, considered the forces between molecules, and established the basic equations of fluid balance and motion, in which there is only one viscosity constant. In 1845, G. G. Stokes improved his hydrodynamic equation of motion from the continuous model, and obtained equations of fluid motion with two viscosity constants (later called Navier–Stokes equation). In 1821, Navier took every molecule as a force center, and derived the equations of equilibrium and motion of elastic solids (published in 1827), in which only contain one elasticity constant. The basic equations of isotropic elasticity with two elastic constants were obtained by A.-L. Cauchy in 1823. Navier's achievements in

Fig. 8.15 Navier (1785–1836)

other aspects of mechanics are as follows: the first (1820) to solve the fourth-order partial differential equation of simply supported rectangular plate with double trigonometric series; the introduction of mechanical work in engineering to measure the efficiency of the machine. In engineering, he changed the tradition of designing and building suspension bridge by experience alone, and adopted theoretical calculation in design. His scientific papers were published in various French scientific journals. Papers on the basic equations of fluid mechanics were published in Vol. 19 (1821) of the Journal of Chemistry. Papers on the equations of equilibrium and motion of elastic solids were published in the Research Report of the French Academy of Sciences, Vol. 7 (1827) (Fig. 8.15).

8.16 Cauchy (1789–1857)

Cauchy (1789–1857) was a French mathematician, physicist, and astronomer, whose greatest contribution in mathematics was that he introduced the concept of limit into calculus and established a logical and clear analysis system based on limit. This is the essence of the history of calculus and Cauchy's great contribution to the development of human science. In 1821, Cauchy put forward the method of limit definition and used inequality to describe the limit process. Nowadays, all calculus textbooks still use the definitions of limit, continuity, derivative, convergence, which were put forward by Cauchy et al. His explanation of calculus was widely adopted by later generations. Cauchy did the most systematic pioneering work on definite integral and defined definite integral as the "limit" of sum. Before the operation of definite integral, he pointed out the existence of integral must be established. He first proved the basic theorem of calculus strictly using the mean value theorem. Through the hard work of Cauchy and later Weierstrass, the basic concepts of mathematical analysis were strictly discussed. He ended the confusion of calculus in the past two hundred years, liberated calculus and its popularization from the complete dependence on geometric concept, motion and intuitive understanding, and made calculus develop into the most basic and largest mathematics subject in modern mathematics. Cauchy's research in other areas is also very rich. The theory of calculus of complex function was founded by him. In algebra, theoretical physics, optics, and elasticity, he also made outstanding contributions. Cauchy's achievements in mathematics are not only brilliant, but also astonishing. There are 27 volumes of Cauchy's complete works and more than 800 works. In the history of mathematics, Cauchy is the second most prolific mathematician after Euler. His glorious name, together with many theorems and rules, are remembered in many textbooks today. As a scholar, he has a quick thinking and outstanding achievements. From his great works and achievements, it is not difficult to imagine how he worked diligently all his life (Fig. 8.16).

Fig. 8.16 Cauchy (1789–1857)

8.17 Saint-Venant (1797–1886)

Adhémar Jean Claude Barré de Saint-Venant (1797–1886) was a French dynamicist whose main research fields are solid mechanics and fluid mechanics, especially in material mechanics and elasticity. In the field of elasticity, he proposed to use the semi inverse method to solve the problems of torsion and bending of the cylinder. The idea of solving is that if the two kinds of external loads at the end of the cylinder are equivalent in statics, the difference of stress field in the two cases outside the end is very small. In 1885, J. V. Boussinesq popularized this idea and called it the Saint-Venant principle. If there is an equilibrium force system (i.e., the resultant force and moment of resultant force are both zero) in a small range of action of the elastomer, the stress caused by the equilibrium force system in the elastomer far away from the action area can be ignored. Saint-Venant principle has been widely used in engineering mechanics for a long time. After 1868, Saint-Venant studied the plastic flow of ductile materials and put forward the basic hypothesis and equation of plastic flow, which he called plastic dynamics. In the field of fluid mechanics, Saint-Venant listed the basic equations of viscous incompressible fluid motion in *The Study of Fluid Dynamics* published in 1843, while the same result of G. G. Stokes was published in 1845. Most of Saint-Venant's results were published in the Journal of the French Academy of Sciences (Fig. 8.17).

Fig. 8.17 Saint–Venant (1797–1886)

8.18 Poiseuille (1799–1869)

Jean-Louis-Marie Poiseuille (1799–1869), a French physiologist, had invented a sphygmomanometer to measure the blood pressure of the dog's aorta since he was a student. He has published a series of papers on the flow of blood in arteries and veins. Among them, the paper *Experimental Study of Liquid Flow in Small Diameter* published from 1840 to 1841 has played an important role in the development of hydrodynamics. He pointed out that the volume flow is proportional to the pressure drop per unit length and to the fourth power of pipe diameter. This law is later known as Poiseuille's law. As a German engineer Hagen got the same result in 1839, W. Ostwald proposed to call the law Hagen–Poiseuille law in 1925. The empirical law of Poiseuille and Hagen is an important experimental proof of the basic theory of viscous fluid motion established by G. G. Stokes in 1845. Nowadays, the flow of viscous fluid in a circular pipe is often called Poiseuille flow (Fig. 8.18).

Fig. 8.18 Poiseuille (1799–1869)

8.19 Darcy (1803–1858)

Henri-Philibert-Gaspard Darcy (1803–1858), a French dynamicist and engineer, was one of the founders of hydrogeology. His experimental results created a science of groundwater dynamics to study the movement of groundwater flow in porous media. In his whole life, he was responsible for the design and construction of canals, railways, highways, bridges, tunnels, and other civil works. After 1845 in France, because of the rapid development of industry and the rapid increase of water consumption, the excavation of deep wells to extract groundwater was very popular and promoted the research of groundwater. Darcy focused on the mechanism of groundwater movement in alluvium. In 1856, through the sand penetration test, it was first proposed that the flow through the sample was directly proportional to the cross-sectional area of the sample and the head difference of the piezometer at both ends of the sample, and inversely proportional to the height of the sample. The penetration law is called Darcy's law in the international academia, which lays the foundation for the experimental research method of water movement in soil, the theory of groundwater movement, and its application in different situations. In 1858, he put forward a famous formula of pipeline resistance loss with German scholar Weisbach (Fig. 8.19).

Fig. 8.19 Darcy (1803–1858)

8.20 Froude (1810–1879)

William Froude (1810–1879), British hydrologist and shipbuilding engineer, carried out the first research on ship hydrodynamics in 1846. It was found that adding a fin bilge keel along the horizontal direction below the waterline on both sides of the ship could reduce the roll of the ship. The device was later adopted by the British navy. In 1868, a series of experiments on ship motion were carried out with ship models, and the data obtained from the experiments were applied to ship construction. The ship resistance is divided into friction resistance and residual resistance (mainly wave-making resistance). When the ratio of velocity to the square root of length of ship and model is the same, the residual resistance per unit displacement is equal. This ratio is often called "Froude number". According to this similarity law, the basis of modern ship model test technology is established, and the accuracy of using ship model test to estimate real ship power is improved, which has a significant impact on ship design and construction. Early aerodynamic scientists used similar techniques to conduct model aircraft experiments in wind tunnels (Fig. 8.20).

Fig. 8.20 Froude (1810–1879)

8.21 Stokes (1819–1903)

George Gabriel Stokes (1819–1903) was a British dynamicist and mathematician, whose main contribution was to study the motion law of viscous fluid. Navier extended L. Euler's equation of motion for fluids from the molecular hypothesis, and obtained the equation of motion with a viscosity constant in 1821. In 1845, Stokes employed the mechanical model of continuous system and Newton's physical law of viscous fluid, gave the basic equations of viscous fluid motion with two viscosity constants in *On the Theory of Internal Friction of Fluid in motion and the Theory of Balance and Motion of Elastomer*. This group of equations is called Navier–Stokes equation later and is the most basic equations in fluid mechanics. In 1851, Stokes put forward the resistance calculation formula when the ball moves slowly in the viscous fluid in *The Research Report of the Effect of Friction on the Pendulum Motion in the Fluid*, indicating that the resistance is proportional to the flow velocity and viscosity coefficient, which is the Stokes formula of the resistance of the ball in the flow. Stokes found the nonlinear characteristics of fluid surface waves, whose velocity depends on wave amplitude, and for the first time dealt with the nonlinear wave problem by perturbation method in 1847. Stokes also studied elastic mechanics and he pointed out that there are two basic resistances in isotropic elastic bodies, namely, the resistance to volume compression and the resistance to shear. He explicitly introduced the shear stiffness and compression stiffness (1845), proving that the elastic P-wave is an irrotational expansion wave and the elastic transverse wave is an isochoric distortion wave (1849). In mathematics, Stokes is famous for a conversion formula (Stokes formula) between line integral and area integral in field theory (Fig. 8.21).

8.22 Helmholtz (1821–1894)

Hermann Ludwig Ferdinand von Helmholtz (1821–1894), German biophysicist and mathematician, was the founder of the law of conservation of energy. He has made great contributions in physiology, optics, electrodynamics, mathematics, thermodynamics, and other fields. After studying the optical structure of the eye, he developed the Younger-Helmholtz theory. His research on muscle activity enriched the earlier theories of Julius Meyer and James Jour, and helped him establish the theory of energy conservation. In terms of electromagnetic theory, he measured the propagation speed of

Fig. 8.21   Stokes (1819–1903)

electromagnetic induction as 314,000 km/s, and deduced from Faraday's electrolytic law that electricity may be particles. As a result of his series of lectures, Maxwell's electromagnetic theory really attracted the attention of European continental physicists, and led his student Hertz to prove the existence of electromagnetic waves in 1887 with experiments and achieve a series of major results. In the field of thermodynamics, he published a paper *Thermodynamics of Chemical Process* in 1882. He distinguished the "binding energy" and "free energy" in chemical reaction, pointing out that the former can only be converted into heat, while the latter can be converted into other forms of energy. From Clausius' equation, he derived the Gibbs–Helmholtz equation. He also studied the formation mechanism of vortex, sea wave, and some meteorological problems in hydrodynamics, and proposed three famous laws of conservation of vorticity (Fig. 8.22).

Fig. 8.22 Helmholtz (1821–1894)

8.23 Kelvin (1824–1907)

Lord Kelvin (1824–1907) was a British physicist and inventor, whose scientific activities are multifaceted. His main contributions to physics are electromagnetism and thermodynamics. He was one of the main founders of thermodynamics and his name is used as a unit of temperature (Kelvin). In 1927, the Seventh International metrology conference took the thermodynamic temperature scale as the most basic temperature scale. Fluid mechanics, especially vortex theory, was one of Kelvin's favorite subjects. Inspired by Helmholtz's work, he found some valuable theorems. The research achievements in electromagnetics theory and engineering application are outstanding. In 1848, he invented the electric image method, which is an effective method to calculate the electrostatic field generated by the charge distribution of a certain shape of conductor. He deeply studied the discharge oscillation characteristics of Leyton bottle, published the paper *The Oscillation Discharge of Leyton Bottle* in 1853, calculated the oscillation frequency, and made a pioneering contribution to the research of electromagnetic oscillation theory. In 1846, the "moving image method of force" of electric force, magnetic force, and electric current was successfully completed. He revealed the similarity between the theory of Fourier heat conduction and

Fig. 8.23 Kelvin (1824–1907)

the theory of potential, discussed Faraday's concept of electric action propagation, and analyzed the oscillating circuit and the resulting alternating current. He had made a series of great contributions to the development of thermodynamics. In 1851, he proposed the second law of thermodynamics: "it is impossible to absorb heat from a single heat source and turn it into useful work without any other impact". This is generally accepted as the second law of thermodynamics. He asserted from the second law of thermodynamics that energy dissipation is a general trend. In 1852, he cooperated with Joule to further study the internal energy of gas, improved the free expansion experiment of Joule gas, carried out the multi-hole plug experiment of gas expansion, and found the Joule–Thomson effect, namely the temperature change phenomenon caused by the adiabatic expansion of gas through the multi-hole plug. In 1856, he predicted a new thermoelectric effect from theoretical research, that is, when the current flows through the conductor with uneven temperature, the conductor not only generates irreversible Joule heat, but also absorbs or emits certain heat. This phenomenon is called Thomson effect (Fig. 8.23).

8.24 Riemann (1826–1866)

Georg Friedrich Bernhard Riemann (1826–1866), a German mathematician and physicist, had made important contributions to mathematical analysis, differential geometry, differential equations, thermology, electromagnetic non-over distance interaction, and shock theory. He introduced the theory of trigonometric series, pointed out the direction of integral theory, laid the foundation of modern analytic number theory, and put forward a series of problems. He first introduced the concept of Riemannian surface, which had a great influence on modern topology. In the field of algebraic function theory, Riemann–Roch theorem is also very important. In differential geometry, Riemann geometry was established after Gauss. His name appears in Riemannian ζ function, Riemannian integral, Riemannian lemma, Riemannian manifold, Riemannian space, Riemannian mapping theorem, Riemannian–Hilbert problem, Cauchy–Riemannian equation, and Riemannian loop matrix. In addition, he had made a great contribution to the partial differential equation and its application in physics. Riemann's work had a direct impact on the development of mathematics in the second half of the nineteenth century. Many outstanding mathematicians demonstrated the theorems that Riemann asserted, and many branches of mathematics made brilliant achievements under the influence of Riemann's thought. Riemann first proposed a new idea and method to study number theory with complex function theory, especially with ζ function, which ushered in a new era of analytic number theory and had a profound impact on the development of single complex function theory. He is one of the most innovative mathematicians in the history of mathematics in the world. Riemann's works are limited in number, but they are all extremely profound. He was good at imagination and creation of concepts (Fig. 8.24).

Fig. 8.24 Riemann (1826–1866)

8.25   Langley (1834–1906)

Langley (1834–1906), American astronomer and pioneer of flight, invented a bolometer in 1881, which was used to precisely measure a tiny amount of heat (a temperature difference of up to one hundred thousandths of a degree), measured by the amount of current generated by heating a blackened white gold wire. In order to commemorate him, the radiation unit of 1 cal per square centimeter is called 1 Langley. During Langley's visit to Whitney hills, California, the instrument was used to measure the solar radiation intensity in the visible and infrared regions of the spectrums. In this process, he extended the knowledge of solar spectrum to far-infrared region for the first time.

Langley had never been to a university, but he was a scholar famous for his tenacious self-study. He had enough ability to engage in astronomy and aeronautics. In 1865, he became an assistant professor of astronomy at Harvard University, and finally obtained the professorship of this subject in several colleges. Langley studied the principles of aerodynamics, showing how birds glide with their wings, and how air can support thin wings of special shapes. His formula for lifting force is still used today. Although Langley's theory is feasible, in practice, due to the structural strength of the materials he used or the defects of the engine, his plane failed to fly (Fig. 8.25).

Fig. 8.25   Langley (1834–1906)

8.26 Mach (1838–1916)

Ernst Mach (1838–1916), Austrian physicist and philosopher, received his doctorate from the University of Vienna in 1860. He had contributed to mechanics, acoustics, thermodynamics, experimental psychology, and philosophy. Mach used schlieren technique to study the flying projectiles. In 1887, he studied the spherical disturbance wave propagating at the speed of sound c emitted by the moving object in the air. When the velocity v of the object is greater than c, the wavefront of the disturbance wave forms a conical envelope surface. The relationship between the angle α, formed by the generatrix of the conical surface and the moving direction of the object v and c is sin α = c / v. In 1907, L. Prandtl called the angle "Mach angle" for the first time. In 1929, in view of the increasing importance of the ratio v/c in aerodynamics, J. Akerlett suggested using the term Mach number. In the late 1930s, Mach number became an important parameter to characterize the state of fluid motion. As a philosopher, Mach was skeptical of many basic ideas of physics at that time. In his important work *Mechanics*, he made a profound criticism on the space–time view, movement view, and material view of classical mechanics. Mach also contributed to the study of the history of mechanics in *Mechanics* (Fig. 8.26).

Fig. 8.26 Mach (1838–1916)

8.27   Reynolds (1842–1912)

O. Reynolds (1842–1912, Ireland), British dynamicist, physicist, and engineer, graduated from Queen's College of Cambridge University in 1867. In 1868, he became the chief engineering professor of Manchester Owen College (later renamed Victoria University). He was elected a member of the Royal Society in 1877. He won the Royal medal in 1888. He was an outstanding experimental scientist. In 1883, he published a classic paper *On the Conditions for Determining the Flow as a Straight or Curved Motion and the Law of Resistance in Parallel Flumes*. In this paper, the experimental results showed that the flow is divided into laminar flow and turbulent flow, and the dimensionless number Re (later called Reynolds number) was proposed as the criterion to distinguish the two flow patterns. Reynolds put forward the lubrication theory of bearings in 1886 and the concept of time average decomposition in 1895, and derived the governing equation of the time average motion of turbulence. He had a wide range of interests and had written many literature in his life, among which nearly 70 papers have a profound impact. The research contents of these papers include mechanics, thermodynamics, electricity, aeronautics, steam engine characteristics, etc. His achievements have been compiled into two volumes of *The Proceedings of Reynolds Mechanics and Physics* (Fig. 8.27).

Fig. 8.27   Reynolds (1842–1912)

8.28 Rayleigh (1842–1919)

Third Baron Rayleigh (1842–1919), was a British physicist and hydrologist, whose initial work was mainly the mathematical research of optical and vibration systems. Later research almost involved all aspects of physics, such as acoustics, wave theory, color vision, electrodynamics, electromagnetics, light scattering, liquid flow, hydrodynamics, gas density, viscosity, capillary action, elasticity, and photography. His persistent and precise experiments led to the establishment of resistance standards, current standards, and electromotive force standards, and later work focused on electrical and magnetic issues. Rayleigh had many achievements in mechanics. He had obtained many important results in the theory of elastic vibration, including the estimation and calculation of the nature frequency of the system. He wrote two volumes of the famous *Theory of Acoustics* (1877–1878), and systematically summarized his achievements in the study of elastic vibration. In 1887, it was first pointed out that there was a surface cover in elastic waves, which played an important role in understanding the mechanism of earthquakes. He also analyzed the convection caused by the temperature difference between the upper and lower parts of the fluid, and introduced the relevant dimensionless number (later called Rayleigh number), which can be used to explain some meteorological phenomena caused by the surface atmospheric convection. In addition, he had studied the propagation of finite-amplitude waves and the resistance of gases to moving objects. In order to explain the long-standing puzzle of why the sky appears blue, he derived the molecular scattering formula, which is called Rayleigh scattering law. In the aspect of experiment, he studied the resolution of grating and diffraction, and was the first one gave a clear definition of the resolution of optical instrument; this work led to a series of basic researches on the optical properties of spectrometer later, which played an important role in the development of spectroscopy. The relationship between the absolute blackbody radiation and the frequency was a common concern in the field of physics in the second half of the nineteenth century. In 1900, Rayleigh put forward a formula about thermal radiation from the perspective of statistical physics, which was later called Rayleigh–Gins formula, the result was in good agreement with the experiment, and provided the conditions for the emergence of quantum theory. Rayleigh paid close attention to the emergence and development of quantum theory and relativity. His research on acoustooptic interaction, mechanical motion mode, nonlinear vibration, and other projects has a profound impact on the development of the whole physics. In 1905, he revised and published the work *Principles of Acoustics*. Up to now, it is not only a classic work for acousticians

Fig. 8.28   Riley (1842–1919)

who study mechanical vibration, but also a very helpful reference for other physicists. Riley donated the Nobel Prize to Cavendish Laboratory and the Cambridge University Library (Fig. 8.28).

8.29   Boussinesq (1842–1929)

Joseph Valentine Boussinesq (1842–1929), French physicist and mathematician, received his doctorate in 1876. Boussinesq had made great contributions to almost every branch of mathematics and physics (except electromagnetics). In the field of fluid mechanics, he mainly studied vortex, wave, resistance of solid to liquid flow, mechanical mechanism of powder medium, and cooling of flowing liquid. In terms of turbulence, the famous eddy viscosity hypothesis was proposed in 1877. In the field of soil mechanics, a new Boussinesq solution of additional stress was proposed. In 1834, J. S. Russell in England observed the solitary wave in the experiment, and reported his results at the meeting of the British Association for Scientific Progress in 1844. Afterwards, he was criticized by the authoritative scholars, such as Airy and Stokes. In 1871, Boussinesq was the first to put forward mathematical theory to support Russell's experimental observation. In 1876, Lord Rayleigh also established a mathematical theory to support Russell's experimental observation. At the end of his paper, Rayleigh admitted that Boussinesq theory was put forward earlier. In 1877, Boussinesq proposed the shallow water long wave approximation and established the famous Boussinesq equation, which has been widely used and extended since then. In 1897, he made great contributions to

Fig. 8.29 Boussinesq (1842–1929)

turbulence and hydrodynamics. It is found that the term turbulence is mostly attributed to Boussinesq. In addition, Boussinesq also proposed the famous Boussinesq approximation for the buoyancy-driven flow in the stratified flow with small density difference and put forward a simple and reliable theory to compute buoyancy. He had also made outstanding contributions in elastic mechanics and geotechnical mechanics. As Boussinesq had contributed to many fields of fluid mechanics, he has been mentioned in many works of fluid mechanics. For example, there are three kinds of Boussinesq approximation, involving shallow water waves, eddy viscosity, and buoyant flow (Fig. 8.29).

8.30 Laval (1845–1913)

Karl Gustaf Patrik De Laval (1845–1913), a Swedish engineer and inventor of single stage impact turbine, proposed the famous Laval nozzle. The concept of impulse turbine was put forward in 1882, and a small impulse turbine was developed in 1887, which confirmed his idea. In 1890, the supersonic jet was successfully realized through the first contraction and then expansion of the pipeline. The impulse steam turbine was manufactured and the famous Laval nozzle was put forward. Today's rocket engines are equipped with this kind of nozzle, and the turbine speed with this nozzle can reach more than

Fig. 8.30   Laval (1845–1913)

30,000 rpm. He put forward the effective way of oil-water separator, through the experiment, the centrifugal separator is the most effective (Fig. 8.30).

8.31   Joukowski (1847–1921)

Joukowski (1847–1921), a Russian aerodynamicist, graduated from the Department of Physics and Mathematics of Moscow University in 1868. In 1882, he received his Ph.D. in Applied Mathematics. His thesis was *On the Permanence of Motion.* In 1902, he guided the construction of the wind tunnel of Moscow University, which was one of the earliest wind tunnels in Europe. Since 1910, he actively participated in the preparation of the Aerodynamics Laboratory of Moscow Institute of Technology. From 1910 to 1912, he taught the course of "theoretical basis of flight", and in 1913, he also taught the course for pilots. In the World War, he was engaged in the study of bombing theory and external ballistics. After the October Revolution, he devoted himself to the founding of the Soviet air force. In December 1918, on his proposal, the Soviet Union established the "Central Institute of aerodynamics and hydrodynamics" and appointed him as its director. On March 17, 1921, he died in Moscow. He had made great contributions in the fields

Fig. 8.31 Rukovsky (1847—921)

of aerodynamics, aviation science, hydraulics, hydrogeography, mechanics, mathematics, astronomy, and so on. His work had a huge impact on the development of the aviation industry, and he was called "father of Russian aviation" by Lenin (Fig. 8.31).

8.32   Lilienthal (1848–1996)

Otto Lilienthal (1848–1896), a German engineer and glider, was one of the world's pioneers in aviation. He was the first to design and manufacture a practical glider. In 1889, he completed the famous book *Bird Flying- the Basis of Aviation*, which discusses the characteristics of bird flying, and points out that the wing should have the same bow section as the bird wing in order to obtain greater lift. Since then, in cooperation with his brother G. Lilienthal, he made a bat-like bow wing glider in 1891, which successfully glided over 30 m, thus confirming the reasonable form of curved wing. Since then, a number of different types of single-wing and double-wing gliders had been built. Lilienthal's glider is equipped with a hanger in the middle, and the pilot hangs on the frame. He can control the direction and speed of the glide by moving his body to master the center of gravity. From 1891 to 1896, he carried out more than 2000 glide tests at the test site near Berlin. He also

Fig. 8.32 Lilienthal (1848–1996)

made detailed records of his experience in gliding, accumulated rich materials, compiled air pressure datasheets, and wrote *Practical Tests in Flight* and other books. He planned to install a steam engine on the glider to realize power flight after fully mastering the stable control, but this wish failed to be realized. In 1896, he lost his life in a flight test. Although Lilienthal did not realize the aircraft power flight, his large number of flight practices and research provided valuable experience for later aircraft researchers, especially the Wright brothers gained many lessons from his experience (Fig. 8.32).

8.33 Lamb (1849–1934)

Horace Lamb (1849–1934) was a British mathematician and mechanist, and was good at fluid mechanics. His research field was wave and its application in earthquake, tide, etc. There is a kind of wave running in thin layer, namely Lamb wave, named after him. Lamb wave is a combination of longitudinal wave and shear wave. From 1921 to 1927, he also studied the air flow on the plane surface. The original name of *Hydrodynamics* (1878) written by lamb was *Fluid Motion Teaching Theory*, which was revised in 1895 and renamed

Fig. 8.33 Lamb (1849–1934)

Hydrodynamics. This book is the representative work and summary of classical fluid mechanics before the end of the nineteenth century. It had been the standard work of fluid mechanics for several decades after its publication, and even now it still has important reference value. Other works of Lamb include *Infinitesimal Analysis* (1897), *Theory of Sound Dynamics* (1910), *Statics* (including hydrostatics and preliminary theory of and elasticity, 1912), *Dynamics* (1914), *Advanced Mechanics* (1920), etc (Fig. 8.33).

8.34 Lorentz (1853–1928)

Hendrik Anton Lorentz (1853–1928), an outstanding Dutch physicist, mathematician, and founder of classical electron theory, received his doctorate in 1875. Lorenz's most important contribution to physics is his electron theory. He used J. C. Maxwell's electromagnetic theory to deal with the reflection and refraction of light on the dielectric interface. He thought that all matter molecules contain electrons, and the particles of cathode rays are electrons. The interaction between ether and matter is attributed to the interaction between ether and electron. This theory successfully explains the

Zeeman effect. Lorentz is the founder of classical electron theory and thinks that electricity has "atomicity" and that electricity itself is composed of tiny entities later called electrons. Lorentz explained the electrical properties of matter on the basis of the concept of electron. The Lorentz force, which is the force acting on the moving charge in the magnetic field, is deduced from the electron theory. He explained the luminescence of an object as the vibration of electrons inside an atom. In this way, when the light source is placed in the magnetic field, the vibration of the electrons in the atoms of the light source will change, which will increase or decrease the vibration frequency of the electrons, resulting in the broadening or splitting of the spectral lines. In 1904, Lorentz proved that when Maxwell's electromagnetic field equations were transformed from one reference system to another by Galileo transformation, the speed of light in vacuum would not be an invariable quantity, so that Maxwell's equations and various electromagnetic effects may be different for the observers of different inertial systems. In order to solve this problem, Lorentz proposed another method called Lorentz transformation. Later, Einstein applied Lorentz transformation to mechanical relations and established special relativity (Fig. 8.34).

Fig. 8.34 Lorentz (1853–1928)

8.35 The Wright Brothers (1867–1912, 1871–1948)

Wilbur Wright (1867–1912) and Orville Wright (1871–1948) were two American inventors and aircraft makers. On December 17, 1903, they completed the first fully controlled flight with external power attached to the aircraft, and continued to stay in the air without landing. So they are credited with "inventing the first airplane in the world". The Wright brothers were well-educated, but none of them got a diploma. In 1892, the brothers opened a bicycle repair shop and began to produce their own brand bicycles in 1896. At the beginning of 1890, the Wright brothers began their mechanical aviation experiments. The image of the Wright brothers in front of the public is always one; they share the achievements and honors of invention. When the Wright brothers finished all the theoretical research, they began to practice. Charlie Taylor, their bicycle assistant, became an important member of the team, and the three worked together to build the first aircraft engine (Fig. 8.35).

Fig. 8.35 Wilbur Wright (1867–1912, left); Orville Wright (1871–1948, right)

8.36 Lanchester (1868–1946)

F. W. Lanchester (1868–1946), British hydrologist, engineer, and a pioneer in aerodynamics, pointed out the principle of aircraft which is heavier than air in his paper in 1891. In 1894, he explained the principle of wing generated lift, put forward the correct calculation method, and solved the lift calculation of two-dimensional wing, preceding M. W. Kutta of Germany (1867–1944) and H. E. Joukowski of Russia (1847–1921). In 1915, he put forward the concepts of attached vortex and free vortex for the calculation of lift force of a wing with finite span (see lift line theory, lift surface theory). Lanchester also got the formula of surface drag, and explained the separation phenomenon and turbulence phenomenon in the boundary layer. Lanchester was also the first person to make a serious and scientific analysis of the role of aircraft in the war. In 1914, he published a series of papers on the application of aircraft and air combat. In 1916, he published the book *Aircraft in War, the Emergence of the Fourth Weapon*. The mathematical equation that he established to describe the changing process of the forces of both sides in combat is called Lanchester equation. Lanchester's main works are *Aerodynamics*, *Aerology* and *Aircraft in War, the Emergence of the Fourth Weapon* (Fig. 8.36)

Fig. 8.36 Lanchester (1868–1946)

8.37   Prandtl (1875–1953)

Ludwig Prandtl (1875–1953), a German mechanic and a world Master of fluid mechanics, received his doctorate in 1900. After 1904, he was hired to establish the Department of Applied Mechanics, the Institute of Aerodynamics Experiment, and the Institute of Fluid Mechanics at the University of Gottingen. Since then, he had been engaged in the research and teaching of aerodynamics. He had made important contributions in boundary layer theory, wind tunnel experiment technology, wing theory, turbulence theory, and so on. Therefore, he is called the father of aerodynamics and modern fluid mechanics. In 1904, Prandtl completed his most famous paper *Fluid Flow under very Small Friction*. In this paper, Prandtl first described the boundary layer and its application in drag reduction and streamline design, described the boundary layer separation, and proposed the concept of stall, which played an epoch-making role. Prandtl's paper attracted the attention of mathematician Klein, who recommended Prandtl as the director of the school of Technical Physics at the University of Gottingen. In 1908, Prandtl and his student Theodor Meyer proposed the theory of supersonic shock wave flow, and Prandt–Meyer expansion wave theory became the theoretical basis for the design of supersonic wind tunnel. From 1913 to 1918, the theory of lifting line and the theory of minimum induced drag were put forward, and then the theory of lifting surface was put forward, which enrich the wing theory.

In 1922, Prandtl and Richard Von Mises founded the International Society of Applied Mathematics and Mechanics. In 1929, he and Adolf Busemann proposed a design method of supersonic nozzle. The Prandtl method is still used in the design of all supersonic wind tunnels and rocket nozzles. The complete theory of supersonic flow was finally completed by Theodore Von Karman, a student of Prandtl.

He is also creative in meteorology. Prandtl and Tijon published *Applied Hydrodynamics and Aerodynamics* in 1931. His monograph *Introduction to Fluid Mechanics* was published in 1942, and the Chinese translation was published in 1974. His mechanical papers were compiled into three volumes of *Complete Works*, which were published in 1961 (Fig. 8.37).

8.38   Karman (1881–1963)

Von. Karman (1881–1963), an American scientist (Hungarian Jew), received a doctorate at the University of Gottingen in 1908. In 1911, the theory of bluff body drag was summed up, that is, the famous "Karman vortex

Fig. 8.37   Prandtl (1875–1953)

street" theory. In 1930, von Karman moved to the United States to guide the design and construction of the Guggenheim Aerodynamic Laboratory and the first wind tunnel of Caltech. During his tenure as director of the laboratory, he also proposed the theory of boundary layer control, and in 1935 he proposed the principle of supersonic drag. In 1938, Von Karman guided the United States to conduct the first supersonic wind tunnel test, invented jet-assisted takeoff, making the United States the first country to use rocket booster on aircraft. Under his guidance, a group of aeronautical engineers from Caltech (including his Chinese students Qian Weichang, Qian Xuesen, Guo yonghuai, etc.) began to engage in jet propulsion and liquid fuel rockets, leading to the establishment of Jet Propulsion Laboratory. The laboratory is the first research institution of the U.S. government for long-range missile and space exploration. The laboratory has many important research results, including the first theory on strange perturbation published by Qian Weichang under Karman's guidance. Therefore, Qian Weichang is internationally recognized as the founder of this field. After 1932, Karman published many papers and research results on supersonic flight. For the first time, he calculated the supersonic resistance of a slender body in a three-dimensional flow field by using the small perturbation linearization theory. He proposed the concept of shock resistance in supersonic flow and the important viewpoint that reducing the relative thickness can reduce

Fig. 8.38 Karman (1881–1963)

shock resistance. In 1939, his student Qian Xuesen established the famous "Karman–Tsien formula" under his guidance. In 1946, von Karman proposed the transonic similarity law, which combined with Prandtl's subsonic similarity law, Qian Xuesen's hypersonic similarity law, and Akley's supersonic similarity law to form a complete theoretical system of compressible aerodynamics. In 1936, when the scientific community generally doubted the rocket propulsion technology, he supported his students to study the subject. In order to improve the performance of aircraft, especially to shorten the takeoff distance from the ground or aircraft carrier, he and Malina first proved in 1940 that they could design a solid rocket motor with stable and durable combustion. A prototype of the takeoff booster rocket was soon developed. This kind of rocket is also the prototype of solid rocket on the long-range missiles of Polaris, Militia, and Poseidon. In 1941, he was involved in the creation of the General Aviation Jet Company that made rocket engines in the United States (Fig. 8.38).

8.39 Taylor (1886–1975)

Geoffrey Ingram Taylor (1886–1975) was a British dynamicist, whose contribution to mechanics is manifold. In terms of fluid mechanics, he clarified the internal structure of shock wave (1910); studied atmospheric turbulence and

Fig. 8.39 Taylor (1886–1975)

turbulent diffusion (1915–1932), put forward the statistical theory of turbulence; obtained the instability conditions of flow between two rotating coaxial shafts (1923); put forward the self-simulation theory of strong explosion in the study of atomic bomb explosion (1946–1950); pointed out that the main role plays in the drop is surface tension rather than viscous force (1959), etc. In the field of solid mechanics, he also contributed to dislocation theory in crystals (1934), plastic flow in perforated sheet (1940), and high-speed loading material test (1946). Taylor's scientific work was characterized by his skillful combination of profound physical insight and advanced mathematical methods, and he was good at designing simple and perfect specialized experiments. In 1970, he made a summary speech on the combination of theory and practice in fluid mechanics, and later published in *Annual Review of Fluid Mechanics* in 1974 (Fig. 8.39).

8.40 Zhou Peiyuan (1902–1993)

Zhou Peiyuan (1902–1993) was a famous hydrologist, theoretical physicist, educator and social activist and one of the founders of modern mechanics and theoretical physics in China. In 1927, he studied at the California Institute of Technology and obtained a doctor's degree. After returning to China in 1929, he became a professor in the Department of Physics of Tsinghua University. Zhou Peiyuan's research field was mainly the theory of gravity in Einstein's general relativity and the theory of turbulence in fluid mechanics, which lays the foundation of turbulence model theory. In the aspect of general relativity, Zhou Peiyuan had been devoted to solving the definite solution of the gravitational field equation and applying it to the research of cosmology. In the field of gravity theory, he put forward the important viewpoint that "harmony condition is physical condition". For the first time in the world, he obtained the result that the speed of light on the earth's surface is identical in the horizontal direction and in the vertical direction with a relative error of 10^{-11}, which may have a significant impact on people's understanding of Einstein's theory of gravity. In turbulence theory, in the early 1930s, Zhou Peiyuan realized that the turbulent flow field is closely related to the boundary conditions. Later, referring to the treatment method of taking quality as integral constant in general relativity, he worked out the differential equation satisfied by Reynolds stress, and hoped to introduce the influence of boundary conditions into the calculation formula of Reynolds stress through the boundary conditions. In 1940, Zhou Peiyuan wrote the first paper on turbulence, which was the first time in the world to put forward the turbulence fluctuation equation, and established the general turbulence theory by solving the correlation function of shear stress and three-dimensional velocity to satisfy the dynamic equation, and laid the foundation of the turbulence model theory. In 1945, Zhou Peiyuan published an important paper *Solutions to the Equations of Velocity Correlation and Turbulence Fluctuation* in the *Quarterly of Applied Mathematics* of the United States, and proposed two methods to solve the turbulence motion, which had a profound impact on the development of turbulence model theory. It is the foundational work of engineering turbulence model theory based on Reynolds stress equation (Fig. 8.40).

8.41 Kolmogorov (1903–1987)

Kolmogorov, Russian mathematician and master of turbulence statistics, was one of the few most influential mathematicians in the twentieth century. His research covered almost all fields of mathematics and he had made many pioneering contributions. In 1928, he obtained the necessary and sufficient conditions for random variable sequence to obey the law of large numbers. In 1929, he obtained the law of iterated logarithm for independent and identically distributed random variable sequence. In 1930, he obtained the very general sufficient conditions for the law of large numbers. In 1931, he published the paper *Analytical Method of Probability Theory*, which laid the foundation of Markov process theory. Markov process is widely used

Fig. 8.40 Zhou Peiyuan (1902–1993)

in physics, chemistry, biology, engineering technology, economic management, and other disciplines. It is still one of the hot spots and focuses of mathematical research in the world. In 1932, he obtained the necessary and sufficient conditions for the infinite separable distribution law of random variables with second-order matrix. In 1933, he published the book *Probability Theory Foundation*, in which he established the axiomatic conclusion of probability theory for the first time in the world on the basis of measure theory and integral theory. This is a great work of epoch-making significance. It is the most glorious page of mathematics in the former Soviet Union in the history of Science. In 1935, he proposed the concept of reversible symmetric Markov process and the necessary and sufficient conditions for its characteristics. This process has become an important model of statistical physics, queuing network, simulated annealing, artificial neural network, and protein structure. From 1936 to 1937, he gave the state distribution of countable state Markov chain. In 1939, he defined and obtained the statistics and distribution function of the maximum deviation between the empirical distribution and the theoretical distribution. In the 1930s–1940s, he and Khinchin developed the Markov process and the stable stochastic process theory, which were applied to the automatic control of artillery and the

Fig. 8.41 Kolmogorov (1903–1987)

production of industry and agriculture, and played an important role in the patriotic war. In 1941, he obtained the prediction and interpolation formula of stationary stochastic process. From 1955 to 1956, he and his student, Y. V. Prokhorov, the Soviet mathematician, initiated the weak limit theory of probability measure in function space. This theory and D-space theory introduced by A. B. Skorohod, the Soviet mathematician, are the epoch-making achievements of weak limit theory (Fig. 8.41).

8.42 Whittle (1907–1996)

Frank Whittle (1907–1996) was a British aviation engineer, inventor, and founder of jet engine. In 1928, he published a paper on gas turbine and jet reaction aircraft and put forward the basic formula of jet thermodynamics. In 1930, he obtained a patent for turbojet engine design. From 1937 to 1944, he was the chief engineer of British Jet Power Co., Ltd. The single rotor turbojet engine developed by Whittle was successfully operated for the first time on April 12, 1937. In May 1941, the Gloucester E-28/39 aircraft with W-1 engine designed by Whittle was successfully tested. Jet fighters such as "Meteor" and "Vampire" used by Britain in the late and post World War II were developed on the basis of such aircraft. In the early 1950s, he developed the world's first turboprop passenger aircraft "Viscount" and the first turbojet "Comet", which made the UK's aviation jet propulsion technology once the world's leading position. Whittle was awarded the rank of brigadier general

Fig. 8.42 Whittle (1907–1996)

and knighthood in 1948. Many professional societies around the world had also given him numerous medals and honorary degrees. Whittle migrated to the United States in 1976 and became a university professor (Fig. 8.42).

8.43 Schlichting (1907–1982)

Hermann Schlichting (1907–1982), a famous German hydrodynamicist, received doctorate from the University of Gottingen under the guidance of Professor Prandtl who is a world-class master of fluid mechanics. His doctoral dissertation was on the problem of wind shadow of aircraft. He had been working in the Institute of Fluid at Munich University of Technology, and had been the director of the Institute of Fluid for a long time. He mainly studied aerodynamics and boundary layer characteristics. For the first time, he studied the development of harmonics in the laminar boundary layer, i.e., Tollmien–Schlichting wave. In 1951, his world-famous German Book *Boundary Layer Theory* was published, in 1955, the English version of *Boundary Layer Theory* was published, and in 2000, the eighth edition was published (Fig. 8.43).

Fig. 8.43 Schlichting (1907–1982)

8.44 Landau (1908–1968)

L. D. Landau (1908–1968), a great Russian theoretical physicist, had published papers on solid physics, nuclear physics, plasma physics, hydrodynamics, astronomy, quantum mechanics, and other disciplines. The complete works of *Theoretical Physics*, which he co-authored with another Russian physicist, Livschitz, are the most perfect works of theoretical physics. The originality of the book and the extensiveness of the materials are rare in the world. Therefore, this great work has gained great reputation, and Landau himself is known as the master of theoretical physics in the world. In 1950, Landau and Ginzburg put forward a theory called "Ginzburg–Landau theory", which can accurately predict the characteristics such as the maximum current of superconductor load. In 1957, Abrakosov, a student of Landau, used this theory to get a classic result in the history of superconductivity theory and materials. This result is the analytical solution of Ginzberg–Landau theory. In 1962, Landau won the Nobel Prize in physics for his research on the superfluidity of liquid helium. Superfluidity is a very strange phenomenon in the eyes of ordinary people: if you inject liquid helium into

Fig. 8.44 Landau (1908–1968)

an open container, liquid helium will "automatically" overflow the container (Fig. 8.44).

8.45 Guo Yonghuai (1909–1968)

Guo Yonghuai (1909–1968) is a famous applied mathematician, aerodynamicist and one of the founders of Chinese modern mechanics. Guo Yonghuai had been engaged in Aeronautical Engineering Research for a long time, discovered the upper critical Mach number, developed the deformation coordinate method in the singular perturbation theory, which is internationally recognized PLK method, advocated the research of hypersonic flow, electromagnetic fluid mechanics and explosion mechanics in China, and trained many excellent mechanical talents. He had taken on the leading role in the scientific research of national defense and made important contributions to the development of missiles, nuclear bombs, and satellites. He led and organized the research work of detonation mechanics, high pressure equation of state, aerodynamics, flight mechanics, structural mechanics and weapon environment experiment science, solved a series of major problems, and was the only scientist who made great contributions to the experimental work of China's nuclear bomb, hydrogen bomb, and satellite. In terms of aero-dynamics, he focused on the study of transonic theory and viscous flow, and published several important articles, such as *Compressible Irrotational Subsonic and Supersonic Mixed Flow and Upper Critical Mach Number* (in coopera-tion with Qian Xuesen), *On the Flow of Incompressible Viscous Fluid around a Plate at Medium Reynolds Number, Reflection of Weak Shock from the Boundary Layer along the Plate*, etc. He solved the major theoretical problems in tran-sonic flow. At the same time, in order to solve the singularity of the boundary layer, he improved the deformation parameters and the deformation coordi-nate method of Poincare and Lethill, and developed the singular perturbation theory. Guo Yonghuai paid attention to the direction of hypersonic flow in the early 1950s, and studied the interference and dissociation effect of hypersonic shock boundary layer.

Guo Yonghuai founded the *Journal of Mechanics* and the *Translation of Mechanics*, and was the editor in chief. He translated and published many famous academic works, such as the *Introduction to Fluid Mechanics*, and successively carried out research on emerging hypersonic aerodynamics, electromagnetic fluid mechanics, and other topics. His achievements have attracted constantly the attention of the international scientific community. At the symposium on interplanetary navigation organized by the Chinese Academy of Sciences, Guo Yonghuai proposed that China should develop its space industry, and expressed many important opinions on such issues as means of delivery and propulsion technology (Fig. 8.45).

8.46   Qian Xuesen (1911–2009)

Qian Xuesen (1911–2009), a famous scientist and aerodynamicist in China, was the founder of China's manned space flight. At the beginning of 1956, Qian Xuesen proposed to the CPC Central Committee and the State Council the proposal on the establishment of China's national defense avia-tion industry. At the same time, Qian Xuesen established China's first rocket and missile research institute, the Fifth Research Institute of the Ministry of National Defense, and served as the first president. He presided over and completed the "establishment of jet and rocket technology" plan; partici-pated in the development of short-range missiles, medium and short-range missiles and China's first man-made earth satellite; directly led the "two bombs combined" test of atomic bombs carried by medium and short-range missiles; participated in the formulation of the launch of China's first inter-stellar aviation planning; developed and established engineering cybernetics

Fig. 8.45 Guo Yonghuai (1909–1968)

and systems. Under the leadership of Qian Xuesen, China's first atomic bomb exploded successfully on October 16, 1964, China's first hydrogen bomb air burst test was successful on June 17, 1967, and China's first man-made satellite was successfully launched on April 24, 1970. He had made a lot of research achievements in aerodynamics, the most outstanding one is to put forward the similarity law of transonic flow, and put forward the concept of hypersonic flow together with Karman, which provides a theoretical basis for aircraft to overcome thermal and acoustic barriers in the early stage, and lays an important theoretical basis for the development of aerodynamics. The formula used in the design of subsonic aircraft is Karman–Tsien formula named after Karman and Qian Xuesen. Qian Xuesen and von Karman carried out the study of compressible boundary layer, revealed the temperature change in this field, and established the Karman–Tsien approximate equation. In cooperation with Guo Yonghuai, the concept of upper and lower critical Mach numbers was first introduced into transonic flow problems. Qian Xuesen's research on the physical, chemical, and mechanical properties of thin gas in 1946 was a pioneering work. In 1953, he formally put forward the concept of physical mechanics, which greatly saved human and material resources, and opened up a new field of high temperature and high pressure. In 1961, his lecture *Notes on Physical Mechanics* was officially published. Qian Xuesen put forward several important concepts in the field of rocket and aerospace: in the 1940s, he put forward and realized the JATO, which shortened the runway distance of the aircraft; in 1949, he put forward

Fig. 8.46   Qian Xuesen (1911–2009)

the concept of rocket passenger aircraft and the idea of nuclear rocket; in 1953, he studied the possibility of interstellar flight; in 1962's *Introduction to Interstellar Navigation*, he put forward that a large jet engine aircraft was used as the first stage vehicle. In the process of forming engineering cybernetics, Qian Xuesen took the engineering practice of designing stability and guidance system as the main research object. In system science, he developed the methodology of system science and open complex system (Fig. 8.46).

8.47   Lu Shijia (1911–1986)

Lu Shijia (1911–1986) is a famous hydrodynamist and educator in China. She was the only female doctor of Prandtl who is the world's master of fluid mechanics. She had been engaged in research and teaching of aerodynamics and aeronautical engineering for a long time. She advocated the research of vortex, separated flow, and turbulence structure. As one of the founders of Beihang University, she founded China's first aerodynamics major and made contributions to the development of Chinese mechanics and the cultivation of scientific and technological talents in the aviation industry. In 1952, Lu Shijia was the first director of the Aerodynamics Office of Beihang and one of the main founders of the first aerodynamics major in China. She made it clear that this major is an engineering specialty for aerospace construction, and its teaching plan should be formulated according to the actual situation in China. She had always been teaching at the front line, teaching theoretical aerodynamics, and other courses for students. Lu Shijia had done a lot of work for the development of magnetohydrodynamics, biological fluid mechanics, separated flow, and vortex motion. In the 1980s, as the vice president of China Aerodynamics Research Association, she initiated and presided over the research on separation flow and vortex motion, and held a national symposium. In 1958, she and other comrades built a complete set of low-speed wind tunnel. At the same time, she actively participated in and organized the staff of the whole teaching office to design and manufacture the supersonic wind tunnel and large mechanical six component balance by themselves. In the 1980s, she also took an active interest in the development of the branch of biological fluid mechanics. Considering the important role of water tunnel experiment in the study of turbulence and drag reduction, she supported middle-aged teachers to build the first water tunnel of Beihang University. In the late 1970s, she retranslated the *Introduction to Fluid Mechanics* (7th edition of German version) written by the famous German professor Plandtl, and made outstanding achievements for China's aviation industry (Fig. 8.47).

Fig. 8.47 Lu Shijia (1911–1986)

8.48   Shen Yuan (1916–2004)

Shen Yuan (1916–2004) was a pioneer and educationalist of Chinese aerospace higher education, a famous aerodynamicist, and one of the founders of Beihang University. In 1945, he received his doctorate from Imperial College of London University. In the early days of Beihang's founding, he devoted a lot of energy to the school's teaching work, from determining the preparation plan to making the teaching plan and syllabus, from organizing a large number of young teachers to learn from Soviet experts to establishing the teaching organization and system. In 1956, he participated in the formulation of the national long-term plan for the development of science and technology, foreseeing the urgency of the demand for talents in the aerospace industry, the rocket and missile industries. Together with other leaders of the University, he took decisive measures to take the lead in creating a whole set of new majors in rocket, missile, and other aspects in colleges and universities across the country. Many graduates of these majors have now become the pillars of Chinese aerospace industry. In 1958, he led teachers and students to design and build China's first medium-sized supersonic wind tunnel in Beihang, which played an important role in teaching and research. At that time, the development of light passenger plane, sounding rocket, drone all gathered his efforts and sweat. In 1975, he personally presided over the import of the third generation of medium-sized electronic computers from abroad, enriched the hardware equipment of the computer major of Beihang, and laid the foundation for the development of the new subject of Beihang. In 1978, the national recruitment system for postgraduates was restored. He put forward the policy of "selecting seedlings, preferring lack to abuse, laying a good foundation, strict requirements, and combining ability training with scientific research tasks". He actively encouraged and organized the reliability research of Beihang, which played an important role in promoting the development of reliability engineering (Fig. 8.48).

8.49   Batchelor (1920–2000)

George Keith Batchelor (1920–2000) was an Australian applied mathematician and hydrologist. He had been a professor of Applied Mathematics at Cambridge University in the UK for a long time. He is the founder of the Department of Applied Mathematics and Theoretical Physics at Cambridge University. In 1956, he established the Journal of Fluid Mechanics and served as the editor in chief for 40 years. As an applied mathematics

Fig. 8.48 Shen Yuan (1916–2004)

worker, he insisted on the basis of experimental data and physical cognition. He established the theory of homogeneous isotropic turbulence and made important contributions to the theory of fluid mechanics. His *Introduction to fluid mechanics* is regarded as a classic book, which is well received by readers (reprinted many times), and becomes the basic book of viscous fluid mechanics. In order to recognize his outstanding contributions in fluid mechanics, the Batchelor prize was established, which was selected by the international mechanical conference every four years (Fig. 8.49).

8.50 Whitcomb (1921–2009)

Richard Whitcomb (1921–2009), an American aerodynamicist, had been working at NASA Langley Research Center for a long time, engaged in the research of aircraft drag reduction and shock wave control technology. In 1952, he put forward the area rule theory for transonic flight of aircraft. It was found that for the aircraft flying at transonic speed, the cross-sectional area of the fuselage in the section where the fuselage and wing are connected should be reduced, so as to reduce the flight resistance. This theory guided the design of transonic aircraft and produced the first batch of supersonic aircraft.

Fig. 8.49 Bacheler (1920–2000)

In 1967, he put forward supercritical airfoil, which delayed the generation of local shock wave, greatly increased the drag divergence Mach number of airfoil, increased the cruising speed of subsonic aircraft and reduced fuel consumption. At present, all kinds of high subsonic aircraft in the world adopt this kind of similar airfoil, which has become one of the core technologies of wing design, and Whitcomb had also won the Collier aviation award. In the mid-1970s, Whitcomb invented winglets, and through a series of wind tunnel tests, it was found that the wing with such winglets could indeed reduce drag. The results of wind tunnel test and flight test show that the winglets can reduce the induced drag of the whole aircraft by 20–35%, which is equivalent to the increase of lift–drag ratio by 5–7% (Fig. 8.50).

8.51 Lighthill (1924–1998)

Michael James Lighthill (1924–1998) is a famous British mathematician and hydrologist. He was mainly engaged in the scientific work in the fields of fluid mechanics, applied mathematics, etc. He had created the new subject fields of aeroacoustics, nonlinear acoustics and biological fluid dynamics, created the theory of aeroacoustics, and made an important contribution to the reduction

Fig. 8.50 Whitcomb (1921–2009)

of jet engine noise. He graduated from Trinity College of Cambridge University in 1941. From 1946 to 1959, he was employed as a senior lecturer at the University of Manchester, then as a professor of Applied Mathematics, and led a research group on fluid mechanics. From 1964 to 1969, he was a professor of Royal Society of Imperial University of Technology and began to study biological fluid mechanics. In 1969, after physicist Dirac, he accepted the position of Lucas mathematics professor in Cambridge, and withdrew from the position in 1980, which was succeeded by Hawking. From 1979 to 1989, he was the president of University of London. He was elected a member of the Royal Society at the age of 29, won 24 honorary doctorates in his life, and was also elected a foreign academician of the authoritative Academy of Sciences (Fig. 8.51).

8.52 Zhuang Fenggan (1925–2010)

Zhuang Fenggan (1925–2010), an aerodynamicist in China, had been engaged in aerodynamics research for a long time. In 1947, he went to the United States to study. He studied at the California Institute of Technology. Under the guidance of H. W. Liepmann, a famous professor of fluid mechanics, he studied aviation engineering and mathematics. In 1950, he received doctor degree. He organized and led the construction of China's

Fig. 8.51 Lighthill (1924–1998)

main aerodynamics experimental base, built a complete set of equipment from low speed to high supersonic speed, and established a backbone team of aerodynamics research. He was one of the main pioneers in this field, and had made outstanding contributions to the development of China's aerospace industry. He obtained the dissipation law of turbulence in the study of the basic characteristics of turbulence in fluid mechanics. He had made outstanding achievements in the research of shock diffraction, thermal protection theory of hypersonic reentry body, and the mechanism and control of vortex formation. He carried out the aerodynamic research of missiles, rockets, and reentry vehicles for a long time. He had made important contributions to the design and construction of large wind tunnels, the design and construction of ramjet test beds, the aerodynamic research and test of launch vehicles, the aerodynamics dominated by unsteady vortices, and computational fluid dynamics. Zhuang Fenggan had carried out extensive research work in many fields of aerodynamics, and had published more than 60 academic papers and reports, including aerodynamic theory, experiment, and testing technology. He was one of the main technical leaders in the construction of aerodynamics research and test base in China. These bases were built in the 1950s and 1970s, respectively, when there was a lack of international assistance and information, and there were many economic and technical difficulties. During his tenure as the leader of Beijing Aerodynamics Research Institute, China Aerodynamics Research and Development Center and the aerodynamics professional group of the National Defense Science and Technology Commission, he had done a lot of pioneering work for the design and construction of China's own aerodynamic test equipment in leading the formulation of the overall plan, the determination of the scheme and the solution of various technical problems. At that time, Qian Xuesen served as the leader of the aerodynamic professional group, Zhuang Fenggan as the deputy leader, and Guo Yonghuai presided over the construction of the experimental base of China Aerodynamic Research and Development Center. All kinds of equipment in the test base played an important role in the development of satellites and missiles in China (Fig. 8.52).

Fig. 8.52 Zhuang Fenggan (1925–2010)

Bibliography

1. L. Prandtl, *Führer durch die Strömungslehre* (Translated by Y. Guo, S. Lu) (Science Press, Beijing, 1987)
2. G.K. Batchelor, *An Introduction to Fluid Dynamics* (Translated by Q. Shen, F. Jia) (Science Press, Beijing, 1987a)
3. G. Li, The Great Scientist's Invention Dispute **11**, 44–45 (2010)
4. S.H. Tsien, Superaerodynamics, mechanics of rarefied gases, J. Aero. Sci. **13**, 653–664 (1946)
5. Z. Zhai, W. Liu, M. Zeng, J. Liu, *Hypersonic Aerodynamics* (National University of Defense Technology Press, Beijing, 2001)
6. W. Zhang, S. Fu, G. Zhang, W. Ren, *Translation of Prandtl Memorial Report - a History of Mechanical Development of Gottingen School*. Academic Monograph of Tsinghua University (Tsinghua University Press, Beijing, 2013)
7. V.K. Theodore, A. Lee, von Karman (Translated by K. Wang) (Xi'an Jiaotong University Press, Xi'an, 2015)
8. X. Qian, *Research on Qian Xuesen's Scientific Thought* (Xi'an Jiaotong University Press, Xi'an, 2008)
9. C. Zhang, *Hydrodynamics* (Higher Education Press, Beijing, 1993)
10. W. Wu, *Fluid Mechanics* (Peking University Press, Beijing, 1983)
11. J.F. Doulas, J.M. Gasiorekand, J.A. Swaffield, *Fluid Mechanics*, 3rd edn. (World Publishing Corporation, Beijing, 2000)
12. W. Xu, *Fluid Mechanics* (National Defense Press, Beijing, 1979)
13. H. Lamb, *Hydrodynamics*, 6th edn. (Cambridge University Press, 1932)
14. G.K. Batchelor, *The Theory of Homogeneous Turbulence* (Cambridge University Press, New York, 1953)
15. H. Schlichting, *Boundary Layer Theory* (McGraw Hill Book Company, New York, 1979)

© Science Press 2021
P. Liu, *A General Theory of Fluid Mechanics*,
https://doi.org/10.1007/978-981-33-6660-2

16. B.E. Launder, D.B. Spalding, *Mathematical Models of Turbulence* (Academic, London, 1972)
17. W. Frost, T.H. Moulden, *Handbook of Turbulence* (Plenum Press, New York, 1977)
18. P. Liu, J. Liu, *Introduction to Modern University* (Beihang University Press, Beijing, 2012)
19. P. Liu, *Principles of Energy Dissipation and Erosion Control in Modern Dam Engineering* (Science Press, Beijing, 2010)
20. X. Zhao, Q. Liao, *Viscous Fluid Mechanics* (China Machine Press, Beijing, 1983)
21. M.M. Stanisic, *The Mathematical Theory of Turbulence* (Springer, New York, 1984)
22. J.O. Hinze, Turbulence (Translated by Y. Huang, D. Yan), vols. I and II (Science Press, Beijing, 1987)
23. Y. Chen, *Turbulence Calculation Model* (University of Science and Technology of China Press, Hefei, 1991)
24. U. Frisch, *Turbulnce* (Cambridge University Press, New York, 1995)
25. M. Chen, *Fundamentals of Viscous Fluid Dynamics* (Higher Education Press, Beijing, 2002)
26. S.A. Orszag, G.S. Patterson, Numerical simulation of three-dimensional homogeneous isotropic turbulence. Phys. Rev. Lett. **28**, 76 (1972)
27. R.D. Moser, P. Moin, The effect of curvature in wall bounded turbulence. JFM **175**, 479 (1987)
28. P. Moin et al., Direct numerical simulation: A tool in turbulence research. Annu. Rev. Fluid Mech. **30**, 539 (1998)
29. L. Kleiser, T.A. Zang, Numerical simulation of transition in wall-bounded shear flows. Annu. Rev. Fluid Mech. **23**, 495 (1991)
30. Z. Wu, *Aerodynamics*, vol. I and II (Tsinghua University Press, Beijing, 2007a)
31. Z. Zhang, G. Cui, C. Xu, *Turbulence Theory and Simulation* (Tsinghua University Press, Beijing, 2005)
32. X. Shi, *Turbulence* (Tianjin University Press, Tianjin, 1994)
33. J.-Z. Wu, H.-Y. Ma, M.-D. Zhou, *Vortical Flows* (Springer, Heidelberg, 2015)
34. Q. Fan, H. Zhang, Y. Guo, X. Wang, W. Lin, *Large Eddy Simulation of Coherent Structure of Plane Free Turbulent Jet* (Journal of Tsinghua University, Beijing, 2001)
35. I.B. Cohen, *Dictionary of Scientific Biography* (Charles Scribner's Sons, New York, 1970)
36. T. Zhu, X. Qin, Y. Sun, *Biography of Foreign Historical Celebrities* (China Social Sciences Press, Beijing, 1984)
37. Newton - A Great Scientist. Sina
38. *World Famous Mathematicians* (Tianjin Engineering Technical Institute, Tianjin, 2014)
39. F. Sun et al., *History of Scientific Development* (Zhengzhou University Press, Zhengzhou, 2006)

40. C. John, A. Michael, *Timelines of Science and Technology* (Translated by H. Zhang) (Heilongjiang Science and Technology Press, Harbin, 2009a)
41. R. Liu, *The History of Science* (Jiangxi University Press, Nanchang, 2009)
42. D. Yuan, J. Wang, Y. Zhao, *A Brief History of Natural Science Development* (China Central Radio and Television University Press, Beijing, 2004a)
43. C. Bai, *Science and China* (Science Press, Beijing, 2016)
44. D.F. Anderson, S. Eberhart, *Understanding Flight* (Translated by L. Han, X. Liu), 2nd edn. (Aviation Industry Press, Beijing, 2011)
45. H. Xu, *Fundamentals of Aerodynamics*, vols. I and II (Beihang University Press, Beijing, 1987)
46. J.D. Anderson Jr., *Fundamentals of Aerodynamics*, 3rd edn., International edn. Mechanical Engineering Series (McGaw-Hill, New York, 2001)
47. S. Martin, *Aerodynamics of Model Aircraft* (Translated by Z. Xiao and D. Ma) (Aviation Industry Press, Beijing, 2007)
48. Z. Chen, F. Liu, G. Bao, *Aerodynamics* (Aviation Industry Press, Beijing, 1993)
49. Y. Qian, *Aerodynamics* (Beihang University Press, Beijing, 2004)
50. Z. Lu, *Aerodynamics* (Beihang University Press, Beijing, 2009)
51. Z. Wu, *Aerodynamics*, vol. I and II (Tsinghua University Press, Beijing, 2007b)
52. E. Gunter, J.G. Michael, *Aircraft Recognition Guide* (Translated by P. Li), 5th edn. (Post and Telecom Press, Beijing, 2014)
53. Rolls-Royce Company, *The Jet Engine* (Rolls-Royce plc, 2005)
54. M.A. Saad, *Compressible Fluid Flow* (Prentice-hall, INC., 1985)
55. L.M. Milne-Thomson, *Theoretical Aerodynamics* (Macmillan and Co, London, 1948)
56. M. Van Dyke, *An Album of Fluid Motion* (PARABOLIC, United States, 1982)
57. J.J. Bertin, M.L. Smith, *Aerodynamics for Engineers*, 2nd edn. (Prentice Hall, Englewood Cliffs, 1989)
58. B.W. McCormick, *Aerodynamics, Aeronautics, and Flight Mechanics*, 2nd edn. (John Wiley & Sons, INC., New York, 1995)
59. Ed. Obert, *Aerodynamic Design of Transport Aircraft* (Delft University Press, 2009)
60. S. Wang, X. Zhang, G. He, T. Liu, Lift enhancement by bats dynamically changing wingspan. J. R. Soc. Interface **12**, 12 (2015)
61. L.R. Jenkinson, P. Simpkin, D. Rhodes, *Civil Jet Aircraft Design* (Translated by G. Li, X. Wu, J. Hua, L. Wang, C. Tang, H. Qiu, Y. Song) (Aviation Industry Press, Beijing, 2001)
62. Z. Guo, *Energy Dissipation and Anti-Scour Principle and Hydraulic Design* (Science Press, 1982)
63. N. Rajaratnam, Hydraulic jumps, in *Advances in Hydrosciens*, vol. 4 (1967), pp. 198–280
64. V.T. Chow, *Open-Channel Hydraulics* (McGRAW-Hill Book Company, New York, 1959)
65. P. Liu, *Basics of Computational Hydraulics* (The Yellow River Water Conservancy Press, Zhengzhou, 2001)

66. K. Mehmed, V. Yevjevich, *Unsteady Flow in Open Channel*, vol. 1 (Water Resources and Electric Power Press, Beijing, 1987)
67. Z. Xia, *Modern Hydraulics*, vol. 4 (Higher Education Press, Beijing, 1992)
68. Department of Hydraulics, Tsinghua University, *Hydraulics*, vol. 2 (People's Education Press, 1981)
69. C. Wu, *Hydraulics*, vol. 1 (Higher Education Press, 1983)
70. Hydraulics Teaching and Research Group, East China University of Water Resources, *Hydraulics*, vol. 2 (East China University of Water Resources, 1978)
71. M. Bernard, *Hydeodynamicque Des Structures Offshore* (Translated by S. Liu) (National Defense Industry Press, Beijing, 2012)
72. C. John, A. Michael, *Timelines of Science and Technology* (Translated by H. Zhang) (Heilongjiang Science and Technology Press, Harbin, 2009b)
73. J.D. Anderson Jr., *Computational Fluid Dynamics–The Basics with Applications* (McGraw-Hill, 1995)
74. P.J. Roache, *Computational Fluid Dynamics* (Hermosa and Publishers, 1976)
75. J.F. Doulas, J.M. Gasiorek, J.A. Swaffield, *Fluid Mechanics*, 3rd edn. (World Publishing Corporation, Beijing, 2000)
76. P.R. Spalart et al., Spectral methods for the Navier-Stokes equations with one infinite and two periodic directions. J. Comput. Phys. **96**, 297 (1991)
77. S.K. Lele, Computational acousitics: a review. in *AIAA, paper* 97-0018 (1997)
78. C. John, A. Michael, *Timelines of Science and Technology* (Translated by H. Zhang) (Heilongjiang Science and Technology Press, Harbin, 2009c)
79. D. Yuan, J. Wang, Y. Zhao, *A Brief History of Natural Science Development* (China Central Radio and Television University Press, Beijing, 2004b)
80. Z. Li, *Wind Tunnel Test Manual* (Aviation Industry Press, Beijing, 2015)
81. D. Yan, *Experimental Fluid Mechanics* (Higher Education Press, Beijing, 1992)
82. Q. Yun, *Experimental Aerodynamics* (National Defense Industry Press, Beijing, 1991)
83. T. Wang, *Aerodynamics Experiment Technology* (Aviation Industry Press, Beijing, 1995)
84. K. He, Low turbulence wind tunnel design. Paper on the 30th Anniversary of Northwestern Polytechnical University, 1987.
85. N. Nagib et al., Flow quality documentation of the national diagnostic facility, in *AIAA paper* 94-2499
86. R.L. Loerke et al., Control of free-steam turbulence by means of honeycombs, a balance between suppression and generation. J. Fluids Eng. (1976)
87. R. Wu, Z. Wang, *Wind Tunnel Design Principle* (Beijing University Press, 1985)
88. R.C. Penkster et al., *Wind Tunnel Experiment Technology*, vol. I (National Defense Industry Press, 1963)
89. G. Li, *Optical Measurement Method for Wind Tunnel Test* (National Defense Industry Press, Beijing, 2008)
90. H. Cheng, *Interference and Correction of Wind Tunnel Experiment* (National Defense Industry Press, Beijing, 2003)

91. D.D. Baals, W.R. Corliss, *Wind tunnels of NASA* (Scientific and Technical Information Branch, National Aeronautics and Space Administration: for sale by the Supt. of Docs., U.S. G.P.O., Washington, D.C., 1981)

92. A. Pope, *Wind-Tunnel Testing* (John Wiley & SONS Inc., New York, 1947)

93. C. Tropea, A.L. Yarin, J.F. Foss, *Handook of experimental fluid mechanics* (Springer, Berlin, 2007)

94. S. Sheng, *Velocity Measurement Technology* (Peking University Press, Beijing, 1987)

95. X. Shen, *Laser Doppler Velocimetry and Its Application* (Tsinghua University Press, Beijing, 2004)

96. Q. Zhan, A. Lu, L. Li, Y. Li, *Digital Image Processing Technology* (Tsinghua University Press, Beijing, 2010)

97. J. Fan, *Flow Visualization and Measurement* (China Machine Press, Beijing, 1997)

98. Committee of Aviation Centennial Activities, *Flying Century - Commemorating the 100th Anniversary of Aircraft Invention* (China Aerospace Press, Beijing, 2003)

99. W. Hu, *General Aircraft* (Aviation Industry Press, Beijing, 2008)

100. D. Liu, G. Chen, *Engine - The Heart of Aircraft* (Aviation Industry Press, Beijing, 2010)

101. S. Li, *Shock Wave and Boundary Layer Dominated Complex Flow* (Science Press, Beijing, 2007)

102. China Encyclopedia (Encyclopedia of China Publishing House, Beijing, 1993)

103. G.K. Batchelor, *An introduction to fluid dynamics* (Translated by Q. Shen, F. Jia) (Science Press, Beijing, 1987b)

104. S.J. Kline, W.C. Reynold, F.H. Schraub, P.W. Runstadler, The strueture of turbulent boundary layers. J. Fluid Mech. **30**, 741–774 (1967)

105. H. Yao, G. He, Spatiotemporal correlation function of pressure in uniform isotropic turbulence, celebrating the 50th anniversary of the establishment of the Chinese society of mechanics and the academic conference of the Chinese society of mechanics (2007)

106. W. Zhang, C. Gao, Z. Ye, Research advances of wing/airfoil transonic buffet. Acta Aeronaut. Astronaut. Sin. **36**(4), 1056–1075 (2015)

107. S.A. Morton, R.M. Cummings, D.B. Kholodar, High resolution turbulence treatment of F/A-18 tail buffet, in *45th AIAA/ASME/ASCE/AHS/ASC Structures, Structural Dynamics & Materials Conference*, 19–22 April 2004, Palm Springs, California (2004)

108. P. Liu, A comparison of heavenly questions and turbulent questions. Mech. Eng. **38**(3), 356–360 (2016)

109. T. Weinkauf, H.-C. Hege, B.R. Noack, M. Schlegel, A. Dillmann, Coherent structures in a transitional flow around a backward-facing step. Phys. Fluids **15**(9) (2003)

110. M. Lighthill, On sound generated aerodynamically I. General theory. Proc. R. Soc. Lond. Ser. A Math. Phys. Sci. **211**(1107), 564–587 (1952)

111. N. Curle, The influence of solid boundaries upon aerodynamic sound. Proc. R. Soc. Lond. Ser A Math. Phys. Sci. **231**(1187), 505–514 (1955)

112. J.E. Ffowcs Williams, D. Hawkings, Sound generation by turbulence and surfaces in arbitrary motion. Philos. Trans. R. Soc. Lond. Ser A Math. Phys. Sci. **264**(1151), 321–342 (1969)

113. A. Powell, Theory of vortex sound. J. Acoust. Soc. Am. **36**, 177 (1964)

114. J.E. Ffowcs Williams, L.H. Hall, Aerodynamic sound generation by turbulence flow in the vicinity of a scattering half plane. J. Fluid Mech **40**(4), 657–670 (1970)

115. F. Farassat, Theory of noise generation from moving bodies with an application to helicopter rotors. NASA TRR-451 (1975)

116. Forbes website cmo.icxo.com

117. PE Daily, http://news.pedaily.cn

118. http://www.sznews.com

119. www.xxsb.com

120. http://www.metacomptech.com

121. http://www.hangkong.com

122. baike.baidu.com

123. www.sina.com.cn

124. http://mil.news.sina.com.cn

125. http://tech.sina.com.cn

126. http://gd.sina.com.cn

127. http://news.xinhuanet.com

128. http://www.afwing.com

129. http://tech.qq.com

130. http://www.ftchinese.com

131. http://life.caijing.com.cn

132. news.hexun.com

133. http://www.newmaker.com

134. http://www.ycfth.com

135. http://www.tooopen.com

136. www.cpc.people.com.cn

137. www.tsinghua.edu.cn

138. www.pku.edu.cn

139. www.buaa.edu.cn

140. tdjxxy.tju.edu.cn

141. www.tongji.edu.cn

142. jpkc.nwpu.edu.cn

143. www.cas.cn

144. sucai.redocn.com

145. baike.sogou.com

146. blog.cntv.cn

147. www.766.com

148. www.sciencenet.cn

149. www.cardc.cn
150. www.avicari.avic.com
151. www.rtri.or.jp

Printed in the United States
by Baker & Taylor Publisher Services